PROBABILITY

and STATISTICS

for ENGINEERS

THIRD EDITION

PROBABILITY

and STATISTICS

for ENGINEERS

IRWIN MILLER

Chairman and Chief Executive Officer
Opinion Research Corporation

JOHN E. FREUND

Arizona State University

Prentice-Hall, Inc., Englewood Cliffs, NJ   07632

Library of Congress Cataloging in Publication Data

Miller, Irwin (date)
    Probability and statistics for engineers.

    Bibliography: p.
    Includes index.
    1. Engineering—Statistical methods.   2. Probabilities.
I. Freund, John E.   II. Title.
TA340.M5 1985   519.2'02462   84–3307
ISBN 0–13–711938–0

Editorial/Production: Nicholas Romanelli
Manufacturing buyer: John Hall
Cover design: Jayne Conte

Printed in the United States of America
10 9 8 7 6 5 4 3 2

ISBN 0-13-711938-0

Prentice-Hall International, *London*
Prentice-Hall of Australia, Pty. Limited, *Sydney*
Editora Prentice-Hall do Brasil, Ltda., *Rio de Janeiro*
Prentice-Hall of Canada, Ltd., *Toronto*
Prentice-Hall of India Private Limited, *New Delhi*
Prentice-Hall of Japan, Inc., *Tokyo*
Prentice-Hall of Southeast Asia Pte. Ltd., *Singapore*
Whitehall Books Limited, Wellington, *New Zealand*

Contents

Preface

This book has been written for an introductory course in probability and statistics for students of engineering and the physical sciences. In the Third Edition, the authors have revised the exercise material, and many new exercises have been added. Exercises are numbered consecutively within each chapter. Each chapter is keynoted by an introductory statement and has a check list of key terms (with page references) at the end. Important formulas, theorems, and rules have been set out from the text; they are enclosed in boxes with a description of each in the margin. Notation and technical terms have been updated to conform with current usage. New material has been added on combinatorial methods, stem-and-leaf plots, and the distinction between "probability" and "confidence." The treatment of statistical inference has been extensively revised, including a more systematic treatment of tests of hypotheses and a discussion of tests of composite null hypotheses.

Throughout, the text has been revised with the aim of improving clarity and bringing the applications up to date. It has been tested extensively in courses for university students as well as by in-plant training of practicing engineers. The authors have found that the material in the book can be covered in a two-semester or three-quarter course consisting of three lectures a week. However, through a choice of topics the book also lends itself as a text for shorter courses with emphasis upon either theory or applications.

Chapters 2, 3, 4, and 6 provide a brief, though rigorous, introduction to the theory of statistics and, together with some of the material in Chapter 15, they are suitable for an introductory semester (or quarter) course on the mathematics of probability and statistics. Chapters 5, 7, 8, 9, and 10 contain conventional material on the elementary methods of statistical inference including a treatment of nonparametric methods. Chapters 11, 12, and 13 comprise an introduction to some of the standard, though more advanced, topics of experimental statistics, while Chapters 14 and 15 deal with special applications which have become increasingly important in recent years.

The mathematical background expected of the reader is a year course in calculus; actually, calculus is required mainly for Chapters 4 and 6 dealing with basic distribution theory in the continuous case, for Chapter 15 dealing with applications to reliability theory, and for the least-squares methods of Chapters 11 and 12. The treatment of probability in Chapter 2 is modern in the sense that it is based on the elementary theory of sets.

The authors would like to express their appreciation and indebtedness to the D. Van Nostrand Company for permission to reproduce the material in Table 2; to the Macmillan Publishing Co., Inc., for permission to reprint Table 4 from *Statistical Methods for Research Workers* by R. A. Fisher; to Professor E. S. Pearson and the *Biometrika* trustees for permission to reproduce the material in Tables 5, 6, and 9; to A. H. Bowker and G. J. Lieberman and Prentice-Hall, Inc., for permission to reproduce Table 8; to Donald B. Owen and Addison-Wesley, Inc., for permission to reproduce part of the table of random numbers from their *Handbook of Statistical Tables*; to Frank J. Massey, Jr., and *Journal of the American Statistical Association* for permission to reproduce the material in Table 10; to D. B. Duncan, H. L. Harter, and *Biometrics* for permission to reproduce Table 12; to the American Society for Testing and Materials for permission to reproduce Table 13; and to the McGraw-Hill Book Company for permission to reproduce the material in Table 14.

The authors would also like to express their appreciation to the editorial staff of Prentice-Hall, Inc., for their courteous cooperation in the production of this book, to Judi Fabian, who helped in the typing of the manuscript, to Rita Ewer for her help with the proofreading, and above all to their wives for not complaining too much about the demands made on their husbands throughout the writing of this book.

IRWIN MILLER AND JOHN E. FREUND

PROBABILITY

and STATISTICS

for ENGINEERS

1

Introduction

Everything dealing with the collection, processing, analysis, and interpretation of numerical data belongs to the domain of statistics. In engineering, this includes such diversified tasks as calculating the average length of the downtimes of a computer, collecting and presenting data on the numbers of persons attending seminars on solar energy, evaluating the effectiveness of commercial products, predicting the reliability of a missile or, studying the vibrations of airplane wings.

In Sections 1.1 and 1.2 we discuss the recent growth of statistics and, in particular, its applications to problems of engineering.

The origin of statistics can be traced to two areas of interest which, on the surface, have little in common: games of chance and what is now called political science. Mid-eighteenth century studies in probability, motivated largely by interest in games of chance, led to the mathematical treatment of errors of measurement and the theory which now forms the foundation of statistics. In the same century, interest in the numerical description of political units (cities, provinces, counties, etc.) led to what is now called **descriptive statistics**. At first, descriptive statistics consisted merely of the presentation of data in tables and charts; nowadays, it includes also the summarization of data by means of numerical descriptions.

In recent decades, the growth of statistics has made itself felt in almost every major phase of human activity, and the most important feature of its growth has been the shift in emphasis from descriptive statistics to **statistical inference**. Statistical inference concerns generalizations based on sample data; it applies to such problems as estimating an engine's average emission of pollutants from trial runs, testing a manufacturer's claim on the basis of measurements performed on samples of his product, and predicting the fidelity of an audio system on the basis of sample data pertaining to the performance of its components.

When one makes a statistical inference, namely, an inference which goes beyond the information contained in a set of data, one must always proceed with caution. One must decide carefully how far one can go in generalizing from a given set of data, whether such generalizations are at all reasonable or justifiable, whether it might be wise to wait until there are more data, and so forth. Indeed, some of the most important problems of statistical inference concern the appraisal of the risks and the consequences to which one might be exposed by making generalizations from sample data. This includes an appraisal of the probabilities of making wrong decisions, the chances of making incorrect predictions, and the possibility of obtaining estimates which do not lie within permissible limits.

In recent years, attempts have been made to treat all these problems within the framework of a unified theory called **decision theory**. Although this theory has many conceptual as well as theoretical advantages, it poses some problems of application that are difficult to overcome. To understand these problems, one must realize that no matter how objectively an experiment or an investigation is planned, it is impossible to eliminate all elements of subjectivity. It is at least partly a subjective decision whether to base an experiment (say, the determination of a specific heat) on 5 measurements, on 12 measurements, or on 25 or more. Also, subjective factors invariably enter the design of equipment, the hiring of personnel, and even one's deciding how to formulate a hypothesis and the alternative against which it is to be tested. (This important problem will be discussed in some detail in

Chapter 7.) An element of subjectivity enters even when we define such terms as "good" or "best" in connection with the description of decision criteria—for instance, in Chapter 11 we shall ask for a straight line which "best" fits a given set of paired data. Above all, subjective judgments are virtually unavoidable when one is asked to put "cash values" on the various risks to which one is exposed. In other words, it is impossible to be completely objective in specifying "rewards" for being right (or close) and "penalties" for being wrong (or not close enough). After all, if scientists are asked to judge the safety of a piece of equipment, how can they possibly put a cash value on the possibility that they might make an error, when such an error may lead to the loss of human lives.

Whether or not statistical inference is viewed within the broader framework of decision theory, it depends heavily on the theory of probability. This is a mathematical theory, but the question of objectivity versus subjectivity arises in its application and in its interpretation. As we shall see in Chapter 2, statements such as "there is a fifty-fifty chance that a given fabric will ignite when exposed to a cigarette ash" or "the probability is 0.05 that a defective part will not be caught in the final inspection" can be looked upon as objective or subjective evaluations of the uncertainties involved.

We shall approach the subject of statistics as a science, developing each statistical idea insofar as possible from its probabilistic foundation, and applying each idea to problems of physical or engineering science as soon as it has been developed. The great majority of the methods we shall use in stating and solving these problems belong to the **classical approach**, because they do not formally take into account the various subjective factors mentioned above; in selected applications we present also the **Bayesian approach** (see Sections 7.3 and 9.2), which accounts formally for at least some of these subjective factors. In any case, we shall endeavor continually to make the reader aware that the subjective factors do exist, and to indicate whenever possible what role they might play in making the final decision. Subjectivity plays an important role in the choice among statistical methods or formulas to be used in a given situation, in deciding on the size of a sample, in specifying the probabilities with which we are willing to risk errors, and so forth. This "bread-and-butter" approach to statistics presents the subject in the form in which it has so successfully contributed to engineering science, as well as to the natural social sciences, in the last 30 years.

1.2 STATISTICS AND ENGINEERING

There are few areas where the impact of the recent growth of statistics has been felt more strongly than in engineering and industrial management. Indeed, it would be difficult to overestimate the contributions statistics has made to problems of production, to the effective use of materials and manpower, to basic research, and to the development of new products. As in the other sciences, statistics has become a vital tool to engineers; in fact, some knowledge of statistics has become a necessity without which they cannot possibly appreciate, understand, or apply much of the work done in their field.

In this text, our attention will be directed largely toward engineering applications, but we shall not hesitate to refer also to other areas to impress upon the reader the great generality of most statistical techniques. Thus, the reader will find that the statistical method which is used to estimate the coefficient of thermal expansion of a metal serves also to estimate the average time it takes a secretary to perform a given task, the average thickness of a pelican egg, or the average I.Q. of an immigrant arriving in the United States. Similarly, the statistical method which is used to compare the strength of two alloys serves also to compare the effectiveness of two teaching methods, the merits of two insect sprays, or the performance of men and women in a current-events test.

In spite of the generality of most statistical techniques, there are also instances where the requirements of different fields have led to the development of special techniques. Thus, problems of economic forecasting led to special methods used in the analysis of series of business data; the pharmacological problem of determining critical dosages led to what is called probit analysis; problems in psychological testing led to factor analysis, and so forth. As far as engineering is concerned, we shall introduce the reader to three areas which have required the development of special techniques. (1) In Chapter 14 we shall present some of the special methods used in problems of **quality assurance**, including the problem of controlling (establishing and maintaining) quality in mass production, the problem of establishing tolerance limits, and problems of sampling inspection. (2) In Chapter 15 we shall present special techniques which have been developed to meet the **reliability** needs of the highly complex products of space-age technology. (3) The third area, **operations research**, is a new technology which is characterized by the application of scientific techniques (including probability theory and statistics) to problems involving the operations of a "system" looked upon as a whole. Thus, it applies to the conduct of a war, the management of a firm, the manufacture of a product, and so forth. This new technology will not be presented in a special chapter, but some of its methods will be illustrated throughout the text in exercises and examples.

1.3 CHECK LIST OF KEY TERMS (with page references)

Bayesian approach to statistics, 3	*Operations research, 4*
Classical approach to statistics, 3	*Quality assurance, 4*
Decision theory, 2	*Reliability, 4*
Descriptive statistics, 2	*Statistical inference, 2*

2

Probability

In the study of probability there are basically three kinds of questions: (1) What do we mean when we say that the probability of an event is, say, 0.50, 0.02, or 0.81 ? (2) How are the numbers we call probabilities determined, or measured in actual practice? (3) What are the mathematical rules which probabilities must obey?

After some mathematical preliminaries in Sections 2.1 and 2.2, we study the first two kinds of questions in Section 2.3 and the third kind of question in Sections 2.4 through 2.7. The related concept of a mathematical expectation is introduced in Section 2.8.

In statistics, a set of all possible outcomes of an experiment is called a **sample space**, owing to the fact that it usually consists of all the things that can happen when one takes a sample. Sample spaces are usually denoted by the letter S. To avoid misunderstandings about the words "experiment" and "outcome" as we have used them here, it should be understood that statisticians use these terms in a very wide sense. An **experiment** may consist of the simple process of noting whether a switch is turned on or off; it may consist of determining the time it takes a car to accelerate to 30 miles per hour; or it may consist of the complicated process of finding the mass of an electron. Thus, the **outcome** of an experiment may be a simple choice between two alternatives; it may be the result of a direct measurement or count; or it may be an answer obtained after extensive measurements and calculations.

When we study the outcomes of an experiment, we usually identify the various possibilities with numbers, points, or some other kinds of symbols. For instance, if four contractors bid on a highway construction job and we let a, b, c, and d denote that it is awarded to Mr. Adam, Mrs. Brown, Mr. Clark, or Ms. Dean, then the sample space for this experiment is the set $S = \{a, b, c, d\}$.

Also, if a government agency must decide where to locate two new nuclear power plants and that (for a certain purpose) it is of interest to indicate how many of them will be located in Texas and how many in California, we can write the sample space as

$$S = \{(0, 0), (1, 0), (0, 1), (2, 0), (1, 1), (0, 2)\}$$

where the first coordinate is the number of plants that will be located in Texas and the second coordinate is the number that will be located in California. Geometrically, this sample space may be pictured as in Figure 2.1, from which it is apparent, for example, that in two of the six possibilities Texas and California will get equally many of the new power plants. The use of points rather than letters or numbers has the advantage that it makes it easier to visualize the various possibilities, and perhaps discover some special features which several of the outcomes may have in common.

Generally, sample spaces are classified according to the number of elements (points) which they contain. In the two preceding examples, the sample spaces had four and six elements, and they are both referred to as **finite**. Other examples of finite sample spaces are the one for the various ways in which a president and a vice-president can be selected from among the 25 members of a union local and the one for the various ways in which a student can answer the 12 questions on a true-false test. As we shall see on page 13, the first of these sample spaces has 600 elements and the other has 4,096.

The following are examples of sample spaces that are not finite. If

Figure 2.1 Sample space for the number of new nuclear power plants to be located in Texas and in California.

persons checking the nitrogen oxide emission of cars are interested in the number of cars they have to inspect before they observe the first one that does not meet government regulations, it could be the first, the second, . . . , the fiftieth, . . . , and for all we know they may have to check thousands of cars before they find one that does not meet government regulations. Not knowing how far they may have to go, it is appropriate in an example like this to take as the sample space the whole set of natural numbers, of which there is a countable infinity. To go one step further, if they were interested in the nitrogen oxide emission of a given car in grams per mile, the sample space would have to consist of all the points on a continuous scale (a certain interval on the line of real numbers), of which there is a continuum.

In general, a sample space is said to be **discrete** if it has finitely many or a countable infinity of elements. If the elements (points) of a sample space constitute a continuum, for example, all the points on a line, all the points on a line segment, or all the points in a plane, the sample space is said to be **continuous**. In the remainder of this chapter we shall consider only discrete and mainly finite sample spaces.

In statistics, any subset of a sample space is called an **event**. By subset we mean any part of a set, including the whole set and, trivially, a set called the **empty set** and denoted by $\varnothing$, which has no elements at all. For instance, with reference to Figure 2.1,

$$C = \{(1, 0), (0, 1)\}$$

is the event that, between them, Texas and California will get one of the two power plants,

$$D = \{(0, 0), (0, 1), (0, 2)\}$$

is the event that Texas will not get either of the two power plants, and

$$E = \{(0, 0), (1, 1)\}$$

is the event that Texas and California will get equally many of the new power plants. Note that events C and E have no elements in common—they are **mutually exclusive events**.

In many probability problems we are interested in events which can be expressed in terms of two or more events by forming **unions**, **intersections**, and **complements**. Although the reader must surely be familiar with these terms, let us review briefly that if A and B are any two sets in a sample space S, their union $A \cup B$ is the subset of S which contains all elements that are either in A, in B, or in both; their intersection $A \cap B$ is the subset of S which contains all elements that are in both A and B; and the complement A' of A is the subset of S which contains all the elements of S that are not in A.

EXAMPLE   With reference to the sample space of Figure 2.1 and the events C, D, and E defined above, list the outcomes comprising each of the following events, and also express the events in words:

(a) $C \cup E$;
(b) $C \cap D$;
(c) D'.

Solution   (a) Since $C \cup E$ contains all the elements that are in C, in E, or in both,

$$C \cup E = \{(1, 0), (0, 1), (0, 0), (1, 1)\}$$

is the event that neither Texas nor California will get both of the new power plants. (b) Since $C \cap D$ contains all the elements that are in both C and D,

$$C \cap D = \{(0, 1)\}$$

is the event that Texas will not get either of the two new power plants and California will get one. (c) Since D' contains all the elements of the sample space that are not in D,

$$D' = \{(1, 0), (1, 1), (2, 0)\}$$

is the event that Texas will get at least one of the new power plants.

Sample spaces and events, particularly relationships among events, are often depicted by means of **Venn diagrams** like those of Figures 2.2 and 2.4.

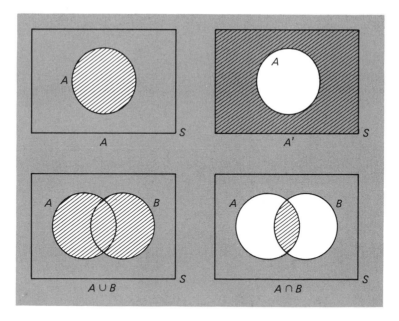

Figure 2.2 Venn diagrams.

In each case the sample space is represented by a rectangle, while events are represented by regions within the rectangle, usually by circles or parts of circles. The shaded regions of the four Venn diagrams of Figure 2.2 represent event A, the complement of event A, the union of events A and B, and the intersection of events A and B.

EXAMPLE If A is the event that a certain student is taking a course in calculus and B is the event that the student is taking a course in applied mechanics, what events are represented by the shaded regions of the four Venn diagrams of Figure 2.2?

> **Solution** The shaded region of the first diagram represents the event that the student is taking a course in calculus; that of the second diagram represents the event that the student is not taking a course in calculus; that of the third diagram represents the event that the student is taking a course in calculus and/or a course in applied mechanics; and that of the fourth diagram represents the event that the student is taking a course in calculus as well as a course in applied mechanics.

Venn diagrams are often used to verify relationships among sets, thus making it unnecessary to give formal proofs based on the algebra of sets. To illustrate, let us show that $(A \cup B)' = A' \cap B'$, which expresses the fact that the complement of the union of two sets equals the intersection of their complements. To begin, note that the shaded region of the first Venn diagram of Figure 2.3 represents the set $(A \cup B)'$ (compare this diagram with the third diagram of Figure 2.2). The crosshatched region of the second Venn diagram of Figure 2.3 was obtained by shading the region representing A' with lines going in one direction and that representing B' with lines going in another direction. Thus, the crosshatched region represents the intersection of A' and B', and it can be seen that it is identical with the shaded region of the first Venn diagram of Figure 2.3.

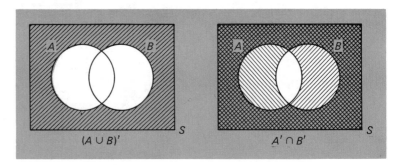

Figure 2.3 Use of Venn diagrams to show that $(A \cup B)' = A' \cap B'$.

When we deal with three events, we draw the circles as in Figure 2.4. In this diagram, the circles divide the sample space into eight regions, numbered 1 through 8, and it is easy to determine whether the corresponding events are parts of A or A', B or B', and C or C'.

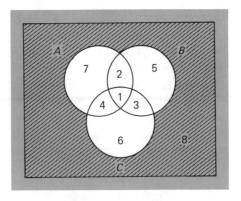

Figure 2.4 Venn diagram.

EXAMPLE If A is the event that a group of efficiency experts will inspect the accounting department of a large firm, B is the event that they will inspect the quality control department, and C is the event that they will inspect the research and development department, express in words what events are represented by the following regions of the Venn diagram of Figure 2.4:

(a) region 2;
(b) regions 1 and 3 together;
(c) regions 3, 5, 6, and 8 together.

Solution (a) Since this region is contained in A and B but not in C, it represents the event that the efficiency experts will inspect the accounting and quality control departments, but not the research and development department. (b) Since this is the region common to B and C, it represents the event that they will inspect the quality control department and the research and development department. (c) Since this is the entire region outside A, it represents the event that they will not inspect the accounting department.

2.2 COUNTING

At times it can be quite difficult, or at least tedious, to determine the number of elements in a finite sample space by direct enumeration. To illustrate, suppose that a consumer testing service rates lawn mowers as being easy, average, or difficult to operate; as being expensive or inexpensive; and as being costly, average, or cheap to repair. In how many different ways can a lawn mower be rated by this testing service?

Clearly, there are many possibilities: a lawn mower may be rated easy to operate, inexpensive, but costly to repair; it may be rated difficult to operate, expensive, and cheap to repair; it may be rated as neither easy nor difficult to operate, inexpensive, with average cost of repairs; and so on. Continuing this way, we may be able to list all 18 possibilities, but the chances are that we will omit at least one or two.

To handle this kind of problem systematically, it helps to draw a **tree diagram** like that of Figure 2.5, where the three alternatives for ease of operation are denoted by E_1, E_2, and E_3; the price is either P_1 or P_2; and the three alternatives for the cost of repairs are denoted by C_1, C_2, and C_3. Following a given path from left to right along the branches of the tree, we obtain a particular rating, namely, a particular element of the sample space, and it can be seen that altogether there are 18 possibilities.

This result could also have been obtained by observing that there are three E-branches, that each E-branch forks into two P-branches, and that each P-branch forks into three C-branches. Thus, there are $3 \cdot 2 \cdot 3 = 18$ com-

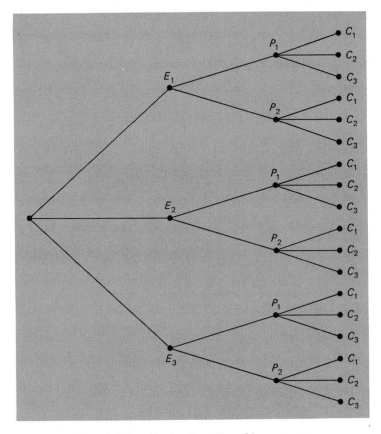

Figure 2.5 Tree diagram for rating of lawn mowers.

binations of branches, or paths. This result is a special case of the following theorem:

Multiplication of choices

> **Theorem 2.1** *If sets $A_1, A_2, \ldots, A_k$ contain, respectively $n_1, n_2, \ldots, n_k$ elements, there are $n_1 \cdot n_2 \cdot \ldots \cdot n_k$ ways of choosing first an element of A_1, then an element of $A_2, \ldots$, and finally an element of A_k.*

In our example we had $n_1 = 3$, $n_2 = 2$, and $n_3 = 3$, and, hence, $3 \cdot 2 \cdot 3 = 18$ possibilities.

EXAMPLE   In how many different ways can a union local with a membership of 25 choose a vice-president and a president?

Solution Since the vice-president can be chosen in 25 ways and, subsequently, the president in 24 ways, there are altogether $25 \cdot 24 = 600$ ways in which the whole choice can be made.

EXAMPLE If a test consists of 12 true-false questions, in how many different ways can a student mark the test paper with one answer to each question?

Solution Since each question can be answered in two ways, there are altogether

$$2 \cdot 2 \cdot 2 \cdot 2 \cdot 2 \cdot 2 \cdot 2 \cdot 2 \cdot 2 \cdot 2 \cdot 2 \cdot 2 = 2^{12} = 4,096 \text{ possibilities}$$

As in the first of these two examples, the rule for the multiplication of choices is often used when several choices are made from one set and we are concerned with the order in which they are made. In general, if r objects are chosen from a set of n distinct objects, any particular arrangement, or order, of these objects is called a **permutation**. For instance, 4 1 2 3 is a permutation of the first four positive integers, and Maine, Vermont, and Connecticut is a permutation, a particular ordered arrangement, of three of the six New England states.

To find a formula for the total number of permutations of r objects selected from a set of n distinct objects, we observe that the first selection is made from the whole set of n objects, the second selection is made from the $n - 1$ objects which remain after the first selection has been made, . . . , and the rth selection is made from the $n - (r - 1) = n - r + 1$ objects which remain after the first $r - 1$ selections have been made. Therefore, by the rule for the multiplication of choices, the total number of permutations of r objects selected from a set of n distinct objects is

$$_nP_r = n(n - 1)(n - 2) \cdot \ldots \cdot (n - r + 1)$$

Since products of consecutive integers arise in many problems relating to permutations or other kinds of special selections, it will be convenient to introduce here the **factorial notation**, where $1! = 1$, $2! = 2 \cdot 1 = 2$, $3! = 3 \cdot 2 \cdot 1 = 6$, $4! = 4 \cdot 3 \cdot 2 \cdot 1 = 24, \ldots$, and in general $n! = n(n - 1)(n - 2) \cdot \ldots \cdot 3 \cdot 2 \cdot 1$. Also, to make various formulas more generally applicable, we let $0! = 1$ by definition.

To express the formula for $_nP_r$ in terms of factorials, we multiply and divide by $(n - r)!$, getting

$$_nP_r = \frac{n(n - 1)(n - 2) \cdot \ldots \cdot (n - r + 1)(n - r)!}{(n - r)!} = \frac{n!}{(n - r)!}$$

To summarize,

> **Theorem 2.2** *The number of permutations of r objects selected from a set of n distinct objects is*
>
> $$_nP_r = n(n - 1)(n - 2) \cdot \vdots \ldots \cdot (n - r + 1)$$
>
> *or, in factorial notation,*
>
> $$_nP_r = \frac{n!}{(n - r)!}$$

EXAMPLE   In how many different ways can one make a first, second, third, and fourth choice among 12 firms leasing construction equipment?

Solution   For $n = 12$ and $r = 4$, the first formula yields

$$_{12}P_4 = 12 \cdot 11 \cdot 10 \cdot 9 = 11{,}880$$

and the second formula yields

$$_{12}P_4 = \frac{12!}{(12 - 4)!} = \frac{12!}{8!} = \frac{12 \cdot 11 \cdot 10 \cdot 9 \cdot 8!}{8!} = 11{,}880$$

EXAMPLE   If a part consists of five different components which can be assembled in any order, in how many different ways can the part be assembled?

Solution   For $n = 5$ and $r = 5$, the first formula yields

$$_5P_5 = 5 \cdot 4 \cdot 3 \cdot 2 \cdot 1 = 120$$

and the second formula yields

$$_5P_5 = \frac{5!}{(5 - 5)!} = \frac{5!}{0!} = 5! = 120$$

The first formula for $_nP_r$ is generally easier to use unless we can refer to a table of factorials or use a calculator which directly yields factorials and/or ratios of factorials.

There are many problems in which we must find the number of ways in which r objects can be selected from a set of n objects, but we do not care about the order in which the selection is made. For instance, we may want to know in how many ways 3 of 20 laboratory assistants can be chosen to assist with an experiment. In general, there are $r!$ permutations of any r objects we select from a set of n distinct objects, and the $_nP_r$ permutations of r objects selected from a set of n objects contain each set of r objects $r!$ times. Therefore, to find the number of ways in which r objects can be selected from a set of n

distinct objects, also called the number of **combinations** of n objects taken r at a time and denoted by $_nC_r$ or $\binom{n}{r}$, we divide $_nP_r$ by $r!$ and get

Number of combinations of n objects taken r at a time

> **Theorem 2.3** *The number of ways in which r objects can be selected from a set of n distinct objects is*
>
> $$\binom{n}{r} = \frac{n(n-1)(n-2) \cdot \ldots \cdot (n-r+1)}{r!}$$
>
> *or, in factorial notation,*
>
> $$\binom{n}{r} = \frac{n!}{r!(n-r)!}$$

EXAMPLE In how many different ways can 3 of 20 laboratory assistants be chosen to assist with an experiment?

Solution For $n = 20$ and $r = 3$, the first formula for $\binom{n}{r}$ yields

$$\binom{20}{3} = \frac{20 \cdot 19 \cdot 18}{3!} = 1,140$$

EXAMPLE In how many different ways can the director of a research laboratory choose two chemists from among seven applicants and three physicists from among nine applicants?

Solution The two chemists can be chosen in $\binom{7}{2} = 21$ ways, the three physicists can be chosen in $\binom{9}{3} = 84$ ways, so that the whole selection can be made in $21 \cdot 84 = 1,764$ ways.

EXERCISES

2.1 A technician has to check the suitability of three solid crystal lasers and two carbon dioxide lasers for a given task.
   (a) Using two coordinates so that (2, 1), for example, represents the event that the technician will find two of the solid crystal lasers and one of the carbon dioxide lasers suitable for the task, draw a diagram similar to that of Figure 2.1 showing the 12 points of the sample space.
   (b) If R is the event that equally many solid crystal lasers and carbon dioxide lasers are suitable for the task, T is the event that none of the carbon dioxide lasers is suitable for the task, and U is the event that fewer solid crystal lasers than carbon dioxide lasers are suitable for the task, express each of these events symbolically by listing its elements.

2.2 With reference to the preceding exercise, which of the three pairs of events, *R* and *T*, *R* and *U*, and *T* and *U*, are mutually exclusive?

2.3 With reference to Exercise 2.1, list the outcomes comprising each of the following events, and also express the events in words:
(a) $R \cup U$;
(b) $R \cap T$;
(c) T'.

2.4 With reference to the sample space of Figure 2.1, express each of the following events in words:
(a) $F = \{(1, 0), (1, 1)\}$;
(b) $G = \{(0, 2), (1, 1), (2, 0)\}$;
(c) $F \cap G$.

2.5 To construct sample spaces for experiments in which we deal with non-numerical data, we often code the various alternatives by assigning them numbers. For instance, if a mechanic is asked whether work on a certain model car is very easy, easy, average, difficult, or very difficult, we might assign these alternatives the codes 1, 2, 3, 4, and 5. If $A = \{3, 4\}$, $B = \{2, 3\}$, and $C = \{4, 5\}$, express each of the following symbolically by listing its elements, and also in words:
(a) $A \cup B$;   (c) $A \cup B'$;
(b) $A \cap B$;   (d) C'.

2.6 With reference to the preceding exercise, which of the three pairs of events, *A* and *B*, *A* and *C*, and *B* and *C*, are mutually exclusive?

2.7 Two professors and three graduate assistants are responsible for the supervision of a physics lab, and at least one professor and one graduate assistant has to be present at all times.
(a) Using two coordinates so that (1, 3), for example, represents the event that one professor and three graduate assistants are present, draw a diagram similar to that of Figure 2.1 showing the points of the corresponding sample space.
(b) Describe in words the events which are represented by $B = \{(1, 3), (2, 3)\}$, $C = \{(1, 1), (2, 2)\}$, and $D = \{(1, 2), (2, 1)\}$.
(c) With reference to part (b), express $C \cup D$ symbolically by listing its elements, and also express this event in words.
(d) With reference to part (b), are *B* and *D* mutually exclusive?

2.8 A salesman of industrial chemicals has four customers in Sacramento, whom he may or may not be able to visit on a 2-day trip to this city. He will not visit any of these customers more than once.
(a) Using two coordinates so that (2, 1), for example, represents the event that he will visit two of his customers on the first day and one on the second day, draw a diagram similar to that of Figure 2.1 showing the points of the corresponding sample space.
(b) List the points of the sample space which constitute the events *X*, *Y*, and *Z* that he will visit all four of his customers, that he will visit more of his customers on the first day than on the second day, and that he will visit at least three of his customers on the second day.

(c) Which of the three pairs of events, X and Y, X and Z, and Y and Z, are mutually exclusive?

2.9 With reference to the preceding exercise, express each of the following events symbolically by listing its elements, and also express it in words:

(a) X';   (c) $X \cap Z$;
(b) $X \cup Y$;   (d) $X' \cap Y$.

2.10 For each of the following experiments decide whether it would be appropriate to use a sample space which is finite, countably infinite, or continuous:

(a) The amount of cosmic radiation to which passengers are exposed during a transcontinental jet flight is measured by means of a suitable counter.
(b) Five of the members of a professional society with 12,600 members are chosen to serve on a nominating committee.
(c) An experiment is conducted to measure the heat of vaporization of water.
(d) A study is made to determine in how many of 450 airplane accidents the main cause is pilot error.
(e) Measurements are made to determine the uranium content of a certain ore.
(f) In a torture test, a watch is dropped from a tall building until it stops running.

2.11 In Figure 2.6, C is the event that an ore contains copper and U is the event that it contains uranium. Explain in words what events are represented by regions 1, 2, 3, and 4.

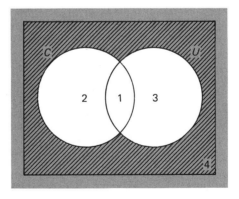

Figure 2.6 Venn diagram for Exercises 2.11 and 2.12.

2.12 With reference to the preceding exercise, what events are represented by
(a) regions 1 and 3 together;
(b) regions 3 and 4 together;
(c) regions 1, 2, and 3 together?

2.13 With reference to Figure 2.4, what events are represented by
(a) region 5;
(b) regions 4 and 6 together;

(c) regions 7 and 8 together;

(d) regions 1, 2, 3, and 5 together?

2.14 With reference to Figure 2.4, what regions or combinations of regions represent the events that the group of efficiency experts will inspect

(a) none of the departments;

(b) the accounting department and the quality control department;

(c) the accounting department and/or the quality control department, but not the research and development department;

(d) the quality control department and accounting department and/or the research and development department?

2.15 Use Venn diagrams to verify that

(a) $(A \cap B)' = A' \cup B'$;

(b) $A \cup (A \cap B) = A$;

(c) $(A \cap B) \cup (A \cap B') = A$;

(d) $A \cup B = (A \cap B) \cup (A \cap B') \cup (A' \cap B)$;

(e) $A \cup (B \cap C) = (A \cup B) \cap (A \cup C)$.

2.16 A building inspector has to check the wiring in a new apartment building either on Monday, Tuesday, Wednesday, or Thursday, and at 8 A.M., 1 A.M., or 2 P.M. Draw a tree diagram which shows the various ways in which he can schedule the inspection of the wiring of the new apartment building.

2.17 If the five finalists in the Miss Universe contest are Miss Spain, Miss U.S.A., Miss Uruguay, Miss Portugal, and Miss Japan, draw a tree diagram which shows the various ways in which the judges can choose the winner and the first runner-up.

2.18 A student can study either 0, 1, or 2 hours for a test in computer programming on any given night. Construct a tree diagram to show that there are 10 different ways in which he can study altogether 6 hours for the test on four consecutive nights.

2.19 If the tone quality of four audio systems is to be rated superior, average, or inferior and we are interested only in how many of the systems get each of these ratings, draw a tree diagram which shows the 15 different possibilities.

2.20 In an optics kit there are six concave lenses, four convex lenses, and three prisms. In how many ways can one choose one of the concave lenses, one of the convex lenses, and one of the prisms?

2.21 A questionnaire sent through the mail as part of a market study consists of eight questions, each of which can be answered in three different ways. In how many different ways can a person answer the eight questions on this questionnaire?

2.22 In a small geology class, each of the four students must write a report on one of eight field trips. In how many different ways can they each choose one of the field trips if

(a) no two students may choose the same field trip;

(b) there is no restriction on their choice?

2.23 If there are nine cars in a race, in how many different ways can they place first, second, and third?

2.24 In how many ways can a television director schedule six different commercials during the six time slots allocated to commercials during the telecast of the first period of a hockey game?

2.25 If among n objects k are alike and the others are all distinct, the number of permutations of these n objects taken all together is $\frac{n!}{k!}$.

(a) How many permutations are there of the letters of the word "class"?

(b) In how many ways can the television director of Exercise 2.24 fill the six time slots allocated to commercials, if she has four different commercials, of which a given one is to be shown three times while each of the others is to be shown once?

2.26 Determine the number of ways in which a manufacturer can choose 2 of 15 locations for a new warehouse.

2.27 If the order does not matter, in how many different ways can 4 of 18 tax returns be chosen for a special audit?

2.28 A carton of 12 transistor batteries contains one that is defective. In how many ways can an inspector choose three of the batteries and
(a) get the one that is defective;
(b) not get the one that is defective?

2.29 With reference to the preceding exercise, suppose that two of the batteries are defective. In how many ways can the inspector choose three of the batteries and get
(a) none of the defective batteries;
(b) one of the defective batteries;
(c) both of the defective batteries?

2.30 A major appliance store carries eight kinds of refrigerators, six kinds of washer-driers, and five kinds of microwave ovens. In how many different ways can two of each kind be chosen for a special sale?

2.3 PROBABILITY

So far we have studied only what is possible in a given situation; now we shall go one step further and judge also what is probable and what is improbable. Historically, the oldest way of measuring uncertainties is the **classical probability concept**, which was developed originally in connection with games of chance. It applies when all possible outcomes are equally likely, in which case we say that

The classical probability concept

> *If there are n equally likely possibilities, of which one must occur and s are regarded as favorable, or as a "success," then the probability of a "success" is given by $\frac{s}{n}$.*

In the application of this rule, the terms "favorable" and "success" are used rather loosely—"favorable" may mean that a television set does not work and "success" may mean that someone catches the flu.

EXAMPLE   What is the probability of drawing an ace from a well-shuffled deck of 52 playing cards?

> *Solution*   There are $s = 4$ aces among the $n = 52$ cards, so we get
>
> $$\frac{s}{n} = \frac{4}{52} = \frac{1}{13}$$

Although equally likely possibilities are found mostly in games of chance, the classical probability concept applies also to a great variety of situations where gambling devices are used to make random selections; say, when offices are assigned to research assistants by lot, when laboratory animals are chosen for an experiment so that each one has the same chance of being selected, or when washing machine parts are chosen for inspection so that each part produced has the same chance of being selected.

EXAMPLE   If 3 of 20 tires are defective and 4 of them are randomly chosen for inspection (that is, each tire has the same chance of being selected), what is the probability that only one of the defective tires will be included?

> *Solution*   There are $\binom{20}{4} = 4{,}845$ equally likely ways of choosing 4 of the 20 tires, so $n = 4{,}845$. The number of favorable outcomes is the number of ways in which one of the defective tires and three of the nondefective tires can be selected, or $s = \binom{3}{1}\binom{17}{3} = 3 \cdot 680 = 2{,}040$. It follows that the probability is
>
> $$\frac{s}{n} = \frac{2{,}040}{4{,}845} = \frac{8}{19}$$

or approximately 0.42.

A major shortcoming of the classical probability concept is its limited applicability, for there are many situations in which the various possibilities cannot all be regarded as equally likely. This would be the case, for example, if we are concerned with the question of whether it will rain the next day, whether a missile launching will be a success, whether a newly designed engine will function for at least 1,000 hours, or whether a certain candidate will win an election.

Among the various probability concepts, most widely held is the **fre-**

quency interpretation, according to which

The frequency interpretation of probability

> *The probability of an event (happening or outcome) is the proportion of the time that events of the same kind will occur in the long run.*

If we say that the probability is 0.78 that a jet from New York to Boston will arrive on time, we mean that such flights arrive on time 78 % of the time. Also, if the Weather Service predicts that there is a 40 % chance for rain (that the probability is 0.40), this means that under the same weather conditions it will rain 40 % of the time. More generally, we say that an event has a probability of, say, 0.90, in the same sense in which we might say that in cold weather our car will start 90 % of the time. We cannot guarantee what will happen on any particular occasion—the car may start and then it may not— but if we kept records over a long period of time, we should find that the proportion of "successes" is very close to 0.90.

In accordance with the frequency interpretation of probability, we estimate the probability of an event by observing what fraction of the time similar events have occurred in the past.

EXAMPLE   If records show that 294 of 300 ceramic insulators tested were able to withstand a certain thermal shock, what is the probability that any one such insulator will be able to withstand the thermal shock?

Solution   Among the insulators tested, $\frac{294}{300} = 0.98$ were able to withstand the thermal shock, and we use this figure as an estimate of the probability.

An alternative point of view which is currently gaining in favor is to interpret probabilities as **personal** or **subjective evaluations**. Such probabilities express the strength of one's belief with regard to the uncertainties that are involved, and they apply especially when there is little or no direct evidence, so that there is no choice but to consider collateral (indirect) evidence, educated guesses, and perhaps intuition and other subjective factors. Subjective probabilities are best determined by referring to risk taking, or betting situations, as will be explained in Exercise 2.59.

2.4   THE AXIOMS OF PROBABILITY

In this section we shall define probabilities mathematically as the values of **additive set functions**. Since the reader is probably most familiar with functions for which the elements of the domain and the range are all numbers, let us first give a very simple example where the elements of the domain are sets while the elements of the range are non-negative integers, namely, a **set function** which assigns to each subset A of a finite sample space S the number of elements in A, written $N(A)$. Suppose that 500 machine parts are inspected before they are shipped, that I denotes that a machine part is improperly assembled, D denotes that it contains one or more defective components, and the distribution of the 500 machine parts among the various categories is as shown in the Venn diagram of Figure 2.7. The numbers in Figure 2.7 are $N(I \cap D') = 20$, $N(I \cap D) = 10$, $N(I' \cap D) = 5$, and $N(I' \cap D') = 465$.

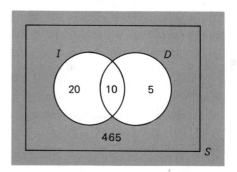

Figure 2.7 Classification of 500 machine parts.

Using these values and the fact that the set function is **additive** (meaning that the number which it assigns to the union of two subsets which have no elements in common is the sum of the numbers assigned to the individual subsets), we can determine the value of $N(X)$ for any other subset X of S. For instance,

$$N(I') = N(I' \cap D) + N(I' \cap D') = 5 + 465 = 470$$
$$N(I \cup D) = N(I \cap D') + N(I \cap D) + N(I' \cap D)$$
$$= 20 + 10 + 5 = 35$$
$$N(I' \cup D) = N(I \cap D) + N(I' \cap D) + N(I' \cap D')$$
$$= 10 + 5 + 465 = 480$$

and

$$N(D) = N(I \cap D) + (I' \cap D) = 10 + 5 = 15$$

Using the concept of an additive set function, let us now explain what we mean by the probability of an event. Given a finite sample space S and an event A is S, we define $P(A)$, the probability of A, to be a value of an additive set function which satisfies the following three conditions:

The axioms of
probability

> **Axiom 1** $0 \leq P(A) \leq 1$ *for each event A in S.*
>
> **Axiom 2** $P(S) = 1.$
>
> **Axiom 3** *If A and B are any mutually exclusive events in S, then* $P(A \cup B)$
> $= P(A) + P(B).$

The first axiom states that probabilities are real numbers on the interval from 0 to 1. The second axiom states that the sample space as a whole is assigned a probability of 1 and this expresses the idea that the probability of a certain event, an event which must happen, is equal to 1. The third axiom states that probability functions must be additive.

Axioms for a mathematical theory require no proof, but if such a theory is to be applied to the physical world, we must show somehow that the axioms are "realistic." Thus, let us show that the three postulates are consistent with the classical probability concept and the frequency interpretation. The situation is more complicated when it comes to subjective probabilities, as is explained in Exercise 2.60.

So far as the first axiom is concerned, fractions of the form $\frac{s}{n}$, where $0 \leq s \leq n$ and n is a positive integer, cannot be negative or exceed 1, and the same is true also for the proportion of the time that an event will occur. To show that the second axiom is consistent with the classical probability concept and the frequency interpretation, we have only to observe that for the whole sample space $P(S) = \frac{n}{n} = 1$ and that an event which is certain to occur must happen 100% of the time. So far as the third axiom is concerned, if $P(A) = \frac{s_1}{n}$, $P(B) = \frac{s_2}{n}$, and A and B are mutually exclusive, then $P(A \cup B) = \frac{s_1 + s_2}{n} = P(A) + P(B)$; also, if one event occurs 36% of the time, another event occurs 41% of the time, and the two events are mutually exclusive, then one or the other will occur $36 + 41 = 77\%$ of the time.

Before we go any further, it is important to stress the point that the axioms of probability do not tell us how to assign probabilities to the various outcomes of an experiment; they merely restrict the ways in which it can be done. In actual practice, probabilities are assigned on the basis of past experience, on the basis of a careful analysis of conditions underlying the experiment, on the basis of subjective evaluations, or on the basis of assumptions —say, the common assumption that all the outcomes are equiprobable.

EXAMPLE If an experiment has the three possible and mutually exclusive outcomes *A*, *B*, and *C*, check in each case whether the assignment of probabilities is permissible:

(a) $P(A) = \frac{1}{3}, P(B) = \frac{1}{3}$, and $P(C) = \frac{1}{3}$;
(b) $P(A) = 0.64, P(B) = 0.38$, and $P(C) = -0.02$;
(c) $P(A) = 0.35, P(B) = 0.52$, and $P(C) = 0.26$;
(d) $P(A) = 0.57, P(B) = 0.24$, and $P(C) = 0.19$.

Solution (a) The assignment of probabilities is permissible because the values are all on the interval from 0 to 1 and their sum is $\frac{1}{3} + \frac{1}{3} + \frac{1}{3} = 1$. (b) The assignment is not permissible because $P(C)$ is negative. (c) The assignment is not permissible because $0.35 + 0.52 + 0.26 = 1.13$, which exceeds 1. (d) The assignment is permissible because the values are all on the interval from 0 to 1 and their sum is $0.57 + 0.24 + 0.19 = 1$.

2.5 SOME ELEMENTARY THEOREMS

With the use of mathematical induction, the third axiom of probability can be extended to include any number of mutually exclusive events; in other words, it can be shown that

Generalization of third axiom of probability

> ***Theorem 2.4*** *If $A_1, A_2, \ldots, A_n$ are mutually exclusive events in a sample space S, then*
>
> $$P(A_1 \cup A_2 \cup \ldots \cup A_n) = P(A_1) + P(A_2) + \ldots + P(A_n)$$

In the next chapter we shall see how the third axiom of probability must be modified so that the axioms apply also to sample spaces which are not finite.

EXAMPLE The probability that a consumer testing service will rate a new antipollution device for cars very poor, poor, fair, good, very good, or excellent are 0.07, 0.12, 0.17, 0.32, 0.21, and 0.11. What are the probabilities that it will rate the device

(a) very poor, poor, fair, or good;
(b) good, very good, or excellent?

Solution Since the possibilities are all mutually exclusive, direct substitution into the formula of Theorem 2.4 yields

$$0.07 + 0.12 + 0.17 + 0.32 = 0.68$$

for part (a) and

$$0.32 + 0.21 + 0.11 = 0.64$$

for part (b).

As it can be shown that a sample space of n points (outcomes) has 2^n subsets, it would seem that the problem of specifying a probability function (namely, a probability for each subset or event) can easily become very tedious. Indeed, for $n = 20$ there are already more than 1 million possible events. Fortunately, this task can be simplified considerably by the use of the following theorem:

Rule of calculating probability of an event

> **Theorem 2.5** *If A is an event in the finite sample space S, then P(A) equals the sum of the probabilities of the individual outcomes comprising A.*

To prove this theorem, let $E_1, E_2, \ldots, E_n$ be the n outcomes comprising event A, so that we can write $A = E_1 \cup E_2 \cup \ldots \cup E_n$. Since the E's are individual outcomes they are mutually exclusive, and by Theorem 2.4 we have

$$P(A) = P(E_1 \cup E_2 \cup \ldots \cup E_n)$$
$$= P(E_1) + P(E_2) + \ldots + P(E_n)$$

which completes the proof.

EXAMPLE With reference to the lawn-mower-rating example on page 11, suppose that the probabilities of the 18 outcomes are as shown in Figure 2.8 (which, except for the probabilities, is identical with Figure 2.5 on page 12). Find $P(E_1)$, $P(P_1)$, $P(C_1)$, $P(E_1 \cap P_1)$, and $P(E_1 \cap C_1)$.

Solution Adding the probabilities of the outcomes comprising the respective events, we get

$$P(E_1) = 0.07 + 0.13 + 0.06 + 0.05 + 0.07 + 0.02 = 0.40$$
$$P(P_1) = 0.07 + 0.13 + 0.06 + 0.07 + 0.14 + 0.07 + 0.02$$
$$+ 0.03 + 0.01 = 0.60$$
$$P(C_1) = 0.07 + 0.05 + 0.07 + 0.08 + 0.02 + 0.01 = 0.30$$

$$P(E_1 \cap P_1) = 0.07 + 0.13 + 0.06 = 0.26$$

and

$$P(E_1 \cap C_1) = 0.07 + 0.05 = 0.12$$

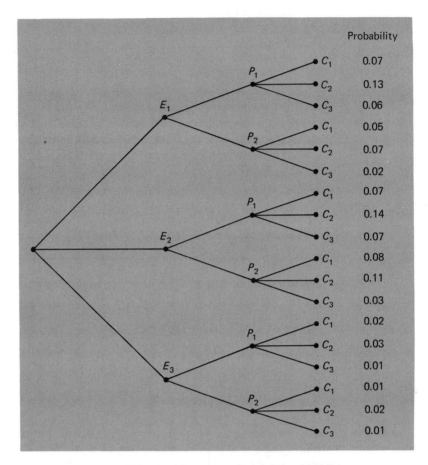

	Probability
C_1	0.07
C_2	0.13
C_3	0.06
C_1	0.05
C_2	0.07
C_3	0.02
C_1	0.07
C_2	0.14
C_3	0.07
C_1	0.08
C_2	0.11
C_3	0.03
C_1	0.02
C_2	0.03
C_3	0.01
C_1	0.01
C_2	0.02
C_3	0.01

Figure 2.8 Ratings of lawn mowers and their probabilities.

In Theorem 2.4 we saw that the third axiom of probability can be extended to include more than two mutually exclusive events. Another useful and important extension of this axiom allows us to find the probability of the union of any two events in S regardless of whether they are mutually exclusive. To motivate the theorem which follows, let us consider the Venn diagram of Figure 2.9, which concerns the job applications of recent engineering school graduates. The letters I and G stand for getting a job in industry or getting a job with the government, and it follows from the Venn diagram that

$$P(I) = 0.18 + 0.12 = 0.30$$
$$P(G) = 0.12 + 0.24 = 0.36$$

and

$$P(I \cup G) = 0.18 + 0.12 + 0.24 = 0.54$$

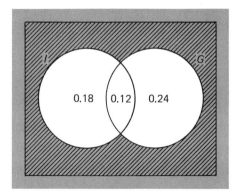

Figure 2.9 Venn diagram for example.

We were able to add the various probabilities because they represent mutually exclusive events.

Had we erroneously used the third axiom of probability to calculate $P(I \cup G)$, we would have obtained $P(I) + P(G) = 0.30 + 0.36$, which exceeds the correct value by 0.12. This error results from adding in $P(I \cap G)$ twice, once in $P(I) = 0.30$ and once in $P(G) = 0.36$, and we could correct for it by subtracting 0.12 from 0.66. Thus, we would get

$$P(I \cup G) = P(I) + P(G) - P(I \cap G)$$
$$= 0.30 + 0.36 - 0.12$$
$$= 0.54$$

and this agrees, as it should, with the result obtained before.

In line with this motivation, let us now state and prove the following theorem:

General addition rule

Theorem 2.6 *If A and B are any events in S, then*
$$P(A \cup B) = P(A) + P(B) - P(A \cap B)$$

To prove this theorem, we make use of the identities of parts (c) and (d) of Exercise 2.15, getting

$$P(A \cup B) = P(A \cap B) + P(A \cap B') + P(A' \cap B)$$
$$= [P(A \cap B) + P(A \cap B')] + [P(A \cap B) + P(A' \cap B)]$$
$$- P(A \cap B)$$
$$= P(A) + P(B) - P(A \cap B)$$

Note that when A and B are mutually exclusive so that $P(A \cap B) = 0$, Theorem 2.6 reduces to the third axiom of probability. For this reason, we sometimes refer to the third axiom of probability as the **special addition rule**.

EXAMPLE   With reference to the lawn-mower-rating example, find the probability that a lawn mower will be rated easy to operate and/or having a high average cost of repairs, namely, $P(E_1 \cup C_1)$.

> **Solution**   Making use of the results obtained on page 25, $P(E_1) = 0.40$, $P(C_1) = 0.30$, and $P(E_1 \cap C_1) = 0.12$, we substitute into the formula of Theorem 2.6 and get
>
> $$P(E_1 \cup C_1) = 0.40 + 0.30 - 0.12$$
> $$= 0.58$$

EXAMPLE   If the probabilities are 0.87, 0.36, and 0.29 that a family, randomly chosen as part of a sample survey in a large metropolitan area, owns a color television set, a black-and-white set, or both, what is the probability that a family in this area will own one or the other or both kinds of sets?

> **Solution**   Substituting these given values into the formula of Theorem 2.6, we get
>
> $$0.87 + 0.36 - 0.29 = 0.94$$

Note that the general addition rule, Theorem 2.6, can be generalized further so that it applies to more than two events (see Exercise 2.55).

Using the axioms of probability, we can derive many other theorems which play important roles in applications. For instance, let us show that

Probability of complement

> **Theorem 2.7**   *If A is any event in S, then $P(A') = 1 - P(A)$.*

To prove this theorem, we make use of the fact that A and A' are mutually exclusive by definition, and that $A \cup A' = S$ (namely, that among them A and A' contain all the elements in S). Hence, we can write

$$P(A) + P(A') = P(A \cup A')$$
$$= P(S)$$
$$= 1$$

so that $P(A') = 1 - P(A)$. As a special case we find that $P(\varnothing) = 1 - P(S) = 0$, since the empty set $\varnothing$ is the complement of S.

EXAMPLE Referring again to the lawn-mower-rating example and the results on page 25, find

(a) the probability that a lawn mower will not be rated easy to operate;

(b) the probability that a lawn mower will be rated as either not being easy to operate or not having a high average repair cost.

Solution (a) $P(E_1') = 1 - P(E_1) = 1 - 0.40 = 0.60$. (b) Since $E_1' \cup C_1' = (E_1 \cap C_1)'$ according to the identity of part (a) of Exercise 2.15, we get $P(E_1' \cup C_1') = P[(E_1 \cap C_1)'] = 1 - P(E_1 \cap C_1) = 1 - 0.12 = 0.88$.

EXERCISES

2.31 If a card is drawn from a well-shuffled deck of 52 playing cards, what is the probability of drawing
(a) a red king;
(b) a 3, 4, 5, or 6;
(c) a black card;
(d) a red ace or a black queen?

2.32 When we roll a pair of balanced dice, what are the probabilities of getting
(a) 7;
(b) 11;
(c) 7 or 11;
(d) 3;
(e) 2 or 12;
(f) 2, 3, or 12?

2.33 A lottery sells tickets numbered from 00001 through 50000. What is the probability of drawing a number that is divisible by 200?

2.34 A car rental agency has 18 compact cars and 12 intermediate-size cars. If four of the cars are randomly selected for a safety check, what is the probability of getting two of each kind?

2.35 In a sample of 446 cars stopped at a roadblock, only 67 of the drivers had their seatbelts fastened. Estimate the probability that a driver stopped on that road will have his or her seatbelt fastened.

2.36 Among 842 armed robberies in a certain city, 143 were never solved. Estimate the probability that an armed robbery in this city will be solved.

2.37 In a group of 160 graduate engineering students, 92 are enrolled in an advanced course in statistics, 63 are enrolled in a course in operations research, and 40 are enrolled in both. How many of these students are not enrolled in either course?

2.38 The personnel manager of a manufacturing plant claims that among the 400 employees 312 got a raise in 1981, 248 got increased pension benefits, 173 got both, and 43 got neither. Explain why this claim should be questioned.

2.39 Among 150 persons interviewed as part of an urban mass transportation study, some live more than 3 miles from the center of the city (*A*), some now regularly drive their own car to work (*B*), and some would gladly switch to

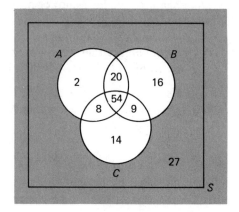

Figure 2.10 Diagram for Exercise 2.39.

public mass transportation if it were available (C). Use the information given in Figure 2.10 to find

(a) $N(A)$;   (f) $N(A \cap B \cap C)$;
(b) $N(B)$;   (g) $N(A \cup B)$;
(c) $N(C)$;   (h) $N(B \cup C)$;
(d) $N(A \cap B)$;   (i) $N(A' \cup B' \cup C)$;
(e) $N(A \cap C)$;   (j) $N[B \cap (A \cup C)]$.

2.40 An experiment has the four possible mutually exclusive outcomes A, B, C, and D. Check whether the following assignments of probability are permissible:
(a) $P(A) = 0.38$, $P(B) = 0.16$, $P(C) = 0.11$, $P(D) = 0.35$;
(b) $P(A) = 0.31$, $P(B) = 0.27$, $P(C) = 0.28$, $P(D) = 0.16$;
(c) $P(A) = 0.32$, $P(B) = 0.27$, $P(C) = -0.06$, $P(D) = 0.47$;
(d) $P(A) = \frac{1}{2}$, $P(B) = \frac{1}{4}$, $P(C) = \frac{1}{8}$, $P(D) = \frac{1}{16}$;
(e) $P(A) = \frac{5}{18}$, $P(B) = \frac{1}{6}$, $P(C) = \frac{1}{3}$, $P(D) = \frac{2}{9}$.

2.41 With reference to Exercise 2.1, suppose that the points $(0, 0)$, $(0, 1)$, $(0, 2)$, $(1, 0)$, $(1, 1)$, $(1, 2)$, $(2, 0)$, $(2, 1)$, $(2, 2)$, $(3, 0)$, $(3, 1)$, and $(3, 2)$ have the probabilities 0.001, 0.002, 0.006, 0.001, 0.017, 0.078, 0.004, 0.069, 0.311, 0.004, 0.092, and 0.415.
(a) Verify that this assignment of probabilities is permissible.
(b) Find the probabilities of events R, T, and U given in part (b) of that exercise.
(c) Calculate the probabilities that zero, one, or two carbon dioxide lasers will be suitable for the given task.

2.42 With reference to Exercise 2.7, suppose that each point (i, j) of the sample space is assigned the probability $\dfrac{15/28}{i+j}$.
(a) Verify that this assignment of probabilities is permissible.
(b) Find the probabilities of events B, C, and D described in part (b) of that exercise.

    (c) Find the probabilities that one, two, or three of the graduate students will be supervising the physics lab.

2.43 Explain why there must be a mistake in each of the following statements:
    (a) The probability that a mineral sample will contain silver is 0.38 and the probability that it will not contain silver is 0.52.
    (b) The probability that a drilling operation will be a success is 0.34 and the probability that it will not be a success is −0.66.
    (c) An air-conditioning repairman claims that the probability is 0.82 that the compressor is all right, 0.64 that the fan motor is all right, and 0.41 that they are both all right.

2.44 Refer to parts (c) and (d) of Exercise 2.15 to show that
    (a) $P(A \cap B) \leq P(A)$;
    (b) $P(A \cup B) \geq P(A)$.

2.45 Explain why there must be a mistake in each of the following statements:
    (a) The probability that a student will get an A in a geology course is 0.32, and the probability that he or she will get either an A or a B is 0.27.
    (b) A company is working on the construction of two shopping centers; the probability that the larger one will be completed on time is 0.35 and the probability that both will be completed on time is 0.42.

2.46 If A and B are mutually exclusive events, $P(A) = 0.29$, and $P(B) = 0.43$, find
    (a) $P(A')$;   (c) $P(A \cap B')$;
    (b) $P(A \cup B)$;   (d) $P(A' \cap B')$.

2.47 If the probabilities that a certain missile will explode during lift-off or have its guidance system fail in flight are 0.0002 and 0.0005, find the probabilities that such a missile will
    (a) not explode during lift-off;
    (b) explode during lift-off or have its guidance system fail in flight;
    (c) neither explode during lift-off nor have its guidance system fail in flight.

2.48 With reference to Exercise 2.39, suppose that the questionnaire filled in by one of the 150 persons is to be double-checked. If it is chosen in such a way that each questionnaire has a probability of $\frac{1}{150}$ of being selected, find the probabilities that the person
    (a) lives more than 3 miles from the center of the city;
    (b) regularly drives his or her car to work;
    (c) does not live more than 3 miles from the center of the city and would not want to switch to public mass transportation if it were available;
    (d) regularly drives his or her car to work but would gladly switch to public mass transportation if it were available.

2.49 A police department needs new tires for its patrol cars and the probabilities are 0.17, 0.22, 0.03, 0.29, 0.21, and 0.08 that it will buy Uniroyal tires, Goodyear tires, Michelin tires, General tires, Goodrich tires, or Armstrong tires. Find the probabilities that it will buy
    (a) Goodyear or Goodrich tires;
    (b) Uniroyal, General, or Goodrich tires;
    (c) Michelin or Armstrong tires;
    (d) Goodyear, General, or Armstrong tires.

2.50 The probabilities that a TV station will receive 0, 1, 2, 3, . . . , 8, or at least 9 complaints after showing a controversial program are, respectively, 0.01, 0.03, 0.07, 0.15, 0.19, 0.18, 0.14, 0.12, 0.09, and 0.02. What are the probabilities that after showing such a program the station will receive
(a) at most 4 complaints;
(b) at least 6 complaints;
(c) from 5 to 8 complaints?

2.51 If each point of the sample space of Figure 2.11 represents an outcome having the probability $\frac{1}{32}$, find

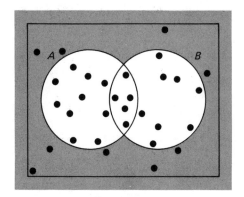

Figure 2.11 Diagram for Exercise 2.51.

(a) $P(A)$;   (d) $P(A \cup B)$;
(b) $P(B)$;   (e) $P(A' \cap B)$;
(c) $P(A \cap B)$;   (f) $P(A' \cap B')$.

2.52 The probability that a person stopping at a gas station will ask to have his tires checked is 0.12, the probability that he will ask to have his oil checked is 0.29, and the probability that he will ask to have them both checked is 0.07.
(a) What is the probability that a person stopping at this gas station will have either his tires or his oil checked?
(b) What is the probability that a person stopping at this gas station will have neither his tires nor his oil checked?

2.53 The probability that a new airport will get an award for its design is 0.16, the probability that it will get an award for the efficient use of materials is 0.24, and the probability that it will get both awards is 0.11.
(a) What is the probability that it will get at least one of the two awards?
(b) What is the probability that it will get only one of two awards?

2.54 Given $P(A) = 0.35$, $P(B) = 0.73$, and $P(A \cap B) = 0.14$, find
(a) $P(A \cup B)$;   (c) $P(A \cap B')$;
(b) $P(A' \cap B)$;   (d) $P(A' \cup B')$.

2.55 It can be shown that for any three events A, B, and C, the probability that at least one of them will occur is given by

$$P(A \cup B \cup C) = P(A) + P(B) + P(C) - P(A \cap B) - P(A \cap C)$$
$$- P(B \cap C) + P(A \cap B \cap C)$$

Verify that this formula holds for the probabilities of Figure 2.12.

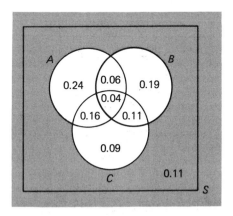

Figure 2.12 Diagram for Exercise 2.55.

2.56 Suppose that in the maintenance of a large medical-records file for insurance purposes the probability of an error in processing is 0.0010, the probability of an error in filing is 0.0009, the probability of an error in retrieving is 0.0012, the probability of an error in processing as well as filing is 0.0002, the probability of an error in processing as well as retrieving is 0.0003, the probability of an error in filing as well as retrieving is 0.0003, and the probability of an error in processing, filing, as well as retrieving is 0.0001. What is the probability of making at least one of these errors?

2.57 If the probability of event A is p, then the **odds** that it will occur are given by the ratio of p to $1 - p$. Odds are usually given as a ratio of two positive integers having no common factor, and if an event is more likely not to occur than to occur, it is customary to give the odds that it will not occur rather than the odds that it will occur. What are the odds for or against the occurrence of an event if its probability is
(a) $\frac{4}{7}$;
(b) 0.05;
(c) 0.80?

2.58 Use the definition of Exercise 2.57 to show that if the odds for the the occurrence of event A are a to b, where a and b are positive integers, then $p = \dfrac{a}{a + b}$.

2.59 The formula of Exercise 2.58 is often used to determine subjective probabilities. For instance, if an applicant for a job "feels" that the odds are 7 to 4 that she will get the job, the subjective probability she assigns to her getting the job is $p = \dfrac{7}{7 + 4} = \dfrac{7}{11}$.

(a) If a businessperson feels that the odds are 3 to 2 that a new venture will succeed (say, by betting $300 against $200 that it will succeed), what subjective probability is he or she assigning to its success?

(b) If a student is willing to bet $30 against $10, but not $40 against $10 that he will get a passing grade in a certain course, what does this tell us about the subjective probability he assigns to his getting a passing grade in the course?

2.60 Subjective probabilities may or may not satisfy the third axiom of probability. When they do, we say that they are **consistent**; when they do not, they ought not to be taken too seriously.

(a) The supplier of delicate optical equipment feels that the odds are 7 to 5 against a shipment arriving late and 11 to 1 against it not arriving at all. Furthermore, he feels that there is a fifty-fifty chance (the odds are 1 to 1) that such a shipment will either arrive late or not at all. Are the corresponding probabilities consistent?

(b) There are two Ferraris in a race, and an expert feels that the odds against their winning are, respectively, 2 to 1 and 3 to 1. Furthermore, she claims that there is a less-than-even chance that either of the two Ferraris will win. Discuss the consistency of these claims.

2.6 CONDITIONAL PROBABILITY

As we have defined probability, it is meaningful to ask for the probability of an event only if we refer to a given sample space S. To ask for the probability that an engineer earns at least $30,000 a year is meaningless unless we specify whether we are referring to all engineers in the western hemisphere, all engineers in the United States, all those in a particular industry, all those affiliated with a university, and so forth. Thus, when we use the symbol $P(A)$ for the probability of A, we really mean the probability of A given some sample space S. Since the choice of S is by no means always evident, and since there are problems in which we are interested in the probabilities of A with respect to more sample spaces than one, the notation $P(A \mid S)$ is used to make it clear that we are referring to a particular sample space S. We read $P(A \mid S)$ as the conditional probability of A relative to S, and every probability is thus a conditional probability. Of course, we use the simplified notation $P(A)$ whenever the choice of S is clearly understood.

To illustrate some of the ideas connected with conditional probabilities, let us consider again the 500 machine parts of which some are improperly assembled and some contain one or more defective components as shown in Figure 2.7. Assuming equal probabilities in the selection of one of the machine parts for inspection, it can be seen that the probability of getting one with one or more defective components is $P(D) = \dfrac{10 + 5}{500} = \dfrac{3}{100}$. To check whether the probability is the same if the choice is restricted to the machine parts that are improperly assembled, we have only to look at the reduced sample space of Figure 2.13 and assume that each of the 30 improperly assembled parts has the same chance of being selected. We thus get

$$P(D|I) = \frac{N(D \cap I)}{N(I)} = \frac{10}{30} = \frac{1}{3}$$

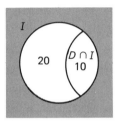

Figure 2.13 Reduced sample space.

and it can be seen that the probability of getting a machine part with one or more defective components has increased from $\frac{3}{100}$ to $\frac{1}{3}$. Note that if we divide the numerator and denominator of the above formula for $P(D|I)$ by $N(S)$, we get

$$P(D|I) = \frac{\dfrac{N(D \cap I)}{N(S)}}{\dfrac{N(I)}{N(S)}} = \frac{P(D \cap I)}{P(I)}$$

where $P(D|I)$ is given by the ratio of $P(D \cap I)$ to $P(I)$.

Looking at this example in another way, note that with respect to the whole sample space S we have

$$P(D \cap I) = \tfrac{10}{500} = \tfrac{1}{50} \quad \text{and} \quad P(D' \cap I) = \tfrac{20}{500} = \tfrac{2}{50}$$

assuming, as before, that each of the 500 machine parts has the same chance of being selected. Thus, the probabilities that the machine part selected will or will not contain one or more defective components, given that it is improperly assembled, should be in the ratio 1 to 2. Since the probabilities of D and D' in the reduced sample space must add up to 1, it follows that

$$P(D|I) = \tfrac{1}{3} \quad \text{and} \quad P(D'|I) = \tfrac{2}{3}$$

and this agrees with the result obtained before. This explains why we had to divide by $P(I)$ when we wrote

$$P(D|I) = \frac{P(D \cap I)}{P(I)}$$

above—division by $P(I)$, or multiplication by $\frac{1}{P(I)}$, takes care of the proportionality factor which makes the sum of the probabilities over the reduced sample space equal to 1.

Following these observations, let us now make the following general definition:

Conditional
probability

> *If A and B are any events in S and $P(B) \neq 0$, the conditional probability of A relative to B is given by*
>
> $$P(A|B) = \frac{P(A \cap B)}{P(B)}$$

EXAMPLE If the probability that a communication system will have high fidelity is 0.81, and the probability that it will have high fidelity and high selectivity is 0.18, what is the probability that a system with high fidelity will also have high selectivity?

> **Solution** If A is the event that a communication system has high selectivity and B is the event that it has high fidelity, we have $P(B) = 0.81$ and $P(A \cap B) = 0.18$, and substitution into the formula yields
>
> $$P(A|B) = \frac{0.18}{0.81} = \frac{2}{9}$$

EXAMPLE If the probability that a research project will be well planned is 0.80 and the probability that it will be well planned and well executed is 0.72, what is the probability that a research project which is well planned will also be well executed?

> **Solution** Substitution into the formula for a conditional probability yields $\frac{0.72}{0.80} = 0.90$.

EXAMPLE Referring to the lawn-mower-rating example, for which the probabilities of the individual outcomes are given in Figure 2.8, use the results obtained on page 25 to find $P(E_1|C_1)$.

Solution Since we had $P(E_1 \cap C_1) = 0.12$ and $P(C_1) = 0.30$, substitution into the formula for a conditional probability yields

$$P(E_1|C_1) = \frac{0.12}{0.30} = 0.40$$

It is of interest to note that the value obtained here for $P(E_1|C_1)$ equals that obtained on page 25 for $P(E_1)$. This means that the probability of a lawn mower being rated easy to repair is the same regardless of whether or not it is given that it has a high average cost of repairs, and we say that E_1 is **independent** of C_1. As the reader will be asked to verify in Exercise 2.65 on page 43, it also follows from the results on page 25 that E_1 is not independent of P_1, namely, that a lawn mower's ease of operation is related to its retail price.

In general, if A and B are any two events in a sample space S, we say that A is independent of B if and only if $P(A|B) = P(A)$, but as it can be shown that B is independent of A whenever A is independent of B, it is customary to say simply that **A and B are independent**.

The following theorem is an immediate consequence of the definition of conditional probability:

General multi-
plication rule

Theorem 2.8 *If A and B are any events in S, then*

$$\begin{aligned}
P(A \cap B) &= P(A) \cdot P(B|A) &&\text{if } P(A) \neq 0 \\
&= P(B) \cdot P(A|B) &&\text{if } P(B) \neq 0
\end{aligned}$$

The second of these rules is obtained directly from the definition of conditional probability by multiplying both sides by $P(B)$; the first is obtained from the second by interchanging the letters A and B.

EXAMPLE The foreman of a group of 20 construction workers wants to get the opinion of 2 of them (to be selected at random) about certain new safety regulations. If 12 of them favor the new regulations and the other 8 are against it, what is the probability that both of the workers chosen by the foreman will be against the new safety regulations?

Solution Assuming equal probabilities for each selection (which is what we mean by the selections being random), the probability that the first worker selected will be against the new safety regulations is $\frac{8}{20}$, and the probability that the second worker selected will be against the new safety regulations given that the first one is against them is $\frac{7}{19}$. Thus, the desired probability is $\frac{8}{20} \cdot \frac{7}{19} = \frac{14}{95}$.

In the special case where A and B are independent, Theorem 2.8 leads to the following result:

Special multi-plication rule

> **Theorem 2.9** *If A and B are independent events, then*
> $$P(A \cap B) = P(A) \cdot P(B)$$

Thus, the probability that two independent events will both occur is simply the product of their probabilities. This rule is sometimes used as the definition of independence; in any case, it may be used to determine whether two given events are independent.

EXAMPLE What is the probability of getting two heads in two flips of a balanced coin?

Solution Since the probability of heads is $\frac{1}{2}$ for each flip and the two flips are independent, the probability is $\frac{1}{2} \cdot \frac{1}{2} = \frac{1}{4}$.

EXAMPLE Two cards are drawn at random from an ordinary deck of 52 playing cards. What is the probability of getting two aces if
    (a) the first card is replaced before the second card is drawn;
    (b) the first card is not replaced before the second card is drawn?

Solution (a) Since there are four aces among the 52 cards, we get $\frac{4}{52} \cdot \frac{4}{52} = \frac{1}{169}$. (b) Since there are only three aces among the 51 cards which remain after one ace has been removed from the deck, we get $\frac{4}{52} \cdot \frac{3}{51} = \frac{1}{221}$.

EXAMPLE If $P(C) = 0.65$, $P(D) = 0.40$, and $P(C \cap D) = 0.24$, are the events C and D independent?

Solution Since $P(C) \cdot P(D) = (0.65)(0.40) = 0.26$ and not 0.24, the two events are not independent.

The special multiplication rule can easily be extended so that it applies to more than two independent events—again, we multiply together all the individual probabilities.

EXAMPLE What is the probability of not rolling any 6's in four rolls of a balanced die?

Solution The probability is $\dfrac{5}{6} \cdot \dfrac{5}{6} \cdot \dfrac{5}{6} \cdot \dfrac{5}{6} = \dfrac{625}{1,296}$.

For three or more dependent events the multiplication rule becomes more complicated, as is illustrated in Exercise 2.77 on page 44.

2.7 BAYES' THEOREM

The general multiplication rules are useful in solving many problems in which the ultimate outcome of an experiment depends on the outcomes of various intermediate stages. Suppose, for instance, that an assembly plant receives its voltage regulators from three different suppliers, 60% from supplier B_1, 30% from supplier B_2, and 10% from supplier B_3. In other words, the probabilities that any one voltage regulator received by the plant comes from these three suppliers are 0.60, 0.30, and 0.10. If 95% of the voltage regulators from B_1, 80% of those from B_2, and 65% of those from B_3 perform according to specifications, what we would like to know is the probability that any one voltage regulator received by the plant will perform according to specifications.

If A denotes the event that a voltage regulator received by the plant performs according to specifications, and B_1, B_2, and B_3 are the events that it comes from the respective suppliers, we can write

$$A = A \cap [B_1 \cup B_2 \cup B_3]$$
$$= (A \cap B_1) \cup (A \cap B_2) \cup (A \cap B_3)$$

and

$$P(A) = P(A \cap B_1) + P(A \cap B_2) + P(A \cap B_3)$$

since $A \cap B_1$, $A \cap B_2$, and $A \cap B_3$ are mutually exclusive. Then, if we apply the second of the general multiplication rules to $P(A \cap B_1)$, $P(A \cap B_2)$, and $P(A \cap B_3)$, we get

$$P(A) = P(B_1) \cdot P(A|B_1) + P(B_2) \cdot P(A|B_2) + P(B_3) \cdot P(A|B_3)$$

and substitution of the given numerical values yields

$$P(A) = (0.60)(0.95) + (0.30)(0.80) + (0.10)(0.65)$$
$$= 0.875$$

for the probability that any one voltage regulator received by the given plant will perform according to specifications.

To visualize this result, we have only to construct a tree diagram like that of Figure 2.14, where the probability of the final outcome is given by the sum of the products of the probabilities corresponding to each branch of the tree.

In the preceding example there were only three alternatives at the intermediate stage, but if there are n mutually exclusive alternatives B_1, B_2, ..., B_n at the intermediate stage, a similar argument will lead to the following result, sometimes called the **rule of elimination** or the **rule of total probability**:

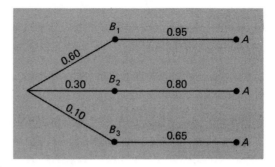

Figure 2.14 Tree diagram for example dealing with three suppliers of voltage regulators.

> **Theorem 2.10** *If $B_1, B_2, \ldots, B_n$ are mutually exclusive events of which one must occur, then*
>
> $$P(A) = \sum_{i=1}^{n} P(B_i) \cdot P(A \mid B_i)$$

To visualize this result, we have only to construct a tree diagram like that of Figure 2.15, where the probability of the final outcome is again given by the sum of the products of the probabilities corresponding to each branch of the tree.

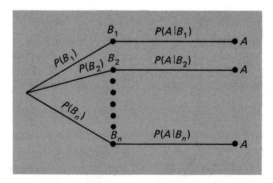

Figure 2.15 Tree diagram for rule of elimination.

To consider a problem which is closely related to the one we have just discussed, suppose we want to know the probability that a particular voltage regulator which is known to perform according to specifications came from supplier B_3. Symbolically, we want to know the value of $P(B_3 \mid A)$, and to find a formula for this probability we first write

$$P(B_3 | A) = \frac{P(A \cap B_3)}{P(A)}$$

Then, substituting $P(B_3) \cdot P(A | B_3)$ for $P(A \cap B_3)$ and $\sum_{i=1}^{3} P(B_i) \cdot P(A | B_i)$ for $P(A)$ in accordance with Theorems 2.8 and 2.10, we get the formula

$$P(B_3 | A) = \frac{P(B_3) \cdot P(A | B_3)}{\sum_{i=1}^{3} P(B_i) \cdot P(A | B_i)}$$

which expresses $P(B_3 | A)$ in terms of given probabilities. Substituting the values from page 39 (or from Figure 2.14), we finally obtain

$$P(B_3 | A) = \frac{(0.10)(0.65)}{(0.60)(0.95) + (0.30)(0.80) + (0.10)(0.65)}$$

$$= 0.074$$

Note that the probability that a voltage regulator is supplied by B_3 decreases from 0.10 to 0.074 once it is known that it performs according to specifications.

The method used to solve the preceding example can easily be generalized to yield the following formula:

Bayes' theorem

Theorem 2.11 *If B_1, B_2, . . . , B_n are mutually exclusive events of which one must occur, then*

$$P(B_r | A) = \frac{P(B_r) \cdot P(A | B_r)}{\sum_{i=1}^{n} P(B_i) \cdot P(A | B_i)}$$

for $r = 1, 2, \ldots$, or n.

Note that the expression in the numerator is the probability of reaching A via the ith branch of the tree, and that the expression in the denominator is the sum of the probabilities of reaching A via the n branches of the tree.

Bayes' theorem provides a formula for finding the probability that the "effect" A was "caused" by the event B_r. For instance, in our example we found the probability that an acceptable voltage regulator was made by supplier B_3. The probabilities $P(B_i)$ are called the "prior" or "a priori" probabilities of the "causes" B_i, and in practice it is often difficult to assign them numerical values. For many years Bayes' theorem was looked upon with suspicion because it was used with the often-erroneous assumption that the prior probabilities are all equal. A good deal of the controversy once surrounding Bayes' theorem has been cleared up with the realization that the probabilities $P(B_3)$ must be determined separately in each case from the nature of the problem, preferably on the basis of past experience. We shall return to this problem as it relates to Bayesian inference in Chapters 7 and 9.

EXAMPLE The four attendants of a gasoline service station are supposed to wash the windshield of each customer's car. Jack, who services 20% of all cars, fails to wash the windshield one time in 20; Tom, who services 60% of all cars, fails to wash the windshield one time in 10; George, who services 15% of all cars, fails to wash the windshield one time in 10; and Peter, who services 5% of all cars, fails to wash the windshield one time in 20. If a customer complains later that her windshield was not washed, what is the probability that her car was serviced by Jack?

Solution Substituting the various probabilities into the formula of Theorem 2.11, we get

$$P(B_1|A) = \frac{(0.20)(0.05)}{(0.20)(0.05) + (0.60)(0.10) + (0.15)(0.10) + (0.05)(0.05)}$$

$$= 0.114$$

and it is of interest to note that although Jack fails to wash the windshield of only 1 car in 20, namely, 5% of the cars, more than 11% of the windshields that are not washed at this service station are his responsibility.

EXERCISES

2.61 With reference to Figure 2.7, find $P(I|D)$ and $P(I|D')$, assuming that originally each of the 500 machine parts has the same chance of being chosen for inspection.

2.62 With reference to the example on page 36, the probability is 0.75 that the research project will be well executed. What is the probability that the project was well planned given that it is well executed?

2.63 With reference to Exercise 2.39 and Figure 2.10, assume that each of 150 persons has the same chance of being selected and find the probabilities that he or she
  (a) lives more than 3 miles from the center of the city given that he or she would gladly switch to public mass transportation;
  (b) regularly drives his or her car to work given that he or she lives more than 3 miles from the center of the city;
  (c) would not want to switch to public mass transportation given that he or she does not regularly drive his or her car to work.

2.64 With reference to Figure 2.12, find
  (a) $P(A|B)$;
  (b) $P(B|C')$;
  (c) $P(A \cap B|C)$;
  (d) $P(B \cup C|A')$;
  (e) $P(A|B \cup C)$;
  (f) $P(A|B \cap C)$;
  (g) $P(A \cap B \cap C|B \cap C)$;
  (h) $P(A \cap B \cap C|B \cup C)$.

2.65 With reference to the lawn-mower-rating example and the probabilities given in Figure 2.8, find
(a) $P(E_1|P_1)$ and compare its value with that of $P(E_1)$;
(b) $P(C_2|P_2)$ and compare its value with that of $P(C_2)$;
(c) $P(E_1|P_1 \cap C_1)$ and compare its value with that of $P(E_1)$.

2.66 With reference to Exercise 2.53, find the probabilities that the airport will get the design award given that
(a) it got the award for the efficient use of materials;
(b) it did not get the award for the efficient use of materials.

2.67 Prove that $P(A|B) = P(A)$ implies that $P(B|A) = P(B)$ provided that $P(A) \neq 0$ and $P(B) \neq 0$.

2.68 If the probabilities are 0.58, 0.25, and 0.19 that a person in a certain income bracket will invest in money market funds, common stocks, or both, find the probabilities that a person in that income bracket
(a) who invests in money market funds will also invest in common stocks;
(b) who invests in common stocks will also invest in money market funds.

2.69 Given $P(A) = 0.50$, $P(B) = 0.30$, and $P(A \cap B) = 0.15$, verify that
(a) $P(A|B) = P(A)$;   (c) $P(B|A) = P(B)$;
(b) $P(A|B') = P(A)$;   (d) $P(B|A') = P(B)$.

2.70 Among the 24 invoices prepared by a billing department, 4 contain errors while the others do not. If we randomly check two of these invoices, what are the probabilities that
(a) both will contain errors;
(b) neither will contain an error?

2.71 Among 60 automobile repair parts loaded on a truck in San Francisco, 45 are destined for Seattle and 15 for Vancouver. If two of the parts are unloaded in Portland by mistake and the "selection" is random, what are the probabilities that
(a) both parts should have gone to Seattle;
(b) both parts should have gone to Vancouver;
(c) one should have gone to Seattle and one to Vancouver?

2.72 For two rolls of a balanced die, find the probabilities of getting
(a) two 4's;
(b) first a 4 and then a number less than 4.

2.73 If events A and B are independent and $P(A) = 0.25$ and $P(B) = 0.40$, find
(a) $P(A \cap B)$;   (c) $P(A \cup B)$;
(b) $P(A|B)$;   (d) $P(A' \cap B')$.

2.74 If $P(A) = 0.80$, $P(C) = 0.35$, and $P(A \cap C) = 0.28$, are events A and C independent?

2.75 If the odds are 5 to 3 that event M will not occur, 2 to 1 that event N will occur, and 4 to 1 that they will not both occur, are the two events M and N independent?

2.76 Find the probabilities of getting
(a) eight heads in a row with a balanced coin;

(b) three 3's and then a 4 or a 5 in four rolls of a balanced die;

(c) five multiple-choice questions answered correctly, if for each question the probability of answering it correctly is $\frac{1}{3}$.

2.77 For three or more events which are not independent, the probability that they will all occur is obtained by multiplying the probability that one of the events will occur times the probability that a second of the events will occur given that the first event has occurred times the probability that a third of the events will occur given that the first two events have occurred, and so on. For instance, for three events we can write

$$P(A \cap B \cap C) = P(A) \cdot P(B|A) \cdot P(C|A \cap B)$$

and we find that the probability of drawing without replacement three aces in a row from an ordinary deck of 52 playing cards is

$$\frac{4}{52} \cdot \frac{3}{51} \cdot \frac{2}{50} = \frac{1}{5,525}$$

(a) If six bullets, of which three are blanks, are randomly inserted into a gun, what is the probability that the first three bullets fired will all be blanks?

(b) In a certain city during the month of May, the probability that a rainy day will be followed by another rainy day is 0.80 and the probability that a sunny day will be followed by a rainy day is 0.60. Assuming that each day is classified as being either rainy or sunny and that the weather on any given day depends only on the weather the day before, find the probability that in the given city a rainy day in May is followed by two more rainy days, then a sunny day, and finally another rainy day.

(c) A department store which bills its charge-account customers once a month has found that if a customer pays promptly one month, the probability is 0.90 that he will also pay promptly the next month; however, if a customer does not pay promptly one month, the probability that he will pay promptly the next month is only 0.50. What is the probability that a customer who has paid promptly one month will not pay promptly the next three months?

(d) If 5 of a company's 12 delivery trucks do not meet emission standards and 4 of the 12 trucks are randomly picked for inspection, what is the probability that none of them meets emission standards?

2.78 At an electronics plant, it is known from past experience that the probability is 0.86 that a new worker who has attended the company's training program will meet the production quota, and that the corresponding probability is 0.35 for a new worker who has not attended the company's training program. If 80% of all new workers attend the training program, what is the probability that a new worker will meet the production quota?

2.79 With reference to the preceding exercise, find the probability that a new worker who meets the production quota will have attended the company's training program.

2.80 A consulting firm rents cars from three agencies, 20% from agency *D*, 20%

from agency E, and 60% from agency F. If 10% of the cars from D, 12% of the cars from E, and 4% of the cars from F have bad tires, what is the probability that the firm will get a car with bad tires?

2.81 With reference to the preceding exercise, what is the probability that a car with bad tires rented by the firm came from agency F?

2.82 Use the information on the tree diagram of Figure 2.16 to determine the value of
(a) $P(A)$;
(b) $P(B|A)$;
(c) $P(B|A')$.

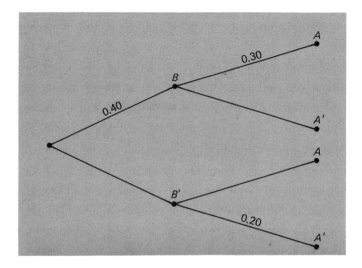

Figure 2.16 Diagram for Exercise 2.82.

2.83 With reference to the example on page 39, find the probabilities that a voltage regulator which performs according to specifications came
(a) from supplier B_1;   (b) from supplier B_2.

2.84 With reference to the example on page 42, find the probabilities that the car of the complaining customer was serviced by
(a) Tom;   (c) Peter.
(b) George;

2.85 Two firms V and W consider bidding on a road-building job which may or may not be awarded depending on the amounts of the bids. Firm V submits a bid and the probability is $\frac{3}{4}$ that it will get the job provided firm W does not bid. The odds are 3 to 1 that W will bid, and if it does, the probability that V will get the job is only $\frac{1}{3}$.
(a) What is the probability that V will get the job?

(b) If V gets the job, what is the probability that W did not oid?

2.86 An explosion in an LNG storage tank in the process of being repaired could have occurred as the result of static electricity, malfunctioning electrical equipment, an open flame in contact with the liner, or purposeful action (industrial sabotage). Interviews with engineers who were analyzing the risks involved led to estimates that such an explosion would occur with probability 0.25 as a result of static electricity, 0.20 as a result of malfunctioning electric equipment, 0.40 as a result of an open flame, and 0.75 as a result of purposeful action. These interviews also yielded subjective estimates of the prior probabilities of these four causes of 0.30, 0.40, 0.15, and 0.15, respectively. What was the most likely cause of the explosion?

2.8 MATHEMATICAL EXPECTATION AND DECISION MAKING

If an insurance agent tells us that in the United States a 40-year-old woman can expect to live 38 more years, this does not mean that anyone really expects a 40-year-old woman to live until her seventy-eighth birthday and then die the next day. Similarly, if we read that a radial tire can be expected to last 40,000 miles, that in the coal mining industry we can expect 41.5 nonfatal injuries per million man-hours, or that the U.S. Patent Office can expect to issue 27,000 certificates of trademarks each year, it must be understood that the word "expect" is not used in its colloquial sense. So far as the first statement is concerned, some 40-year-old women will live another 15 years, some will live another 30 years, some will live another 50 years, . . . , and the life expectance of "38 more years" will have to be interpreted as an average, namely, as a **mathematical expectation**.

Originally, the concept of a mathematical expectation arose in connection with games of chance, and in its simplest form it is the product of the amount a player stands to win and the probability that he or she will win.

EXAMPLE What is our mathematical expectation if we stand to win $8 if and only if a balanced coin comes up heads?

Solution The probability of heads is $\frac{1}{2}$ and our mathematical expectation is $8 \cdot \frac{1}{2} = \$4$.

EXAMPLE What is our mathematical expectation if we buy one of 1,000 raffle tickets for a grand prize of $500.

Solution The probability of winning the grand prize is $\frac{1}{1,000}$ and the mathematical expectation is $500 \cdot \frac{1}{1,000} = \$0.50$.

In both of these examples there was a single prize, but in each case there were two possible payoffs—$8 or $0 in the first example and $500 or $0 in the other. Indeed, in the second example we can argue that 999 of the tickets will pay $0 and one of the tickets will pay $500; altogether, the 1,000 tickets will thus pay $500, or on the average $0.50 per ticket, and this is the mathematical expectation.

To generalize the concept of a mathematical expectation, let us consider the following example:

EXAMPLE A distributor makes a profit of $20 on an item if it is shipped from the factory in perfect condition and arrives on time, but it is reduced by $2 if it does not arrive on time, and by $12 regardless of whether it arrives on time if it is not shipped from the factory in perfect condition. If 70% of such items are shipped in perfect condition and arrive on time, 10% are shipped in perfect condition but do not arrive on time, and 20% are not shipped in perfect condition, what is the distributor's expected profit per item?

> **Solution** We can argue that the distributor will make the $20 profit 70% of the time, the $18 profit 10% of the time, and the $8 profit 20% of the time, so that the average profit per item (the expected profit) is
>
> $$20(0.70) + 18(0.10) + 8(0.20) = \$17.40$$
>
> This result is the sum of the products obtained by multiplying each amount by the corresponding proportion or probability.

Generalizing from this example, let us now make the following definition:

Mathematical expectation

> *If the probabilities of obtaining the amounts* $a_1, a_2, \ldots,$ *or* a_k *are* $p_1, p_2,$ *$\ldots,$ and p_k, then the mathematical expectation is*
> $$E = a_1 p_1 + a_2 p_2 + \ldots + a_k p_k$$

As far as the a's are concerned, it is important to keep in mind that they are positive when they represent profits, winnings, or gains and that they are negative when they represent losses, deficits, or penalties.

EXAMPLE An engineering firm is faced with the task of preparing a proposal for a research contract. The cost of preparing the proposal is $5,000 and the probabilities for potential gross profits of $50,000, $30,000, $10,000, or $0 are 0.20, 0.50, 0.20, and 0.10, provided that the proposal is accepted. If the probability is 0.30 that the firm's proposal will be accepted, what is its expected net profit?

> **Solution** The probability that the firm will make a net profit of $45,000 ($50,000 minus the cost of the proposal) is $(0.30)(0.20) = 0.06$. Similarly, the probabilities that the firm will make net profits of $25,000 or $5,000 are $(0.30)(0.50) = 0.15$ and $(0.30)(0.20) = 0.06$, while the

probability of a $5,000 loss is $(0.30)(0.10) + 0.70 = 0.73$, allowing for the possibility that the proposal will not be accepted. Thus, the expected net profit is

$$45,000(0.06) + 25,000(0.15) + 5,000(0.06) - 5,000(0.73) = \$3,100$$

Whether or not it is wise to risk $5,000 to make an expected profit of $3,100 is not a question of statistics. If a company has little capital and can make only few such proposals, the 0.73 chance of losing $5,000 may well make the venture unattractive. On the other hand, if a company is well capitalized and prepares many such proposals, the risk may be worth taking, for it promises an expected return of 62% on the original investment.

The following is a typical inventory problem in which expected values are used to determine optimum conditions:

EXAMPLE It is known from past experience that the daily demand for a perishable product is as shown in the following table:

Number of orders	3	4	5	6	7	8	9
Probability	0.05	0.12	0.20	0.24	0.17	0.14	0.08

If each item costs $35 (including the cost of carrying it in stock), it sells for $50 provided that it is in stock, and it represents a total loss if it remains in stock at the end of a day, how many items should be stocked each day so as to maximize the expected profit?

Solution If three items are stocked, the profit is $150 - 105 = \$45$, since there is a probability of 1 that there will be a demand for three or more. If four items are stocked, there is a probability of 0.05 that three items will be sold, a probability of 0.95 that there will be a demand for four or more, and the expected profit is

$$150(0.05) + 200(0.95) - 140 = \$57.50$$

Similarly, if five items are stocked, there is a probability of 0.05 that three items will be sold, a probability of 0.12 that four items will be sold, a probability of 0.83 that there will be a demand for five or more, and the expected profit is

$$150(0.05) + 200(0.12) + 250(0.83) - 175 = \$64.00$$

Continuing in this way (see Exercise 2.94), it can be shown that if six items are stocked the expected profit is $60.50, if seven items are stocked the expected profit is $45.00, and that the expected profit is a maximum when the number of items stocked is five.

In recent years, mathematical expectations have been playing an increasingly important role in scientific decision making, as it is generally considered rational to select whichever alternative has the most promising mathematical expectation: the one which maximizes the expected profit, minimizes the expected cost, maximizes tax advantages, minimizes the expected loss, and so forth.

In applying the methods of decision making illustrated in this section, there are essentially two practical limitations: first, we must be able to assign "cash values" to the various outcomes, and second, we must have adequate information concerning all relevant probabilities. The problem of assigning cash values to the various outcomes can be much more difficult than it was in our examples. For instance, if a government agency were faced with the decision of whether to require crash barriers for interstate highway construction, it would be extremely difficult to assign a cash value to the savings in life and reductions in injuries that might result.

The specification of probabilities concerning the various outcomes can be equally difficult. In our examples, the probabilities were presumably based on past experience, and, hence, are really only estimates (or approximations). This may be questioned with respect to the example dealing with the preparation of the research proposal, where it would be extremely difficult even to estimate the probability that a given proposal will be accepted. This would have to depend not only on the quality of the proposal itself, but on such unknown factors as the number and quality of competing proposals as well as the policy of the sponsoring agency. The estimate of 0.30 in our example was purely subjective, being based on some executive's "feeling" about the situation measured against the background of his personal experience. While such estimates of probabilities are not without value, they do leave open the possibility that different individuals (acting in the same situation and having the same information) would arrive at different "optimum" decisions.

EXERCISES

2.87 If a service club sells 4,000 raffle tickets for a cash prize of $800, what is the mathematical expectation of a person who buys one of the tickets?

2.88 A charitable organization raises funds by selling 2,000 raffle tickets for a first prize worth $500 and a second prize worth $100. What is the mathematical expectation of a person who buys one of the tickets?

2.89 What is our mathematical expectation if we win $10 if a balanced coin comes up heads and lose $10 if it comes up tails?

2.90 A game between two players is fair if each player has the same mathematical expectation. If someone gives us $5 each time we roll a 1 or a 2 with a balanced die, how much must we pay him each time we roll a 3, 4, 5, or 6 to make the game fair?

2.91 When the American League and National League champions are evenly matched, the probabilities that a World Series will end in 4, 5, 6, or 7 games are, respectively, $\frac{1}{8}$, $\frac{1}{4}$, $\frac{5}{16}$, and $\frac{5}{16}$. What is the expected length of a World Series when the two teams are evenly matched?

2.92 It costs $60 to test a certain component of a machine. If a defective component is installed, it costs $1,200 to repair the resulting damage to the machine. Is it more profitable to install the component without testing if it is known that

(a) 3% of all components produced are defective;

(b) 5% of all components produced are defective;

(c) 8% of all components produced are defective?

2.93 The two finalists in a golf tournament play 18 holes, with the winner getting $50,000 and the runner-up getting $30,000. What are the two players' mathematical expectations if

(a) they are evenly matched;

(b) one of the two players should be favored by odds of 3 to 1?

2.94 With reference to the example on page 48, show that for six and seven items stocked, the expected profits are $60.50 and $45.00.

2.95 To handle a liability suit, a lawyer has to decide whether to charge a straight fee of $3,000 or a contingent fee of $12,000, which she will get only if her client wins. What does the lawyer think about her client's chances if she feels that

(a) taking the straight fee will give her a higher mathematical expectation;

(b) taking the contingent fee will give her a higher mathematical expectation?

2.96 A union wage negotiator feels that the probabilities are 0.40, 0.30, 0.20, and 0.10 that the union members will get a $1.50 an hour raise, a $1.00 an hour raise, a 50-cent an hour raise, or no raise at all. What is their expected raise?

2.97 An importer is offered a shipment of machine tools for $140,000, and the probabilities that he will be able to sell them for $180,000, $170,000, or $150,000 are 0.32, 0.55, and 0.13. What is the importer's expected gross profit?

2.98 The manufacturer of a new battery additive has to decide whether to sell his product for $0.80 a can, or for $1.20 with a "double-your-money-back-if-not-satisfied" guaranty. How does he feel about the chances that a person will ask for double his money back if

(a) he decides to sell the product for $0.80;

(b) he decides to sell the product for $1.20 with the guarantee;

(c) he cannot make up his mind?

2.99 A contractor has to choose between two jobs. The first job promises a profit of $240,000 with a probability of 0.75 or a loss of $60,000 (due to strikes and other delays) with a probability of 0.25; the second job promises a profit of $360,000 with a probability of 0.50 or a loss of $90,000 with a probability of 0.50.

(a) Which job should the contractor choose if he wants to maximize his expected profit?

(b) Which job would the contractor probably choose if his business is in bad shape and he will go broke unless he can make a profit of at least $300,000 on his next job?

2.100 A merchant can buy an item for $2.10 and sell it for $4.50. The probabilities for a demand of 0, 1, 2, 3, 4, or "5 or more" items are, respectively, 0.05, 0.15, 0.30, 0.25, 0.15, and 0.10. Calculate the expected profit resulting from stocking 0, 1, 2, 3, 4, or 5 items and determine how many the merchant should stock so as to maximize his expected profit.

2.101 Mr. Brown and Mr. Jones are betting on repeated flips of a balanced coin. At the beginning Mr. Brown has m dollars, Mr. Jones has n dollars; at each flip the loser pays the winner one dollar, and they continue playing until one of them has lost all the money with which he began. To find p, the probability that Mr. Brown will win Mr. Jones' n dollars before he loses his m dollars, set up an equation in m, n, and p which makes use of the fact that in a fair game each player's mathematical expectation is the same. What is the probability that Mr. Brown will win if he begins with $12 and Mr. Jones begins with $4?

2.102 The management of a mining company must decide whether to continue an operation at a certain location. If they continue and are successful, this will be worth $1,000,000; if they continue and are not successful, this will entail a loss of $600,000; if they do not continue but would have been successful, this will entail a loss of $400,000 (for competitive reasons); and if they do not continue and would not have been successful anyhow, this will be worth $100,000 to the company (because funds allocated to the operation remain unspent). What decision would maximize the company's expected profit if the probabilities for and against success are 0.40 and 0.60?

2.103 With reference to the preceding exercise, what decision would maximize the company's expected profit if the probabilities for and against success are 0.20 and 0.80?

2.104 With reference to Exercise 2.102, suppose that it is possible to continue the operation for two months at a cost of $300,000, after which it will be possible to know for certain whether the operation will be a success. Would it be worthwhile to spend these additional funds?

2.105 A power company is faced with the choice of whether to construct a light-water reactor (LWR) or a fossil-fuel power plant (FF). The LWR plant will cost $300 per kilowatt to construct, and the FF plant will cost $150 per kilowatt. Due to uncertainties concerning fuel availability and the impact of future air- and water-quality regulations, the useful operating life of each plant is unknown, but the following probabilities have been estimated:

Useful life (years)	10	20	30	40
Probabilities $\begin{cases} LWR\ plant \\ FF\ plant \end{cases}$	0.05 0.10	0.25 0.50	0.50 0.30	0.20 0.10

(a) Which plant can be expected to cost less per year of useful life?

(b) What relative fuel cost per kilowatt would favor construction of the LWR plant?

(c) A further uncertainty concerning safety regulations makes it a fifty-fifty bet that $50 per kilowatt will be added to the actual construction cost of the LWR plant. What effect does this have on these results?

2.9 CHECK LIST OF KEY TERMS *(with page references)*

Additive set function, 21
Axioms of probability, 23
Bayes' theorem, 41
Classical probability concept, 19
Combination, 15
Complement, 8
Conditional probability, 36
Consistency of probabilities, 34
Continuous sample space, 7
Discrete sample space, 7
Empty set, 7
Event, 7
Experiment, 6
Factorial notation, 13
Finite sample space, 6
Frequency interpretation, 21
General addition rule, 27
General multiplication rule, 37

Independent events, 37
Intersection, 8
Mathematical expectation, 46
Multiplication of choices, 12
Mutually exclusive events, 8
Odds, 33
Outcome, 6
Permutation, 13
Rule of elimination, 39
Rule of total probability, 39
Sample space, 6
Set function, 22
Special addition rule, 28
Special multiplication rule, 38
Subjective probability, 21
Tree diagram, 11
Union, 8
Venn diagram, 8

3

Probability Distributions

In most statistical problems we are concerned with one number, or a few numbers, that are associated with the outcomes of experiments. In the inspection of a manufactured product we may be interested only in the number of defectives; in the analysis of a road test we may be interested only in the average speed and the average fuel consumption; and in the study of the performance of a rotary switch we may be interested only in its lubrication, the current, and the humidity. All these numbers are associated with situations involving an element of chance—in other words, they are values of random variables.

In the study of random variables we are usually interested in their probability distributions, namely, in the probabilities with which they take on the various values in their range. The introduction to random variables and probability distributions in Section 3.1 is followed by a discussion of various special probability distributions in Sections 3.2, 3.3, 3.7, 3.8, and 3.9, and descriptions of the most important features of probability distributions in Sections 3.4 and 3.5.

To be more explicit about the concept of a random variable, let us refer again to the lawn-mower-rating example on page 11 and the corresponding probabilities shown in Figure 2.8. Now let us suppose that we refer to E_1 (easy to operate), P_2 (inexpensive), and C_3 (low average cost of repairs) as preferred ratings, and that we are interested only in the number of preferred ratings a lawn mower will get. To find the probabilities that a lawn mower will get 0, 1, 2, or 3 preferred ratings, let us refer to Figure 3.1, which is like Figure 2.8 except that we indicate for each outcome the number of preferred ratings. Adding the respective probabilities, we find that for zero preferred ratings the probability is $0.07 + 0.14 + 0.02 + 0.03 = 0.26$, for one preferred rating

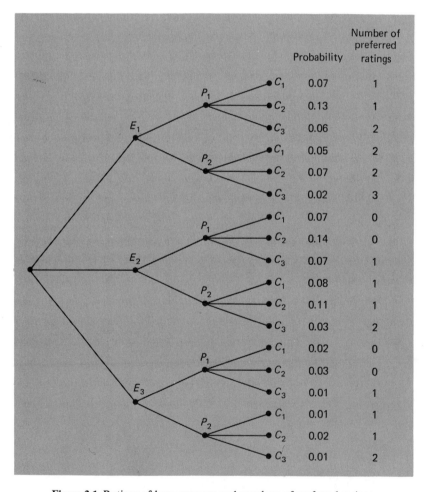

Figure 3.1 Ratings of lawn mowers and numbers of preferred ratings.

the probability is $0.07 + 0.13 + 0.07 + 0.08 + 0.11 + 0.01 + 0.01 + 0.02 = 0.50$, for two preferred ratings the probability is $0.06 + 0.05 + 0.07 + 0.03 + 0.01 = 0.22$, and for three preferred ratings the probability is 0.02. These results may be summarized as in the following table, where x denotes the number of preferred ratings:

x	0	1	2	3
Probability	0.26	0.50	0.22	0.02

The numbers 0, 1, 2, and 3 in this table are values of a **random variable**—the number of preferred ratings. Corresponding to each outcome in the sample space there is one and only one value x of this random variable, so that the random variable may be thought of as a function defined over the elements of the sample space. This is how we define random variables in general; they are functions defined over the elements of a sample space.

To find the probability that a random variable will take on any one value within its range, we proceed as in the above example; indeed, the table which we obtained displays another function, called the **probability distribution** of the random variable. To denote the values of a probability distribution we shall use such symbols as $f(x)$, $g(x)$, $\varphi(y)$, $h(z)$, and so on.

Random variables are usually classified according to the number of values which they can assume. In this chapter we shall limit our discussion to **discrete random variables**, which can take on only a finite number, or a countable infinity of values; **continuous random variables** will be taken up in Chapter 4.

Whenever possible, we try to express probability distributions by means of equations; otherwise we must give a table which actually exhibits the correspondence between the values of the random variable and the associated probabilities. For instance,

$$f(x) = \tfrac{1}{6} \quad \text{for } x = 1, 2, 3, 4, 5, 6$$

gives the probability distribution for the number of points we roll with a balanced die.

Of course, not every function defined for the values of a random variable can serve as a probability distribution. Since the values of probability distributions are probabilities and one value of a random value must always occur, it follows that if $f(x)$ is a value of a probability distribution, then

$$f(x) \geq 0 \quad \text{for all } x$$

and

$$\sum_{\text{all } x} f(x) = 1$$

EXAMPLE Check whether the following can serve as probability distributions:

(a) $f(x) = \dfrac{x-2}{2}$ for $x = 1, 2, 3, 4$;

(b) $h(x) = \dfrac{x^2}{25}$ for $x = 0, 1, 2, 3, 4$.

> ***Solution*** (a) This function cannot serve as a probability distribution because $f(1)$ is negative. (b) This function cannot serve as a probability distribution because the sum of the five probabilities is $\frac{6}{5}$ and not 1.

It is often helpful to visualize probability distributions by means of graphs like those of Figure 3.2. The one on the left is called a **histogram**; the heights of the rectangles are proportional to the corresponding probabilities and the bases touch so that there are no gaps between the rectangles representing the successive values of the random variable. The one on the right is called a **bar chart**; the heights of the rectangles are also proportional to the corresponding probabilities, but they are narrow and their width is of no significance.

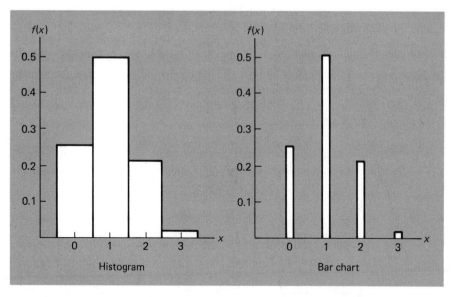

Figure 3.2 Graphs of the probability distribution of the number of preferred ratings.

As we shall see later in this chapter, there are many problems in which we are interested not only in the probability $f(x)$ that the value of a random variable is x, but also in the probability $F(x)$ that the value of a random variable is less than or equal to x. We refer to the function which assigns a value $F(x)$ to each x within the range of a random variable as the **distribution function** of the random variable. Referring again to the lawn-mower-rating example and basing our calculations on the table on page 55, we get

x	0	1	2	3
$F(x)$	0.26	0.76	0.98	1.00

for the distribution function of the number of preferred ratings.

3.2 THE BINOMIAL DISTRIBUTION

Many statistical problems deal with situations referred to as **repeated trials**. For example, we may want to know the probability that one of five rivets will rupture in a tensile test, the probability that 9 of 10 vacuum tubes will last at least 1,000 hours, the probability that 45 of 300 drivers stopped at a road block will be wearing seatbelts, or the probability that 66 of 200 television viewers (interviewed by a rating service) will recall what products were advertised on a given program. To borrow from the language of games of chance, we might say that in each of these examples we are interested in the probability of getting x **successes** in n **trials**, or in other words, x successes and $n - x$ failures in n attempts.

In the problems we shall study in this section, we shall always make the following assumptions:

1. **There are only two possible outcomes for each trial** (arbitrarily called "success" and "failure," without inferring that a success is necessarily desirable).
2. **The probability of a success is the same for each trial.**
3. **There are n trials, where n is a constant.**
4. **The n trials are independent.**

Trials satisfying these assumptions are referred to as **Bernoulli trials**; if the assumptions cannot be met, the theory we shall develop here does not apply.

To solve problems which do meet the conditions listed in the preceding paragraph, we use a formula obtained in the following way: If p and $1 - p$ are the probabilities of success and failure on any one trial, then the probability of getting x successes and $n - x$ failures *in some specific order* is $p^x(1 - p)^{n-x}$; clearly, in this product of p's and $(1 - p)$'s there is one factor p for each success, one factor $1 - p$ for each failure, and the x factors p and $n - x$ factors $1 - p$ are all multiplied together by virtue of the generalized multiplication rule for more than two independent events. Since this probability applies to any point of the sample space which represents x successes and $n - x$ failures (in any specific order), we have only to count how many points of this kind there are, and then multiply $p^x(1-p)^{n-x}$ by this number. Clearly, the number of ways in which we can select the x trials on which there is to be a success is $\binom{n}{x}$, the number of combinations of x objects selected from a set of n objects, and we thus arrive at the following result:

Binomial distribution

$$b(x; n, p) = \binom{n}{x} p^x(1-p)^{n-x} \qquad \text{for } x = 0, 1, 2, \ldots, n$$

This probability distribution is called the **binomial distribution** because for $x = 0, 1, 2, \ldots$, and n, the values of the probabilities are the successive terms of the binomial expansion of $[(1-p) + p]^n$; for the same reason, the combinatorial quantities $\binom{n}{x}$ are referred to as **binomial coefficients**. Actually, the above equation defines a family of probability distributions with each member characterized by given values of the **parameters** n and p.

EXAMPLE It has been claimed that in 60% of all solar heat installations the utility bill is reduced by at least one-third. Accordingly, what are the probabilities that the utility bill will be reduced by at least one-third in

(a) four of five installations;
(b) at least four of five installations?

Solution (a) Substituting $x = 4$, $n = 5$, and $p = 0.60$ into the formula for the binomial distribution, we get

$$b(4; 5, 0.60) = \binom{5}{4}(0.60)^4(1 - 0.60)^{5-4}$$
$$= 0.259$$

(b) Substituting $x = 5$, $n = 5$, and $p = 0.60$ into the formula for the binomial distribution, we get

$$b(5; 5, 0.60) = \binom{5}{5}(0.60)^5(1 - 0.60)^{5-5}$$
$$= 0.078$$

and the answer is $b(4; 5, 0.60) + b(5; 5, 0.60) = 0.259 + 0.078 = 0.337$.

If n is large, the calculation of binomial probabilities can become quite tedious; in case it may be desirable to use numerical approximations or refer to special tables. Values of the binomial distribution and the corresponding cumulative distribution function have been tabulated for $n = 2$

to $n = 49$ by the National Bureau of Standards and for $n = 50$ to $n = 100$ by H. G. Romig (see the Bibliography at the end of the book). Table 1 at the end of this book gives the values of

$$B(x; n, p) = \sum_{k=0}^{x} b(k; n, p) \qquad \text{for } x = 0, 1, 2, \ldots, n$$

for $n = 2$ to $n = 20$ and $p = 0.05, 0.10, 0.15, 0.20, \ldots, 0.50$. We tabulated the **cumulative probabilities** rather than the values of $b(x; n, p)$, because the values of $B(x; n, p)$ are the ones needed more often in statistical applications. Note, however, that the values of $b(x; n, p)$ can be obtained with the use of Table 1 by making use of the identities in parts (c) and (d) of Theorem 3.1 below. Also note that values of $B(x; n, p)$ are not given in Table 1 for p greater than 0.50, but they can be obtained from this table with the use of the identity given in part (b) of Theorem 3.1.

Binomial identities

Theorem 3.1

(a) $b(x; n, p) = b(n - x; n, 1 - p)$

(b) $B(x; n, p) = 1 - B(n - x - 1; n, 1 - p)$

(c) $b(x; n, p) = B(x; n, p) - B(x - 1; n, p)$

(d) $b(x; n, p) = B(n - x; n, 1 - p) - B(n - x - 1; n, 1 - p)$

The first of these identities can be verified by substituting on both sides the corresponding expressions given by the equation defining the binomial distribution; the second identity follows from (a) and the definition of a distribution function; the third identity follows directly from the definition of a distribution function; and the fourth identity follows by substitution from parts (a) and (c). The reader will be asked to verify these identities in detail in Exercises 3.7 and 3.8 on page 65.

The examples which follow illustrate the direct use of Table 1, the use of part (c) of Theorem 3.1, and the use of parts (b) and (d) of Theorem 3.1:

EXAMPLE If the probability is 0.05 that a certain wide-flange column will fail under a given axial load, what are the probabilities that among 16 such columns

  (a) at most two will fail;
  (b) at least four will fail?

  Solution (a) Table 1 shows that $B(2; 16, 0.05) = 0.9571$. (b) Since

$$\sum_{x=4}^{16} b(x; 16, 0.05) = 1 - B(3; 16, 0.05)$$

  Table 1 yields $1 - 0.9930 = 0.0070$.

EXAMPLE If the probability is 0.20 that any one person will dislike the taste of a new toothpaste, what is the probability that 5 of 18 randomly selected persons will dislike it?

Solution First using part (c) of Theorem 3.1 and then looking up the required probabilities in Table 1, we get

$$b(5; 18, 0.20) = B(5; 18, 0.20) - B(4; 18, 0.20)$$
$$= 0.8671 - 0.7164$$
$$= 0.1507$$

EXAMPLE Somebody claims that 75% of all industrial accidents can be prevented by paying strict attention to safety regulations. Assuming that the claim is true, what are the probabilities that

(a) fewer than 16 of 20 industrial accidents can be prevented by paying strict attention to safety regulations;

(b) 12 of 15 industrial accidents can be prevented by paying strict attention to safety regulations?

Solution (a) First using part (b) of Theorem 3.1 and then looking up the required probabilities in Table 1, we get

$$B(15; 20, 0.75) = 1 - B(4; 20, 0.25)$$
$$= 1 - 0.4158$$
$$= 0.5842$$

(b) First using part (d) of Theorem 3.1 and then looking up the required probabilities in Table 1, we get

$$b(12; 15, 0.75) = B(3; 15, 0.25) - B(2; 15, 0.25)$$
$$= 0.4613 - 0.2361$$
$$= 0.2252$$

The following example illustrates the use of the binomial distribution in a problem of decision making:

EXAMPLE A washing machine manufacturer claims that only 10% of his machines require repairs within the warranty period of 12 months. If five of 20 of his machines required repairs within the first year, does this tend to support or refute the claim?

Solution Let us first find the probability that five or more of 20 of the washing machines will require repairs within a year when the probability that any one will require repairs within a year is 0.10. Using Table 1,

we get

$$\sum_{x=5}^{20} b(x; 20, 0.10) = 1 - B(4; 20, 0.10)$$

$$= 1 - 0.9568$$

$$= 0.0432$$

Since this value is very small, it would seem reasonable to reject the washing machine manufacturer's claim.

Important information about the shape of binomial distributions is shown in Figures 3.3 and 3.4. First, if $p = 0.50$ the equation for the binomial distribution is

$$b(x; n, 0.50) = \binom{n}{x}(0.5)^n$$

Figure 3.3 Symmetrical binomial distribution with $n = 5$ and $p = 0.50$.

and since $\binom{n}{n-x} = \binom{n}{x}$ it follows that $b(x; n, 0.50) = b(n - x; n, 0.50)$. This means that the histograms of such binomial distributions are **symmetrical**, as is illustrated in Figure 3.3. Note, however, that if p is less than 0.50, it is more likely that x will be small rather than large compared to n, and that the opposite is true if p is greater than 0.50. This is illustrated by Figure 3.4, showing the histograms of binomial distributions with $n = 5$ and $p = 0.20$ and $p = 0.80$. A probability distribution which has a histogram like either of those of Figure 3.4 is said to be **skewed**; it is said to be **positively skewed** if the tail is on the right, and it is said to be **negatively skewed** if the tail is on the left.

Figure 3.4 Skewed binomial distributions with $n = 5$ and $p = 0.20$ and $p = 0.80$.

3.3 THE HYPERGEOMETRIC DISTRIBUTION

Suppose that we are interested in the number of defectives in a sample of n units drawn from a lot containing N units, of which a are defective. If the sample is drawn in such a way that at each successive drawing whatever units are left in the lot have the same chance of being selected, the probability that the first drawing will yield a defective unit is $\frac{a}{N}$, but for the second drawing it is $\frac{a-1}{N-1}$ or $\frac{a}{N-1}$ depending on whether or not the first unit drawn was defective. Thus, the trials are not independent, the fourth assumption underlying the binomial distribution is not met, and the binomial distribution does not apply. Note that the binomial distribution would apply if we **sample with replacement**, namely, if each unit selected for the sample is replaced before the next one is drawn.

To solve the problem of **sampling without replacement** (that is, as we originally formulated the problem), let us proceed as follows: The x successes (defectives) can be chosen in $\binom{a}{x}$ ways, the $n - x$ failures (nondefectives) can be chosen in $\binom{N-a}{n-x}$ ways, and, hence, x successes and $n - x$ failures can be chosen in $\binom{a}{x}\binom{N-a}{n-x}$ ways. Also, n objects can be chosen from a set of N objects in $\binom{N}{n}$, and if we consider all these possibilities as equally likely, it follows that for sampling without replacement the probability of getting "x successes in n trials" is

Hypergeometric distribution

$$h(x; n, a, N) = \frac{\binom{a}{x}\binom{N-a}{n-x}}{\binom{N}{n}} \quad \text{for } x = 0, 1, \ldots, n$$

where x cannot exceed a and $n - x$ cannot exceed $N - a$. This equation defines the **hypergeometric distribution**, whose parameters are the sample size n, the lot size (or population size) N, and the number of "successes" in the lot a.

EXAMPLE A shipment of 20 tape recorders contains 5 that are defective. If 10 of them are randomly chosen for inspection, what is the probability that 2 of the 10 will be defective?

> **Solution** Substituting $x = 2, n = 10, a = 5$, and $N = 20$ into the formula for the hypergeometric distribution, we get

$$h(2; 10, 5, 20) = \frac{\binom{5}{2}\binom{15}{8}}{\binom{20}{10}} = \frac{10 \cdot 6{,}435}{184{,}756} = 0.348$$

> Probabilities like this are usually calculated using a table of binomial coefficients or a table of the logarithms of factorials, which can be found in most handbooks of mathematical tables.

In the preceding example, n was large compared to N, and if we had made the mistake of using the binomial distribution with $n = 10$ and $p = \frac{5}{20} = 0.25$ to calculate the probability of two defectives, the result would have been 0.282, which is much too small. However, when n is small compared to N, the composition of the lot is not seriously affected by drawing the sample, and the binomial distribution with the parameters n and $p = \frac{a}{N}$ will yield a good approximation.

EXAMPLE Repeat the preceding example for a lot of 100 tape recorders of which 25 are defective, by using
   (a) the formula for the hypergeometric distribution;

(b) the formula for the binomial distribution as an approximation.

Solution (a) Substituting $x = 2$, $n = 10$, $a = 25$, and $N = 100$ into the formula for the hypergeometric distribution, we get

$$h(2; 10, 25, 100) = \frac{\binom{25}{2}\binom{75}{8}}{\binom{100}{10}} = 0.292$$

(b) Substituting $x = 2$, $n = 10$, and $p = \frac{25}{100} = 0.25$ into the formula for the binomial distribution, we get

$$b(2; 10, 0.25) = \binom{10}{2}(0.25)^2(1 - 0.25)^{10-2}$$

$$= 0.282$$

Observe that the difference between the two values is only 0.010. In general, it can be shown that $h(x; n, a, N)$ approaches $b(x; n, p)$ with $p = \dfrac{a}{N}$ when $N \longrightarrow \infty$, and a good rule of thumb is to use the binomial distribution as an approximation to the hypergeometric distribution if $n \leq \dfrac{N}{10}$.

Although we have introduced the hypergeometric distribution in connection with a problem of sampling inspection, it has many other applications. For instance, it can be used to find the probability that three of 12 housewives prefer Brand A detergent to Brand B, if they are selected from among 200 housewives among whom 40 actually prefer Brand A to Brand B. Also, it can be used in connection with a problem of selecting industrial diamonds, some of which have superior qualities and some do not, in connection with a problem of sampling income tax returns, where a among N returns filed contain questionable deductions, and so on.

EXERCISES

3.1 Suppose that a probability of $\frac{1}{12}$ is assigned to each point of the sample space of part (a) of Exercise 2.1 on page 15. Find the probability distribution of the total number of lasers that are suitable for the given task.

3.2 An experiment consists of four tosses of a coin. Denoting the outcomes *HHTH*, *THTT*, ..., and assuming that all 16 outcomes are equally likely, find the probability distribution for the total number of heads.

3.3 Determine whether the following can be probability distributions of a random variable which can take on only the values 1, 2, 3, and 4:
(a) $f(1) = 0.26$, $f(2) = 0.26$, $f(3) = 0.26$, and $f(4) = 0.26$;
(b) $f(1) = 0.15$, $f(2) = 0.28$, $f(3) = 0.29$, and $f(4) = 0.28$;
(c) $f(1) = 0.33$, $f(2) = 0.37$; $f(3) = -0.03$, and $f(4) = 0.33$.

3.4 Check whether the following can define probability distributions, and explain your answers:

(a) $f(x) = \frac{x}{15}$ for $x = 0, 1, 2, 3, 4, 5$;

(b) $f(x) = \frac{5 - x^2}{6}$ for $x = 0, 1, 2, 3$;

(c) $f(x) = \frac{1}{4}$ for $x = 3, 4, 5, 6$;

(d) $f(x) = \frac{x + 1}{25}$ for $x = 1, 2, 3, 4, 5$.

3.5 Given that $f(x) = \frac{k}{2^x}$ is a probability distribution for a random variable which can take on the values $x = 0, 1, 2, 3$, and 4, find k.

3.6 With reference to the preceding exercise, find an expression for the values $F(x)$ of the distribution function of the random variable.

3.7 Prove parts (a) and (b) of Theorem 3.1 on page 59.

3.8 Prove parts (c) and (d) of Theorem 3.1 on page 59.

3.9 Prove that

$$\frac{b(x + 1; n, p)}{b(x; n, p)} = \frac{p(n - x)}{(1 - p)(x + 1)}$$

for $x = 0, 1, 2, \ldots, n - 1$, and use this recursion formula to calculate the values of the binomial distribution with $n = 6$ and $p = 0.30$. Also, verify the results by means of Table 1.

3.10 Use Table 1 to find

(a) $B(8; 16, 0.40)$;

(b) $b(8; 16, 0.40)$;

(c) $B(9; 12, 0.60)$;

(d) $b(9; 12, 0.60)$;

(e) $\sum_{k=6}^{20} b(k; 20, 0.15)$;

(f) $\sum_{k=6}^{9} b(k; 9, 0.70)$.

3.11 Use Table 1 to find

(a) $B(7; 19, 0.45)$;

(b) $b(7; 19, 0.45)$;

(c) $B(8; 10, 0.95)$;

(d) $b(8; 10, 0.95)$;

(e) $\sum_{k=4}^{10} b(k; 10, 0.35)$;

(f) $\sum_{k=2}^{4} b(k; 9, 0.30)$.

3.12 Rework the decision problem on page 60, supposing that only 3 of the 20 washing machines required repairs within the first year.

3.13 In a given city, medical expenses are given as the reason for 75% of all personal bankruptcies. Use the formula for the binomial distribution to find the probability that medical expenses will be given as the reason for two of the next four personal bankruptcies filed in that city.

3.14 If the probability is 0.40 that steam will condense in a thin-walled aluminum tube at 10 atm pressure, use the formula for the binomial distribution to find the probability that under the stated conditions steam will condense in 4 of 12 such tubes.

3.15 If the probability is 0.20 that a downtime of an automated production process will exceed 2 minutes, find the probability that three of eight downtimes of the process will exceed 2 minutes, using
(a) the formula for the binomial distribution;
(b) Table 1.

3.16 Suppose that a civil service examination is designed so that 70% of all persons with an I.Q. of 90 can pass. Use Table 1 to find the probabilities that among 15 persons with an I.Q. of 90 who take the test
(a) at least 12 will pass;
(b) at most six will pass;
(c) 10 will pass.

3.17 The probability that the noise level of a wide-band amplifier will exceed 2 dB is 0.05. Use Table 1 to find the probabilities that among 12 such amplifiers the noise level of
(a) one will exceed 2 dB;
(b) at most two will exceed 2 dB;
(c) two or more will exceed 2 dB.

3.18 If the probability that a fluorescent light has a useful life of at least 500 hours is 0.85, find the probabilities that among 20 such lights
(a) 18 will have a useful life of at least 500 hours;
(b) at least 15 will have a useful life of at least 500 hours;
(c) at least two will not have a useful life of at least 500 hours.

3.19 An agricultural cooperative claims that 90% of the watermelons shipped out are ripe and ready to eat. Find the probabilities that among 18 watermelons shipped out
(a) all 18 are ripe and ready to eat;
(b) at least 16 are ripe and ready to eat;
(c) at most 14 are ripe and ready to eat.

3.20 A quality control engineer wants to check whether (in accordance with specifications) 95% of the electronic components shipped by his company are in good working condition. To this end, he randomly selects 15 from each large lot ready to be shipped and passes it if they are all in good working condition; otherwise, each of the components in the lot is checked. Find the probabilities that the quality control engineer will commit the error of
(a) holding a lot for further inspection even though 95% of the components are in good working condition;
(b) letting a lot pass through without further inspection even though only 90% of the components are in good working condition;
(c) letting a lot pass through without further inspection even though only 80% of the components are in good working condition.

3.21 A food processor claims that at most 10% of his jars of instant coffee contain less coffee than claimed on the label. To test this claim, 16 jars of his instant coffee are randomly selected and the contents are weighed; his claim is accepted if fewer than 3 of the jars contain less coffee than claimed on the label. Find the probabilities that the food processor's claim will be accepted when the actual percentage of his jars containing less coffee than claimed on

the label is
(a) 5%;   (c) 15%;
(b) 10%;   (d) 20%.

3.22 A study shows that a computer firm answers 70% of all inquiries within 6 days. Find the probabilities that the firm will answer 0, 1, 2, . . . , or 10 of 10 inquiries within 6 days, and draw a histogram of this probability distribution.

3.23 What is the probability that an IRS auditor will catch only 2 income tax returns with illegitimate deductions, if she randomly selected 6 returns from among 18 returns of which 8 contain illegitimate deductions?

3.24 Among the 12 solar collectors on display at a trade show, 9 are flat-plate collectors and the others are concentrating collectors. If a person visiting the show randomly selects four of the solar collectors to check out, what is the probability that three of them will be flat-plate collectors?

3.25 A quality control engineer inspects a random sample of 3 batteries from each lot of 24 car batteries that is ready to be shipped. If such a lot contains six batteries with slight defects, what are the probabilities that the inspector's sample will contain
(a) none of the batteries with defects;
(b) only one of the batteries with defects;
(c) at least two of the batteries with defects?

3.26 If 6 of 18 new buildings in a city violate the building code, what is the probability that a building inspector, who randomly selects 4 of the new buildings for inspection, will catch
(a) none of the new buildings that violate the building code;
(b) one of the new buildings that violate the building code;
(c) two of the new buildings that violate the building code;
(d) at least three of the new buildings that violate the building code?

3.27 In 15 experiments studying the electrical behavior of single cells, 11 use microelectrodes made of metal and the other 4 use microelectrodes made from glass tubing. If two of the experiments are to be terminated for financial reasons, and they are selected at random, what are the probabilities that
(a) neither uses microelectrodes made from glass tubing;
(b) only one uses microelectrodes made from glass tubing;
(c) both use microelectrodes made from glass tubing?

3.28 Among the 16 cities which a professional society is considering for its next three annual conventions, 7 are in the western part of the United States. To avoid arguments, the selection is left to chance. If none of the cities can be chosen more than once, what are the probabilities that
(a) none of the conventions will be held in the western part of the United States;
(b) all of the conventions will be held in the western part of the United States?

3.29 A shipment of 120 burglar alarms contains 5 that are defective. If three of these alarms are randomly selected and shipped to a customer, find the probability that the customer will get one bad unit by using

(a) the formula for the hypergeometric distribution;

(b) the formula for the binomial distribution as an approximation.

3.30 Among the 300 employees of a company, 240 are union members while the others are not. If eight of the employees are chosen by lot to serve on the committee which administers the pension fund, find the probability that five of them will be union members while the others are not, by using

(a) the formula for the hypergeometric distribution;

(b) the formula for the binomial distribution as an approximation.

3.4 THE MEAN AND THE VARIANCE OF A PROBABILITY DISTRIBUTION

Besides the binomial and hypergeometric distributions, there are many other probability distributions that have important engineering applications. However, before we go any further, let us discuss some general characteristics of probability distributions.

One such characteristic, that of the symmetry or skewness of a probability distribution, was illustrated by means of Figures 3.3 and 3.4; two others will be presented here with reference to Figure 3.5, which shows the histograms of two binomial distributions. One of these binomial distributions has the parameters $n = 4$ and $p = \frac{1}{2}$, and the other has the parameters $n = 16$ and $p = \frac{1}{2}$. Essentially, these two probability distributions differ in two respects.

Figure 3.5 Histograms of two binomial distributions.

The first probability distribution is centered about $x = 2$ while the other (whose histogram is shaded) is centered about $x = 8$, and we say that the two distributions differ in their **location**. Another distinction is that the histogram of the second distribution is more spread out, and we say that the two distributions differ in **variation**. To make such comparisons more specific, we shall introduce in this section two of the most important statistical measures, describing, respectively, the location and the variation of a probability distribution—the **mean** and the **variance**.

The mean of a probability distribution is simply the mathematical expectation of a corresponding random variable. If a random variable takes on the values $x_1, x_2, \ldots$, or x_k, with the probabilities $f(x_1), f(x_2), \ldots$, and $f(x_k)$, its mathematical expectation is

$$x_1 \cdot f(x_1) + x_2 \cdot f(x_2) + \ldots + x_k \cdot f(x_k)$$

and, using the $\sum$ notation, we can thus write

Mean of probability distribution

$$\mu = \sum_{\text{all } x} x \cdot f(x)$$

where the **mean** is denoted by the Greek letter μ (*mu*). The mean of a probability distribution measures its center in the sense of an average, or better, in the sense of a center of gravity. Note that the above formula for μ is, in fact, that for the **first moment about the origin** of a discrete system of masses $f(x)$ arranged on a straight line at distances x from the origin. We do not have to divide here by $\sum_{\text{all } x} f(x)$, as we do in the usual formula for the x-coordinate of the center of gravity, since this sum equals 1 by definition.

EXAMPLE   Find the mean of the probability distribution of the number of heads obtained in three flips of a balanced coin.

Solution   The probabilities for 0, 1, 2, or 3 heads are $\frac{1}{8}, \frac{3}{8}, \frac{3}{8}$, and $\frac{1}{8}$, as can easily be verified by counting equally likely possibilities or by using the formula for the binomial distribution with $n = 3$ and $p = \frac{1}{2}$. Thus,

$$\mu = 0 \cdot \tfrac{1}{8} + 1 \cdot \tfrac{3}{8} + 2 \cdot \tfrac{3}{8} + 3 \cdot \tfrac{1}{8} = \tfrac{3}{2}$$

EXAMPLE   With reference to the lawn-mower-rating example and the probabilities given on page 55, find the mean of the probability distribution of the number of preferred ratings.

Solution   Substituting $x = 0, 1, 2$, and 3, and the corresponding probabilities, into the formula for μ, we get

$$\mu = 0(0.26) + 1(0.50) + 2(0.22) + 3(0.02)$$
$$= 1.00$$

Returning to the second probability distribution of Figure 3.5, we could find its mean by calculating all the necessary probabilities (or by looking them up in Table 1) and substituting them into the formula for μ. However, if we reflect for a moment, we might argue that there is a fifty-fifty chance for a success on each trial, there are 16 trials, and it would seem reasonable to expect 8 heads and 8 tails (in the sense of a mathematical expectation). Similarly, we might argue that if a binomial distribution has the parameters $n = 200$ and $p = 0.20$, we can expect a success 20% of the time and, hence, on the average $200(0.20) = 40$ successes in 200 trials. These two values are, indeed, correct, and it can be shown in general that

Mean of binomial distribution

$$\mu = n \cdot p$$

for the mean of a binomial distribution. To prove this formula, we substitute the expression which defines $b(x; n, p)$ into the formula for μ, and we get

$$\mu = \sum_{x=0}^{n} x \cdot \frac{n!}{x!(n-x)!} p^x (1-p)^{n-x}$$

Then, making use of the fact that $\dfrac{x}{x!} = \dfrac{1}{(x-1)!}$ and $n! = n(n-1)!$, and factoring out n and p, we obtain

$$\mu = np \sum_{x=1}^{n} \frac{(n-1)!}{(x-1)!(n-x)!} p^{x-1}(1-p)^{n-x}$$

where the summation starts with $x = 1$ since the original summand is zero for $x = 0$. If we now let $y = x - 1$ and $m = n - 1$, we obtain

$$\mu = np \sum_{y=0}^{m} \frac{m!}{y!(m-y)!} p^y (1-p)^{m-y}$$

and this last sum can easily be recognized as that of all the terms of the binomial distribution with the parameters m and p. Hence, this sum equals 1 and it follows that $\mu = np$.

EXAMPLE   Find the mean of the probability distribution of the number of heads obtained in three flips of a balanced coin.

Solution   For a binomial distribution with $n = 3$ and $p = \frac{1}{2}$, we get $\mu = 3 \cdot \frac{1}{2} = \frac{3}{2}$, and this agrees with the result obtained on page 69.

The formula $\mu = np$ applies, of course, only to binomial distributions, but for other special distributions, we can, similarly, express the mean in terms of their parameters. For instance, for the mean of the hypergeometric distribution with the parameters n, a, and N, we can write

***Mean of
hypergeometric
distribution***

$$\mu = n \cdot \frac{a}{N}$$

In Exercise 3.43 on page 79, the reader will be asked to derive this formula by a method similar to the one we used to derive the formula for the mean of a binomial distribution.

EXAMPLE   With reference to the example on page 63, where 5 of 20 tape recorders were defective, find the mean of the probability distribution of the number of defectives in a sample of 10 randomly chosen for inspection.

> ***Solution***   Substituting $n = 10$, $a = 5$, and $N = 20$ into the above formula for μ, we get
>
> $$\mu = 10 \cdot \tfrac{5}{20} = 2.5$$

In other words, if we inspect 10 of the tape recorders, we can expect 2.5 defectives, where "expect" is to be interpreted in the sense of a mathematical expectation.

To study the second of the two properties of probability distributions mentioned on page 69, their variation, let us refer again to the two probability distributions of Figure 3.5. For the one where $n = 4$ there is a high probability of getting values close to the mean, but for the one where $n = 16$ there is a high probability of getting values scattered over considerable distances away from the mean. Using this property, it may seem reasonable to measure the variation of a probability distribution with the quantity

$$\sum_{\text{all } x} (x - \mu) \cdot f(x)$$

namely, the average amount by which the values of the random variable deviate from the mean. Unfortunately,

$$\sum_{\text{all } x} (x - \mu) \cdot f(x) = \sum_{\text{all } x} x \cdot f(x) - \sum_{\text{all } x} \mu \cdot f(x)$$

$$= \mu - \mu \cdot \sum_{\text{all } x} f(x) = \mu - \mu = 0$$

so that this expression is always equal to zero. However, since we are really interested in the magnitude of the deviations $x - \mu$ and not in their signs, it suggests itself that we average the absolute values of these deviations from the mean. This would, indeed, provide a measure of variation, but on purely theoretical grounds we prefer to work instead with the squares of the deviations from the mean. These quantities are also nonnegative, and their average is indicative of the spread or dispersion of a probability distribution. We thus define the **variance** of a probability distribution with values $f(x)$ as

Variance of probability distribution

$$\sigma^2 = \sum_{\text{all } x} (x - \mu)^2 \cdot f(x)$$

where σ is the lowercase Greek letter for s. This measure is not in the same units (or dimension) as the values of the random variable, but we can adjust for this by taking the square root, thus defining the **standard deviation** as

Standard deviation of probability distribution

$$\sigma = \sqrt{\sum_{\text{all } x} (x - \mu)^2 \cdot f(x)}$$

EXAMPLE Compare the standard deviations of the two probability distributions of Figure 3.5.

Solution   Since $\mu = 4 \cdot \frac{1}{2} = 2$ for the binomial distribution with $n = 4$ and $p = \frac{1}{2}$, we find that the variance of this probability distribution is

$$\sigma^2 = (0 - 2)^2 \cdot \tfrac{1}{16} + (1 - 2)^2 \cdot \tfrac{4}{16} + (2 - 2)^2 \cdot \tfrac{6}{16}$$
$$+ (3 - 2)^2 \cdot \tfrac{4}{16} + (4 - 2)^2 \cdot \tfrac{1}{16}$$
$$= 1$$

and, hence, that its standard deviation is $\sigma = 1$. Similarly, it can be shown that for the other distribution $\sigma = 2$, and we find that the second (shaded) distribution with the greater spread also has the greater standard deviation.

Given any probability distribution, we can always calculate σ^2 by substituting the corresponding probabilities $f(x)$ into the formula which defines the variance. As in the case of the mean, however, this work can be simplified to a considerable extent when we deal with special kinds of distributions. For instance, it can be shown that the variance of the binomial distribution with the parameters n and p is given by the formula

Variance of binomial distribution

$$\sigma^2 = n \cdot p \cdot (1 - p)$$

and that the variance of the hypergeometric distribution with the parameters n, a, and N is given by the formula

Variance of hypergeometric distribution

$$\sigma^2 = \frac{n \cdot a \cdot (N - a) \cdot (N - n)}{N^2 \cdot (N - 1)}$$

EXAMPLE   Verify the result stated in the preceding example, that $\sigma = 2$ for the binomial distribution with $n = 16$ and $p = \frac{1}{2}$.

Solution   Substituting $n = 16$ and $p = \frac{1}{2}$ into the formula for the variance of a binomial distribution, we get

$$\sigma^2 = 16 \cdot \tfrac{1}{2} \cdot \tfrac{1}{2} = 4$$

and, hence, $\sigma = \sqrt{4} = 2$.

EXAMPLE   With reference to the example on page 63, where 5 of 20 tape recorders were defective, find the standard deviation of the probability distribution of the number of defectives in a sample of 10 randomly chosen for inspection.

Solution   Substituting $n = 10$, $a = 5$, and $N = 20$ into the formula for the variance of a hypergeometric distribution, we get

$$\sigma^2 = \frac{10 \cdot 5 \cdot (20 - 5) \cdot (20 - 10)}{20^2 \cdot 19} = \frac{75}{76}$$

and, hence, $\sigma = \sqrt{75/76} = 0.99$.

When we first defined the variance of a probability distribution, it may have occurred to the reader that the formula looked exactly like the one which we use in physics to define second moments, or moments of inertia. Indeed, it is customary in statistics to define the **kth moment about the origin** as

$$\mu'_k = \sum_{\text{all } x} x^k \cdot f(x)$$

and the **kth moment about the mean** as

$$\mu_k = \sum_{\text{all } x} (x - \mu)^k \cdot f(x)$$

Thus, the mean μ is the first moment about the origin, and the variance σ^2 is the second moment about the mean. Higher moments are often used in statistics to give further descriptions of probability distributions. For instance, the third moment about the mean (divided by σ^3 to make this measure independent of the scale of measurement) is used to describe the symmetry or skewness of a distribution; the fourth moment about the mean (divided by σ^4) is, similarly, used to describe its "peakedness," or **kurtosis**. To determine moments about the mean, it is usually easiest to express them in terms of moments about the origin and then to calculate the necessary moments about the origin. For the second moment about the mean we thus have the important formula

Computing formula for the variance

$$\sigma^2 = \mu_2' - \mu^2$$

which the reader will be asked to prove in part (a) of Exercise 3.50 on page 79; part (b) of that exercise pertains to a similar formula for expressing μ_3 in terms of moments about the origin.

EXAMPLE   Use the above computing formula to determine the variance of the probability distribution of the number of points rolled with a balanced die.

Solution   Since $f(x) = \frac{1}{6}$ for $x = 1, 2, 3, 4, 5,$ and 6, we get

$$\mu = 1 \cdot \tfrac{1}{6} + 2 \cdot \tfrac{1}{6} + 3 \cdot \tfrac{1}{6} + 4 \cdot \tfrac{1}{6} + 5 \cdot \tfrac{1}{6} + 6 \cdot \tfrac{1}{6}$$
$$= \tfrac{7}{2}$$
$$\mu_2' = 1^2 \cdot \tfrac{1}{6} + 2^2 \cdot \tfrac{1}{6} + 3^2 \cdot \tfrac{1}{6} + 4^2 \cdot \tfrac{1}{6} + 5^2 \cdot \tfrac{1}{6} + 6^2 \cdot \tfrac{1}{6}$$
$$= \tfrac{91}{6}$$

and, hence, $\sigma^2 = \tfrac{91}{6} - (\tfrac{7}{2})^2 = \tfrac{35}{12}$.

3.5 CHEBYSHEV'S THEOREM

Earlier in this chapter we used examples to show how the standard deviation measures the variation of a probability distribution, that is, how it controls the concentration of probability in the neighborhood of the mean. If σ is small, there is a high probability of getting values close to the mean; if σ is large there is a correspondingly higher probability for getting values farther away from the mean. Formally, this idea is expressed by the following theorem:

Chebyshev's theorem

Theorem 3.2 *If a probability distribution has the mean μ and the standard deviation σ, the probability of getting a value which deviates from μ by at least $k\sigma$ is at most $\dfrac{1}{k^2}$.*

Symbolically,

$$P(|x - \mu| \geq k\sigma) \leq \frac{1}{k^2}$$

where $P(|x - \mu| \geq k\sigma)$ is the probability associated with the set of outcomes for which x, the value of a random variable having the given probability distribution, is such that $|x - \mu| \geq k\sigma$.

Thus, the probability that a random variable will take on a value which deviates (differs) from the mean by at least 2 standard deviations is at most $\frac{1}{4}$, the probability that it will take on a value which deviates from the mean by at least 5 standard deviations is at most $\frac{1}{25}$, and the probability that it will take on a value which deviates from the mean by 10 standard deviations or more is less than or equal to $\frac{1}{100}$.

To prove this theorem, consider any probability distribution with the values $f(x)$, the mean μ, and the variance σ^2. Dividing the sum defining the variance into three parts as indicated in Figure 3.6, we have

$$\sigma^2 = \sum_{\text{all } x} (x - \mu)^2 f(x)$$
$$= \sum_{R_1} (x - \mu)^2 f(x) + \sum_{R_2} (x - \mu)^2 f(x) + \sum_{R_3} (x - \mu)^2 f(x)$$

where R_1 is the region for which $x \leq \mu - k\sigma$, R_2 is the region for which $\mu - k\sigma < x < \mu + k\sigma$, and R_3 is the region for which $x \geq \mu + k\sigma$. Since $(x - \mu)^2 f(x)$ cannot be negative, the above sum over R_2 is non-negative, and without it the sum of the summations over R_1 and R_3 is less than or equal to

Figure 3.6 Diagram for proof of Chebyshev's theorem.

σ^2, that is,

$$\sigma^2 \geq \sum_{R_1} (x - \mu)^2 f(x) + \sum_{R_3} (x - \mu)^2 f(x)$$

But $x - \mu \leq -k\sigma$ in the region R_1 and $x - \mu \geq k\sigma$ in the region R_3, so that in either case $|x - \mu| \geq k\sigma$; hence, in both regions $(x - \mu)^2 \geq k^2\sigma^2$. If we now replace $(x - \mu)^2$ in each sum by $k^2\sigma^2$, a number less than or equal to $(x - \mu)^2$, we obtain the inequality

$$\sigma^2 \geq \sum_{R_1} k^2\sigma^2 f(x) + \sum_{R_3} k^2\sigma^2 f(x)$$

or

$$\frac{1}{k^2} \geq \sum_{R_1} f(x) + \sum_{R_3} f(x)$$

Since $\sum_{R_1} f(x) + \sum_{R_3} f(x)$ represents the probability that x is in the region $R_1 \cup R_3$, namely, that $|x - \mu| \geq k\sigma$, this completes the proof of Theorem 3.2.

To obtain an alternative form of Chevyshev's theorem, note that the event $|x - \mu| < k\sigma$ is the complement of the event $|x - \mu| \geq k\sigma$; hence, the probability of getting a value which deviates from μ by less than $k\sigma$ is at least $1 - \frac{1}{k^2}$.

EXAMPLE   The number of customers who visit a Buick dealer's showroom on a Saturday morning is a random variable with $\mu = 18$ and $\sigma = 2.5$. With what probability can we assert that there will be between 8 and 28 customers?

Solution   Since $k = \dfrac{28 - 18}{2.5} = \dfrac{18 - 8}{2.5} = 4$, the probability is at least $1 - \dfrac{1}{4^2} = \dfrac{15}{16}$.

Theoretically speaking, the most important feature of Chevyshev's theorem is that it applies to any probability distribution for which μ and σ exist. However, so far as applications are concerned, this generality is also its greatest weakness—it only provides an upper limit (often a very poor one) to the probability of getting a value which deviates from the mean by k standard deviations or more. For instance, we can assert in general that the probability of getting a value which differs from the mean by at least 2 standard deviations is at most 0.25, whereas the corresponding exact probability for the binomial distribution with $n = 16$ and $p = \frac{1}{2}$ is only 0.0768— "at most 0.25" is correct, but it does not tell us that the actual probability may be as small as 0.0768.

An important result is obtained if we apply Chebyshev's theorem to the binomial distribution. To illustrate this result, consider the following example:

EXAMPLE Show that for 40,000 flips of a balanced coin the probability is at least 0.99 that the proportion of heads will fall between 0.475 and 0.525.

Solution Since $\mu = 40{,}000 \cdot \frac{1}{2} = 20{,}000$, $\sigma = \sqrt{40{,}000 \cdot \frac{1}{2} \cdot \frac{1}{2}} = 100$, and $1 - \frac{1}{k^2} = 0.99$ yields $k = 10$, the alternative form of Chebyshev's theorem tells us that the probability is at least 0.99 that we will get between $20{,}000 - 10(100) = 19{,}000$ and $20{,}000 + 10(100) = 21{,}000$ heads. Hence, the probability is at least 0.99 that the proportion of heads will fall between $\frac{19{,}000}{40{,}000} = 0.475$ and $\frac{21{,}000}{40{,}000} = 0.525$.

Correspondingly, the reader will be asked to show in Exercise 3.48 that for 1,000,000 flips of a balanced coin the probability is at least 0.99 that the proportion of heads will fall between 0.495 and 0.505, and these results suggest that when n is large, the chances are that the proportion of heads will be very close to $p = \frac{1}{2}$. When formulated for any binomial distribution with the parameters n and p, this result is referred to as the **law of large numbers**.

EXERCISES

3.31 Suppose that the probabilities are 0.4, 0.3, 0.2, and 0.1 that there will be 0, 1, 2, or 3 power failures in a certain city during the month of July. Use the formulas which define μ and σ^2 to find
(a) the mean of this probability distribution;
(b) the variance of this probability distribution.

3.32 Use the computing formula for σ^2 to rework part (b) of the preceding exercise.

3.33 The following table gives the probabilities that a certain computer will malfunction 0, 1, 2, 3, 4, 5, or 6 times on any one day:

Number of malfunctions	x	0	1	2	3	4	5	6
Probability	$f(x)$	0.17	0.29	0.27	0.16	0.07	0.03	0.01

Use the formulas which define μ and σ to find
(a) the mean of this probability distribution;
(b) the standard deviation of this probability distribution.

3.34 Use the computing formula for σ^2 to rework part (b) of the preceding exercise.

3.35 Find the mean and the variance of the uniform probability distribution given by

$$f(x) = \frac{1}{n} \qquad \text{for } x = 1, 2, 3, \ldots, n$$

[*Hint*: The sum of the first n positive integers is $\frac{1}{2}n(n + 1)$, and the sum of their squares is $\frac{1}{6}n(n + 1)(2n + 1)$.]

3.36 Find the mean and the variance of the binomial distribution with $n = 4$ and $p = 0.70$ by using

(a) Table 1 and the formulas defining μ and σ^2;

(b) the special formulas for the mean and the variance of a binomial distribution.

3.37 As can easily be verified by means of the formula for the binomial distribution (or by listing all 32 possibilities), the probabilities of getting 0, 1, 2, 3, 4, or 5 heads in five flips of a balanced coin are $\frac{1}{32}$, $\frac{5}{32}$, $\frac{10}{32}$, $\frac{10}{32}$, $\frac{5}{32}$, and $\frac{1}{32}$. Find the mean of this probability distribution using

(a) the formula which defines μ;

(b) the special formula for the mean of a binomial distribution.

3.38 With reference to the preceding exercise, find the variance of the probability distribution using

(a) the formula which defines σ^2;

(b) the computing formula for σ^2;

(c) the special formula for the variance of a binomial distribution.

3.39 If 95% of certain radial tires last at least 30,000 miles, find the mean and the standard deviation of the distribution of the number of these tires, among 20 selected at random, that last at least 30,000 miles, using

(a) Table 1, the formula which defines μ, and the computing formula for σ^2;

(b) the special formulas for the mean and the variance of a binomial distribution.

3.40 Find the mean and the standard deviation of the distribution of each of the following random variables (having binomial distributions):

(a) the number of heads obtained in 676 flips of a balanced coin;

(b) the number of 4's obtained in 720 rolls of a balanced die;

(c) the number of defectives in a sample of 600 parts made by a machine, when the probability is 0.04 that any one of the parts is defective;

(d) the number of students among 800 interviewed who do not like the food served at the university cafeteria, when the probability is 0.65 that any one of them does not like the food.

3.41 With reference to Exercise 3.27, find the mean and the variance of the distribution of the number of microelectrodes made from glass tubing, using

(a) the probabilities obtained in that exercise;

(b) the special formulas for the mean and the variance of a hypergeometric distribution.

3.42 Find the mean and the standard deviation of the hypergeometric distribution with the parameters $n = 3$, $a = 4$, and $N = 8$

(a) by first calculating the necessary probabilities and then using the formulas which define μ and σ;

(b) by using the special formulas for the mean and the variance of a hypergeometric distribution.

3.43 Prove the formula for the mean of the hypergeometric distribution with the parameters n, a, and N, namely, $\mu = n \cdot \dfrac{a}{N}$. [*Hint*: Make use of the identity

$$\sum_{r=0}^{k} \binom{m}{r}\binom{s}{k-r} = \binom{m+s}{k}$$

which can be obtained by equating the coefficients of x^k in $(1 + x)^m(1 + x)^s$ and in $(1 + x)^{m+s}$.]

3.44 Construct a table showing the upper limits provided by Chebyshev's theorem for the probabilities of obtaining values differing from the mean by at least $1, 2$, and 3 standard deviations, and also the corresponding probabilities for the binomial distribution with $n = 16$ and $p = \frac{1}{2}$.

3.45 If a student answers the 144 questions of a true-false test by flipping a balanced coin—heads is "true" and tails is "false"—what does Chebyshev's theorem with $k = 4$ tell us about the number of correct answers he or she will get?

3.46 What does Chebyshev's theorem tell us about the probability of getting at most 30 or at least 105 sixes in 405 rolls of a balanced die?

3.47 The daily number of orders filled by the parts department of a repair shop is a random variable with $\mu = 142$ and $\sigma = 12$. According to Chebyshev's theorem, with what probability can we assert that on any one day it will fill between 82 and 202 orders?

3.48 Show that for 1,000,000 flips of a balanced coin the probability is at least 0.99 that the proportion of heads will fall between 0.495 and 0.505.

3.49 How many times do we have to flip a balanced coin to be able to assert with a probability of at most 0.01 that the difference between the proportion of tails and 0.50 will be at least 0.04?

3.50 Prove that
(a) $\sigma^2 = \mu_2' - \mu^2$;   (b) $\mu_3 = \mu_3' - 3\mu_2' \cdot \mu + 2\mu^3$.

3.6 THE POISSON APPROXIMATION TO THE BINOMIAL DISTRIBUTION

When n is large and p is small, binomial probabilities are often approximated by means of the formula

Poisson distribution

$$f(x; \lambda) = \frac{\lambda^x e^{-\lambda}}{x!} \qquad \text{for } x = 0, 1, 2, \dots$$

with λ (*lambda*) equal to the product np. Before we justify this approximation, let us point out that $x = 0, 1, 2, \dots$ means that there is a countable infinity of possibilities, and this requires that we modify the third axiom of probability given on page 23. In its place we substitute

Axiom 3' *If* $A_1, A_2, A_3, \ldots$ *is a finite or infinite sequence of mutually exclusive events in S, then*

$$P(A_1 \cup A_2 \cup A_3 \cup \ldots) = P(A_1) + P(A_2) + P(A_3) + \ldots$$

while the other postulates remain unchanged. To verify that $P(S) = 1$ for the formula given above, we make use of Axiom 3' and write

$$\sum_{x=0}^{\infty} f(x; \lambda) = \sum_{x=0}^{\infty} \frac{e^{-\lambda}\lambda^x}{x!} = e^{-\lambda} \sum_{x=0}^{\infty} \frac{\lambda^x}{x!}$$

Since the infinite series in the expression on the right is the Maclaurin's series for e^{λ}, it follows that

$$\sum_{x=0}^{\infty} f(x; \lambda) = e^{-\lambda} \cdot e^{\lambda} = 1$$

Let us now show that when $n \to \infty$ and $p \to 0$, while $np = \lambda$ remains constant, the limiting form of the binomial distribution is as given in the beginning of the preceding paragraph. First let us substitute $\dfrac{\lambda}{n}$ for p into the formula for the binomial distribution and simplify the resulting expression; thus, we get

$$b(x; n, p) = \frac{n!}{x!(n-x)!} \left(\frac{\lambda}{n}\right)^x \left(1 - \frac{\lambda}{n}\right)^{n-x}$$

$$= \frac{n(n-1)(n-2) \cdot \ldots \cdot (n-x+1)}{x! \, n^x} (\lambda)^x \left(1 - \frac{\lambda}{n}\right)^{n-x}$$

$$= \frac{\left(1 - \frac{1}{n}\right)\left(1 - \frac{2}{n}\right) \cdot \ldots \cdot \left(1 - \frac{x-1}{n}\right)}{x!} (\lambda)^x \left(1 - \frac{\lambda}{n}\right)^{n-x}$$

If we now let $n \to \infty$, we find that

$$\left(1 - \frac{1}{n}\right)\left(1 - \frac{2}{n}\right) \cdot \ldots \cdot \left(1 - \frac{x-1}{n}\right) \longrightarrow 1$$

that

$$\left(1 - \frac{\lambda}{n}\right)^{n-x} = \left[\left(1 - \frac{\lambda}{n}\right)^{n/\lambda}\right]^{\lambda} \left(1 - \frac{\lambda}{n}\right)^{-x} \longrightarrow e^{-\lambda}$$

and, hence, that the binomial distribution approaches

$$\frac{\lambda^x e^{-\lambda}}{x!} \qquad \text{for } x = 0, 1, 2, \ldots$$

This completes our proof; the distribution at which we arrived is called the **Poisson distribution**, as we already indicated on page 79.

Since the Poisson distribution has many important applications besides approximating binomial probabilities (see Section 3.7), it has been extensively tabulated. Table 2 at the end of the book gives the values of the probabilities $F(x; \lambda) = \sum_{k=0}^{x} f(k; \lambda)$ for values of λ in varying increments from 0.02 to 25, and its use is very similar to that of Table 1.

The following are some examples illustrating the Poisson approximation of binomial probabilities. An acceptable rule of thumb is to use this approximation if $n \geq 20$ and $p \leq 0.05$; if $n \geq 100$, the approximation is generally excellent so long as $np \leq 10$.

EXAMPLE   It is known that 5% of the books bound at a certain bindery have defective bindings. Find the probability that 2 of 100 books bound by this bindery will have defective bindings, using

(a) the formula for the binomial distribution;

(b) the Poisson approximation to the binomial distribution.

Solution   (a) Substituting $x = 2$, $n = 100$, and $p = 0.05$ into the formula for the binomial distribution, we get

$$b(2; 100, 0.05) = \binom{100}{2}(0.05)^2(0.95)^{98} = 0.081$$

(b) Substituting $x = 2$ and $\lambda = 100(0.05) = 5$ into the formula for the Poisson distribution, we get

$$f(2; 5) = \frac{5^2 \cdot e^{-5}}{2!} = 0.084$$

It is of interest to note that the difference between the two values we obtained (the error we would make by using the Poisson approximation) is only 0.003. [Had we used Table 2 instead of looking up the value of e^{-5} in a mathematical table, we would have obtained $f(2; 5) = F(2; 5) - F(1; 5) = 0.125 - 0.040 = 0.085$.]

EXAMPLE   A fire insurance company has 3,840 policyholders. If the probability is $\frac{1}{1,200}$ that any one of the policyholders will file at least one claim in any given year, find the probabilities that $0, 1, 2, 3, 4, \ldots$ of the policyholders will file at least one claim in a given year.

Solution   For practical reasons we cannot use the formula for the binomial distribution, but use of the Poisson distribution with

$$\lambda = 3{,}840 \cdot \frac{1}{1{,}200} = 3.2$$

Table 2, and the identity $f(x; \lambda) = F(x; \lambda) - F(x - 1; \lambda)$ yields the results shown in the histogram of Figure 3.7.

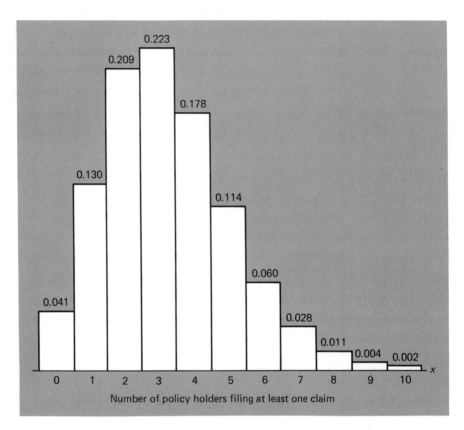

Figure 3.7 Histogram of Poisson distribution with $\lambda = 3.2$.

Using a method similar to that employed on page 70 to derive the formula for the mean of the binomial distribution, we can show that the mean and the variance of the Poisson distribution with the parameter λ are given by

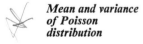

Mean and variance of Poisson distribution

$$\mu = \lambda \text{ and } \sigma^2 = \lambda$$

The first of these results should really have been expected, for in our justification of the Poisson approximation to the binomial distribution we let $\lambda = np$. For the variance we can write $\sigma^2 = np(1 - p) = \lambda(1 - p)$, which approaches λ as $p \longrightarrow 0$.

3.7 POISSON PROCESSES

In general, a **random process** is a physical process which is wholly or in part controlled by some sort of chance mechanism. It may be a sequence of repeated flips of a coin, measurements of the quality of manufactured products coming off an assembly line, the vibrations of airplane wings, the noise in a radio signal, or any one of numerous other phenomena. What characterizes such processes is their time dependence, namely, the fact that certain events do or do not take place (depending on chance) at regular intervals of time or throughout continuous intervals of time.

In this section we shall be concerned with processes taking place over continuous intervals of time, such as the occurrence of imperfections on a continuously produced bolt of cloth, the recording of radiation by means of a Geiger counter, the arrival of telephone calls at a switchboard, or the passing by cars of an electronic checking device. We will now show that the mathematical model which we can use to describe many situations like these is that of the Poisson distribution. To find the probability of x successes during a time interval of length T, we divide the interval into n equal parts of length Δt, so that $T = n \cdot \Delta t$, and we assume that:

1. The probability of a success during a very small interval of time Δt is given by $\alpha \cdot \Delta t$.
2. The probability of more than one success during such a small time interval Δt is negligible.
3. The probability of a success during such a time interval does not depend on what happened prior to that time.

This means that the assumptions underlying the binomial distribution are satisfied, and the probability of x successes in the time interval T is given by the binomial probability $b(x; n, p)$ with $n = \dfrac{T}{\Delta t}$ and $p = \alpha \cdot \Delta t$. Then, following the argument on page 80, we find that when $n \longrightarrow \infty$ the probability of x successes during the time interval T is given by the corresponding Poisson probability with the parameter $\lambda = n \cdot p = \dfrac{T}{\Delta t} \cdot (\alpha \cdot \Delta t) = \alpha T$. Since λ is the mean of this Poisson distribution, note that α is the average (mean) number of successes per unit time.

EXAMPLE If a bank receives on the average $\lambda = 6$ bad checks per day, what are the probabilities that it will receive

    (a) four bad checks on any given day;
    (b) 10 bad checks on any two consecutive days?

    Solution (a) Substituting $x = 4$ and $\lambda = 6$ into the formula for the Poisson distribution, we get

$$f(4; 6) = \frac{6^4 \cdot e^{-6}}{4!} = \frac{1,296(0.0025)}{24} = 0.135$$

(b) Here we want to find $f(10; 12)$, and we write

$$f(10; 12) = F(10; 12) - F(9; 12)$$
$$= 0.347 - 0.242$$
$$= 0.105$$

where the values of $F(10; 12)$ and $F(9; 12)$ were obtained from Table 2.

EXAMPLE   In the inspection of tin plate by a continuous electrolytic process, 0.2 imperfection is spotted on the average per minute. Find the probabilities of spotting
(a) one imperfection in 3 minutes;
(b) at least two imperfections in 5 minutes;
(c) at most one imperfection in 15 minutes.

Solution   Referring in each case to Table 2, we get

$$F(1; 0.6) - F(0; 0.6) = 0.878 - 0.549$$
$$= 0.329$$

for part (a),

$$1 - F(1; 1.0) = 1 - 0.736$$
$$= 0.264$$

for part (b), and

$$F(1; 3.0) = 0.199$$

for part (c).

The Poisson distribution has many important applications in **queueing** problems, where we may be interested, for example, in the number of customers arriving for service at a cafeteria, the number of ships or trucks arriving to be unloaded at a receiving dock, the number of aircraft arriving at an airport, and so forth. Thus, if on the average 0.3 customer arrive per minute at a cafeteria, then the probability that exactly 3 customers will arrive during a 5-minute span is

$$F(3; 1.5) - F(2; 1.5) = 0.934 - 0.809 = 0.125$$

and if on the average 3 trucks arrive per hour to be unloaded at a warehouse, then the probability that at most 20 will arrive during an 8-hour day is

$$F(20; 24) = 0.243$$

Also, if on the average 14.5 planes arrive per day at a private airport, then the probability that the number of arrivals on any one day will be anywhere from 12 to 15, inclusive, is

$$F(15; 14.5) - F(11; 14.5) = 0.619 - 0.220 = 0.399$$

3.8 THE GEOMETRIC DISTRIBUTION

On page 7 we indicated that a countably infinite sample space would be needed if we are interested in the number of cars persons have to inspect until they find one whose nitrogen oxide emission does not meet government standards. To treat this kind of problem in general, suppose that in a sequence of trials we are interested in the number of the trial on which the first success occurs, and that all but the third of the assumptions underlying the binomial distribution are satisfied; in other words, n is not fixed.

Clearly, if the first success is to come on the xth trial, it has to be preceded by $x - 1$ failures, and if the probability of a success is p, the probability of $x - 1$ failures on $x - 1$ trials is $(1 - p)^{x-1}$. Then, if we multiply this expression by the probability p of a success on the xth trial, we find that the probability of getting the first success on the xth trial is given by

Geometric distribution

$$g(x; p) = p(1 - p)^{x-1} \quad \text{for } x = 1, 2, 3, 4, \ldots$$

This probability distribution is called the **geometric distribution**, and as the reader will be asked to verify in Exercise 3.72 on page 90, its mean is

Mean of geometric distribution

$$\mu = \frac{1}{p}$$

EXAMPLE If the probability is 0.20 that a burglar will get caught on any given job, what is the probability that he will get caught for the first time on his fourth job?

> ***Solution*** Substituting $x = 4$ and $p = 0.20$ into the formula for the geometric distribution, we get
>
> $$g(4; 0.20) = (0.20)(1 - 0.20)^{4-1}$$
> $$= 0.102$$

EXAMPLE If the probability is 0.05 that a certain kind of measuring device will show excessive drift, what is the probability that the sixth of these measuring devices tested will be the first to show excessive drift?

Solution Substituting $x = 6$ and $p = 0.05$ into the formula for the geometric distribution, we get

$$g(6; 0.05) = (0.05)(1 - 0.05)^{6-1}$$
$$= 0.039$$

In **queueing theory**, the geometric distribution has important applications in connection with the number of units (customers, trucks, airplanes, etc.) that are being served or are waiting to be served at any given time. Suppose, for instance, that the arrival of cars at a gas station with one pump is a Poisson process with a **mean arrival rate** (average number of arrivals per unit time) of α cars per hour, and that the cars are served at a **mean service rate** of β cars per hour. Since the arrivals constitute a Poisson process, the probability of more than one arrival during a small time interval Δt is negligible, and we shall assume that the same is true also for the completions of service.† Furthermore, we shall assume that $\alpha < \beta$, namely, that the mean number of arrivals per unit time is less than the mean number of services that can be completed per unit time.

To arrive at an expression for $\pi(k)$, the probability that there are k cars in the gas station at any given time being served or waiting to be served, we shall first obtain a set of recursion relations expressing $\pi(k)$ in terms of $\pi(k - 1)$ and $\pi(k - 2)$. To illustrate how these relations are obtained, let us find the probability that there are no cars in the gas station (that is, $k = 0$) at time $t + \Delta t$. If $k = 0$ at time $t + \Delta t$, the only values that k could have had at time t are 0 and 1 by the assumption of Poisson arrivals. If $k = 0$ at time t, then no cars arrived in the interval from t to $t + \Delta t$ (with probability $1 - \alpha \cdot \Delta t$), but if $k = 1$ at time t, then a service was completed in this interval (with probability $\beta \cdot \Delta t$). Using the rule of elimination (see page 40), we thus conclude that

$$\pi(0) = \pi(0)(1 - \alpha \cdot \Delta t) + \pi(1)\beta \cdot \Delta t$$

or

$$\pi(1) = \frac{\alpha}{\beta} \cdot \pi(0)$$

In general, the situation is typified by the the tree diagram of Figure 3.8, and the rule of elimination leads to the following recursion relation:

$$\pi(k) = \pi(k - 1)\alpha \cdot \Delta t + \pi(k)[1 - (\alpha + \beta)\Delta t] + \pi(k + 1)\beta \cdot \Delta t$$

for $k \geq 1$. In Exercise 3.71 on page 90, the reader will be asked to verify that a solution of this set of equations is given by the geometric distribution

---

†In fact, it will be assumed that so long as the service station is not empty, completions of service constitute a Poisson process.

$$\pi(k) = \left(1 - \frac{\alpha}{\beta}\right)\left(\frac{\alpha}{\beta}\right)^{k} \qquad \text{for } k = 0, 1, 2, \ldots$$

which may be obtained from the expression on page 85 by letting $p = 1 - \dfrac{\alpha}{\beta}$ and $x = k + 1$.

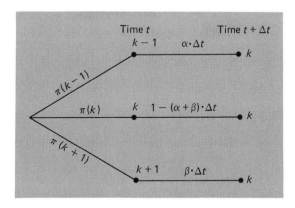

Figure 3.8 Tree diagram for queueing problem

This distribution reveals many important properties of queues. Note first that since $\alpha < \beta$ by assumption, we have

$$\lim_{k \to \infty} \pi(k) = 0$$

so that the probability is 1 that the waiting line remains finite. (In fact, it can be shown that if $\alpha \geq \beta$ there is a nonzero probability that the waiting line will eventually grow beyond all bounds.) The mean number of cars in the station (being served and waiting to be served) is given by the mean of the above distribution, which is

$$\frac{1}{p} - 1 = \frac{1}{1 - \dfrac{\alpha}{\beta}} - 1 = \frac{\alpha}{\beta - \alpha}$$

in accordance with the formula for μ on page 85. (We subtracted the 1 because the values of the random variable started with $k = 0$ instead of $x = 1$.)

Also, the distribution of the number of cars in the queue waiting to be served (thus, excluding the car being served) is given by $\pi(k + 1)$ for $k \geq 1$ and $\pi(0) + \pi(1)$ for $k = 0$. Thus, the mean number of cars in the queue is

$$\left(1 - \frac{\alpha}{\beta}\right) \sum_{k=1}^{\infty} k \left(\frac{\alpha}{\beta}\right)^{k+1} = \frac{\alpha^2}{\beta(\beta - \alpha)}$$

To find the mean time a customer spends waiting in the queue, it is necessary first to find the distribution of waiting times, and then to find its mean. We shall omit this work, stating only that the resulting mean waiting time is

$$\frac{\alpha}{\beta(\beta - \alpha)}.$$

EXERCISES

3.51 Prove that for the Poisson distribution

$$\frac{f(x + 1; \lambda)}{f(x; \lambda)} = \frac{\lambda}{x + 1}$$

for $x = 0, 1, 2, \ldots$.

3.52 Use the recursion formula of Exercise 1 to calculate the values of the Poisson distribution with $\lambda = 3$ for $x = 0, 1, 2, \ldots$, and 9, and draw the histogram of this distribution. Verify your results by referring to Table 2.

3.53 Use Table 2 to find
(a) $F(4; 7)$;
(b) $f(4; 7)$;
(c) $\sum\limits_{k=6}^{19} f(4; 8)$.

3.54 Use Table 2 to find
(a) $F(9; 12)$;
(b) $f(9; 12)$;
(c) $\sum\limits_{k=3}^{12} f(k; 7.5)$.

3.55 Use the Poisson distribution to approximate the binomial probability $b(3; 100, 0.03)$.

3.56 In a given city, 6% of all drivers get at least one parking ticket per year. Use the Poisson approximation to the binomial distribution to determine the probabilities that among 80 drivers (randomly chosen in this city)
(a) four will get at least one parking ticket in any given year;
(b) at least three will get at least one parking ticket in any given year;
(c) anywhere from three to six, inclusive, will get at least one parking ticket in any given year.

3.57 Records show that the probability is 0.00004 that a car will have a flat tire while driving through a certain tunnel. Use the formula for the Poisson distribution to approximate the probability that at least two of 10,000 cars passing through the tunnel with have a flat tire.

3.58 If 0.8% of the fuses delivered to an arsenal are defective, use the Poisson approximation to determine the probability that four fuses will be defective in a random sample of 400.

3.59 The number of weekly breakdowns of a computer is a random variable having a Poisson distribution with $\lambda = 0.3$. What is the probability that the computer will operate without a breakdown for two consecutive weeks?

3.60 The number of gamma rays emitted per second by a certain radioactive substance is a random variable having the Poisson distribution with $\lambda = 5.8$. If a recording instrument becomes inoperative when there are more than 12 rays per second, what is the probability that this instrument becomes inoperative during any given second?

3.61 Given that the switchboard of a consultant's office receives on the average 0.6 call per minute, find the probabilities that
(a) in a given minute there will be at least one call;
(b) in a 4-minute interval there will be at least three calls.

3.62 At a checkout counter customers arrive at an average rate of 1.5 per minute. Find the probabilities that
(a) at most four will arrive in any given minute;
(b) at least three will arrive during an interval of 2 minutes;
(c) at most 15 will arrive during an interval of 6 minutes.

3.63 A company rents out time on a computer for periods of t hours, for which it receives $600 an hour. The number of times the computer breaks down during t hours is a random variable having the Poisson distribution with $\lambda = (0.8)t$, and if the computer breaks down x times during t hours it costs $50x^2$ dollars to fix it. How should the company select t in order to maximize its expected profit?

3.64 The determination of geometric probabilities is often simplified by making use of the identity

$$g(x; p) = \frac{1}{x} \cdot b(1; x, p)$$

and looking up $b(1; x, p)$ in a table of binomial probabilities. Verify this identity and use it (and Table 1) to evaluate
(a) $g(12; 0.10)$;   (b) $g(10; 0.30)$.

3.65 An expert shot hits a target 95% of the time. What is the probability that he will miss the target for the first time on the fifteenth shot?

3.66 In a "torture test" a light switch is turned on and off until it fails. If the probability that the switch will fail any time it is turned on or off is 0.001, what is the probability that the switch will fail *after* it has been turned on or off 1,200 times. Assume that the conditions underlying the geometric distribution are met. [*Hint*: Use the formula for the value of an infinite geometric progression and logarithms.]

3.67 A pool company's records show that the probability is 0.20 that one of its new pools will require repairs within a year. What is the probability that the sixth pool it builds in a given year will be the first one to require repairs within a year?

3.68 In the example on page 84 we considered customers arriving at a cafeteria at an average rate of 0.3 per minute. If they are being served at an average rate of 0.5 per minute, find
(a) the average number of customers being served or waiting to be served at any given time;
(b) the average number of customers waiting to be served at any given time;
(c) the average time a customer spends waiting in line.

3.69 The arrival of trucks at a receiving dock is a Poisson process with a mean arrival rate of two per hour. The trucks can be unloaded at a mean rate of three per hour, and, so long as the receiving dock is not empty, completions of service constitute a Poisson process.
(a) What is the average number of trucks being unloaded and waiting to be unloaded?
(b) What is the mean number of trucks in the queue?
(c) What is the mean time a truck spends waiting in the queue?
(d) What is the probability that there are no trucks waiting to be unloaded?

3.70 With reference to Exercise 3.69, suppose that the cost of keeping a truck in the system is $15 per hour. If it were possible to increase the mean loading rate to 3.5 trucks per hour at a cost of $100 per day, would this be worthwhile?

3.71 Verify (by substitution) that the geometric distribution is a solution of the recursion relation given on page 86.

3.72 Differentiating with respect to p both sides of the equation

$$\sum_{x=1}^{\infty} p(1-p)^{x-1} = 1$$

show that the geometric distribution

$$f(x) = p(1-p)^{x-1} \qquad \text{for } x = 1, 2, 3, \ldots$$

has the mean $1/p$.

3.73 The following is an alternative way of introducing the Poisson distribution. Suppose that $f(x, t)$ is the probability of getting x successes during a time interval of length t when (1) the probability of a success during a very small time interval from t to $t + \Delta t$ is $\alpha \cdot \Delta t$, (2) the probability of more than one success occurring during such a time interval is negligible, and (3) the probability of a success during such a time interval does not depend on what happened prior to time t.
(a) Show that

$$f(x, t + \Delta t) = f(x, t)[1 - \alpha \Delta t] + f(x-1, t)\alpha \Delta t$$

(b) Using the result of part (a), show that

$$\frac{d[f(x, t)]}{dt} = \alpha[f(x-1, t) - f(x, t)]$$

(c) Verify by substitution that the differential equation obtained in (b) is satisfied by the formula for the Poisson distribution with $\lambda = \alpha t$.

3.74 Use the formulas defining μ and σ^2 to show that the mean and the variance of the Poisson distribution are both equal to λ.

3.9 THE MULTINOMIAL DISTRIBUTION

An immediate generalization of the binomial distribution arises when each trial can have more than two possible outcomes. This happens, for example, when a manufactured product is classified as superior, average, or poor, when a student's performance is judged by giving him an A, B, C, D, or F, or when an experiment is judged successful, unsuccessful, or inconclusive. To treat this kind of problem in general, let us consider the case where there are n independent trials, with each trial permitting k mutually exclusive outcomes whose respective probabilities are $p_1, p_2, \ldots, p_k$ (with $\sum_{i=1}^{k} p_i = 1$). Referring to the outcomes as being of the first kind, the second kind, $\ldots$, and the kth kind, we shall be interested in the probability $f(x_1, x_2, \ldots, x_k)$ of getting x_1 outcomes of the first kind, x_2 outcomes of the second kind, $\ldots$, and x_k outcomes of the kth kind, with $\sum_{i=1}^{k} x_i = n$. Using arguments similar to those which we employed in deriving the equation for the binomial distribution in Section 3.2, it can be shown that the desired probability is given by

Multinomial distribution

$$f(x_1, x_2, \ldots, x_k) = \frac{n!}{x_1! x_2! \cdot \ldots \cdot x_k!} p_1^{x_1} \cdot p_2^{x_2} \cdot \ldots \cdot p_k^{x_k}$$

for $x_i = 0, 1, \ldots, n$ for each i, subject to the restriction $\sum_{i=1}^{k} x_i = n$. The **joint probability distribution** whose values are given by these probabilities is called the **multinomial distribution**; it owes its name to the fact that for the various values of the x_i the probabilities are given by the corresponding terms of the multinomial expansion of $(p_1 + p_2 + \ldots + p_k)^n$.

EXAMPLE The probabilities that the light bulb of a certain kind of slide projector will last fewer than 40 hours of continuous use, anywhere from 40 to 80 hours of continuous use, or more than 80 hours of continuous use, are 0.30, 0.50, and 0.20. Find the probability that among eight such bulbs two will last fewer than 40 hours, five will last anywhere from 40 to 80 hours, and one will last more than 80 hours.

Solution Substituting $n = 8$, $x_1 = 2$, $x_2 = 5$, $x_3 = 1$, $p_1 = 0.30$, $p_2 = 0.50$, and $p_3 = 0.20$ into the formula, we get

$$f(2, 5, 1) = \frac{8!}{2!\, 5!\, 1!} (0.30)^2 (0.50)^5 (0.20)^1$$

$$= 0.0945$$

EXERCISES

3.75 Suppose that the probabilities are, respectively, 0.40, 0.40, and 0.20 that in city driving a certain kind of imported car will average less than 22 miles per gallon, anywhere from 22 to 25 miles per gallon, or more than 25 miles per gallon. Find the probability that among 12 such cars tested, four will average less than 22 miles per gallon, six will average anywhere from 22 to 25 miles per gallon, and two will average more than 25 miles per gallon.

3.76 As can easily be shown, the probabilities of getting 0, 1, or 2 heads with a pair of balanced coins are $\frac{1}{4}$, $\frac{1}{2}$, and $\frac{1}{4}$. What is the probability of getting two tails twice, one head and one tail three times, and two heads once in six tosses of a pair of balanced coins?

3.77 Suppose that the probabilities are, respectively, 0.60, 0.20, 0.10, and 0.10 that an income tax form will be filled in correctly, that it will contain an error favoring the taxpayer, that it will contain an error favoring the government, or that it will contain both kinds of errors. Find the probability that among 10 of the income tax forms randomly selected for audit five will be correct, three will contain an error favoring the taxpayer, one will contain an error favoring the government, and one will contain both kinds of errors.

3.78 Using the same sort of reasoning as in the derivation of the formula for the hypergeometric distribution, we can derive a formula which is analogous to the multinomial distribution but applies to sampling without replacement. If a set of N objects contains a_1 objects of the first kind, a_2 objects of the second kind, ..., and a_k objects of the kth kind, so that $a_1 + a_2 + \cdots + a_k = N$, the number of ways in which we can select x_1 objects of the first kind, x_2 objects of the second kind, ..., and x_k objects of the kth kind is given by the product of the number of ways in which we can select x_1 of the a_1 objects of the first kind, x_2 of the a_2 objects of the second kind, ..., and x_k of the a_k objects of the kth kind. Thus, the probability of getting that many objects of each kind is simply this product divided by the total number of ways in which $x_1 + x_2 + \cdots + x_k = n$ objects can be selected from the whole set of N objects.

(a) Write a formula for the probability of thus obtaining x_1 objects of the first kind, x_2 objects of the second kind, ..., and x_k objects of the kth kind.

(b) If 20 defective glass bricks include 10 that have cracks but no discoloration, 7 that are discolored but have no cracks, and 3 that have cracks and discoloration, what is the probability that among 6 of the bricks chosen at random for further checks 3 will have cracks only, 2 will only be discolored, and 1 will have cracks as well as discoloration?

3.10   SIMULATION

In recent years, simulation techniques have been applied to many problems in the various sciences, and if the processes which are being simulated involve an element of chance, these techniques are referred to as **Monte Carlo**

methods. Very often, the use of Monte Carlo simulation eliminates the cost of building and operating expensive equipment; it is used, for instance, in the study of collisions of photons with electrons, the scattering of neutrons, and similar complicated phenomena. Monte Carlo methods are also useful in situations where direct experimentation is impossible—say, in studies of the spread of cholera epidemics, which, of course, are not induced experimentally on human populations. In addition, Monte Carlo techniques are sometimes applied to the solution of mathematical problems which actually cannot be solved by direct means, or where a direct solution is too costly or requires too much time.

A classical example of the use of Monte Carlo methods in the solution of a problem of pure mathematics is the determination of π (the ratio of the circumference of a circle to its diameter) by probabilistic means. Early in the eighteenth century George de Buffon, a French naturalist, proved that if a very fine needle of length a is thrown at random on a board ruled with equidistant parallel lines, the probability that the needle will intersect one of the lines is $2a/\pi b$, where b is the distance between the parallel lines. What is remarkable about this fact is that it involves the constant $\pi = 3.1415926\ldots$, which in elementary geometry is approximated by the circumferences of regular polygons enclosed in a circle of radius $\frac{1}{2}$. Buffon's result implies that if such a needle is actually tossed a great many times, the proportion of the time it crosses one of the lines gives an estimate of $2a/\pi b$ and, hence, an estimate of π since a and b are known. Early experiments of this kind yielded an estimate of 3.1519 (based on 5,000 trials) and an estimate of 3.155 (based on 3,204 trials) in the middle of the nineteenth century.

Although Monte Carlo methods are sometimes based on actual gambling devices (for example, the needle tossing in the estimation of π), it is usually expedient to use so-called tables of **random digits** or **random numbers.** Tables of random numbers consist of many pages on which the digits 0, 1, 2, $\ldots$, and 9 are set down in a "random" fashion, much as they would appear if they were generated one at a time by a gambling device giving each digit an equal probability of being selected. Actually, we could construct such tables ourselves—say, by repeatedly drawing numbered slips out of a hat or by using a perfectly constructed spinner—but in practice such tables are usually generated by means of electronic computers. For instance, in one such method the computer begins with a 4-digit number, say, 3571, and squares it, getting 12752041. The middle four digits, in this case 7520, are looked upon as a set of 4 random digits. (If necessary, a 0 is added on the left to make the square have eight digits.) Then 7520 is squared, yielding 56550400, and 5504 is looked upon as the next set of 4 random digits. Continuing in this way, we obtain a table of pseudo-random digits which are quite satisfactory for most practical purposes.

Although tables of random numbers are constructed so that the digits can be looked upon as values of a random variable having the discrete uniform distribution $f(x) = \frac{1}{10}$ for $x = 0, 1, 2, \ldots$, or 9, they can be used to simulate values of any discrete random variable, and even continuous random variables.

To illustrate the use of a table of random numbers, let us show how we can play "heads or tails" without actually flipping a coin. Letting 0, 2, 4, 6, and 8 represent heads and 1, 3, 5, 7, and 9 represent tails, we might arbitrarily choose the 4th column of the first page of Table 7, start at the top and go down the page. Thus, we get 6, 2, 7, 5, 5, 0, 1, 8, 6, 3, ..., and we interpret this as head, head, tail, tail, tail, head, tail, head, head, tail, ...

Repeated flips of any number of coins, say, three coins, can be simulated in the same way. If we arbitrarily choose the 9th, 10th, and 11th columns of the second page of Table 7, start at the top and go down the page, we get 480, 280, 085, 265, 303, 288, 295, 388, 127, 222, ..., and, counting the number of even digits in each case, we interpret this as 3, 3, 2, 2, 1, 3, 1, 2, 1, 3, ..., heads. If we did not want to count even digits, we could make use of the fact that the probabilities of getting 0, 1, 2, or 3 heads are $\frac{1}{8}$, $\frac{3}{8}$, $\frac{3}{8}$, and $\frac{1}{8}$, and use the following scheme:

Number of heads	Probability	Random digits
0	$\frac{1}{8}$	0
1	$\frac{3}{8}$	1, 2, 3
2	$\frac{3}{8}$	4, 5, 6
3	$\frac{1}{8}$	7

Ignoring the digits 8 and 9 wherever they may occur, we would thus interpret the random digits 4, 8, 1, 1, 1, 9, 3, 3, 3, 7, 6, 3, 6, 2, 6, 5, 8, 9, 3, 1, 0, and 6 (in the 5th row of the third page of Table 7) as getting 2, 1, 1, 1, 1, 1, 1, 3, 2, 1, 2, 1, 2, 2, 1, 1, 0, and 2 heads in 18 tosses of three coins.

Of the two methods used in the preceding example, the first has the disadvantage that we have to count how many even digits there are in each case, and the second has the disadvantage that we have to ignore 8 and 9. To avoid such waste of effort and time, we could have used the following scheme:

Number of heads	Probability	Cumulative probability	Random numbers
0	0.125	0.125	000–124
1	0.375	0.500	125–499
2	0.375	0.875	500–874
3	0.125	1.000	875–999

Here we used three-digit random numbers because the probabilities are given to three decimals, and we allocated 125 (or one-eighth) of the 1,000 random numbers from 000 to 999 to 0 heads, 375 (or three-eighths) to 1 head, 375 (or three-eighths) to 2 heads, and 125 (or one-eighth) to 3 heads. The column of "or less" **cumulative probabilities** was added to facilitate the assignment of the random numbers. Observe that in each case the last random digit is one less than the number formed by the three decimal digits of the corresponding cumulative probability.

With this scheme, if we arbitrarily use the 22nd, 23rd, and 24th columns of the first page of Table 7, starting with the 6th row and going down the page, we get 197, 565, 157, 520, 946, 951, 948, 568, 586, and 089, and we interpret this as 1, 2, 1, 2, 3, 3, 3, 2, 2, and 0 heads.

The method we have illustrated here with reference to a game of chance, can be used to simulate observations of any random variable with a given probability distribution.

EXAMPLE Suppose that the probabilities are 0.082, 0.205, 0.256, 0.214, 0.134, 0.067, 0.028, 0.010, 0.003, and 0.001 that 0, 1, 2, 3, . . . , or 9 cars will arrive at a toll booth of a turnpike during any 1 minute interval in the early afternoon.

    (a) Distribute the three-digit random numbers from 000 to 999 among the 10 values of this random variable, so that they can be used to simulate the arrival of cars at the toll booth.

    (b) Use the 5th, 6th, and 7th columns of the fourth page of Table 7, starting with the 11th row and going down the page, to simulate the arrival of cars at the toll booth during 20 one-minute intervals in the early afternoon.

Solution (a) Calculating the cumulative probabilities and following the suggestion given above, we arrive at the following scheme:

Number of cars	Probability	Cumulative probability	Random numbers
0	0.082	0.082	000–081
1	0.205	0.287	082–286
2	0.256	0.543	287–542
3	0.214	0.757	543–756
4	0.134	0.891	757–890
5	0.067	0.958	891–957
6	0.028	0.986	958–985
7	0.010	0.996	986–995
8	0.003	0.999	996–998
9	0.001	1.000	999

(b) Following the instructions, we get the random numbers 036, 417, 962, 458, 778, 541, 869, 379, 973, 553, 325, 674, 907, 710, 709, 499, 493,

384, 346, and 301, and this means that 0, 2, 6, 2, 4, 2, 4, 2, 6, 3, 2, 3, 5, 3, 3, 2, 2, 2, 2, and 2 cars arrived at the toll booth during the 20 one-minute intervals.

EXERCISES

3.79 Letting any five digits represent heads and the other five digits represent tails, use random numbers to simulate 100 flips of a balanced coin.

3.80 Using the digits 1, 2, 3, 4, 5, and 6 to represent the corresponding faces of a die (and omitting 0, 7, 8, and 9), simulate 120 rolls of a balanced die.

3.81 The probabilities that a salesman of computer software will make 0, 1, 2, 3, 4, or 5 sales on any one day are 0.14, 0.28, 0.27, 0.18, 0.09, and 0.04.
(a) Distribute the two-digit random numbers from 00 to 99 among the six values of this random variable, so that the corresponding random numbers can be used to simulate the salesman's sales.
(b) Use the result of part (a) to simulate the salesman's sales on 25 days.

3.82 With reference to the example on page 81 and the probabilities shown in Figure 3.7, distribute the three-digit random numbers from 000 to 999 among the 11 values of the random variable, and simulate the fire insurance company's experience (that is, the number of policyholders who file at least one claim) over a period of 20 years.

3.83 Suppose that the probabilities are 0.2466, 0.3452, 0.2417, 0.1128, 0.0395, 0.0111, 0.0026, and 0.0005 that there will be 0, 1, 2, 3, 4, 5, 6, or 7 polluting spills in the Great Lakes on any one day.
(a) Distribute the four-digit random numbers from 0000 to 9999 to the eight values of this random variable, so that the corresponding random numbers can be used to simulate daily polluting spills in the Great Lakes.
(b) Use the result of part (a) to simulate the numbers of polluting spills in the Great Lakes on 30 days.

3.84 Depending on the availability of parts, a company can manufacture 3, 4, 5, or 6 units of a certain item per week with corresponding probabilities of 0.10, 0.40, 0.30, and 0.20. The probabilities that there will be a weekly demand for 0, 1, 2, 3, ..., or 8 units are, respectively, 0.05, 0.10, 0.30, 0.30, 0.10, 0.05, 0.05, 0.04, and 0.01. If a unit is sold during the week that it is made, it will yield a profit of $100; this profit is reduced by $20 for each week that a unit has to be stored. Use random numbers to simulate the operations of this company for 50 consecutive weeks and estimate its expected weekly profit.

3.11   CHECK LIST OF KEY TERMS (with page references)

Bar chart, 56

Bernoulli trials, 57

Binomial coefficient, 58

Binomial distribution, 58

Chebychev's theorem, 74

Cumulative probability, 59

Discrete random variable, 55

Distribution function, 57

4

Probability Densities

Continuous sample spaces and continuous random variables arise when we deal with quantities that are measured on a continuous scale—for instance, when we measure the speed of a car, the amount of alcohol in a person's blood the efficiency of a solar collector, or the amount of tar in a cigarette.

In this chapter we shall learn how to determine, and work with, probabilities relating to continuous sample spaces and continuous random variables. The introduction to probability densities in Section 4.1 is followed by a discussion of the normal distribution in Sections 4.2 and 4.3, and various other special probability densities in Sections 4.4 through 4.9. Problems involving more than one continuous random variable are discussed in Section 4.10.

When we first introduced the concept of a random variable in Chapter 3, we presented it as a real-valued function defined over the sample space of an experiment, and we illustrated this idea with the random variable giving the number of preferred ratings which a lawn mower received, assigning the numbers 0, 1, 2, or 3 (whichever was appropriate) to the 18 possible outcomes of the experiment. In the continuous case, where random variables can assume values on a continuous scale, the procedure is very much the same. The outcomes of an experiment are represented by the points on a line segment or a line, and the value of a random variable is a number appropriately assigned to each point by means of some rule or equation. When the value of a random variable is given directly by a measurement or observation, we usually do not bother to differentiate between the value of the random variable, the measurement which we obtain, and the outcome of the experiment, the corresponding point on the real axis. Thus, if an experiment consists of determining what force is required to break a given tensile-test specimen, the result itself, say, 138.4 pounds, is the value of the random variable with which we are concerned. There is no real need in that case to add that the sample space of the experiment consists of all (or part of) the points on the positive real axis. In general, we shall write $P(a \leq x \leq b)$ for the probability associated with the points of the sample space for which the value of a random variable falls on the interval from a to b.

The problem of defining probabilities in connection with continuous sample spaces and continuous random variables involves some complications. To illustrate the nature of these complications, let us consider the following situation: Suppose we want to know the probability that if an accident occurs on a freeway whose length is 200 miles, it will happen at some given location or, perhaps, some particular stretch of the road. The outcomes of this experiment can be looked upon as a continuum of points, namely, those on the continuous interval from 0 to 200. Suppose the probability that the accident occurs on any interval of length L is $L/200$, with L measured in miles. Note that this arbitrary assignment of probability is consistent with Axioms 1 and 2 on page 23, since the probabilities are all nonnegative and less than or equal to 1, and $P(S) = \frac{200}{200} = 1$. Of course, we are considering so far only events represented by intervals which form part of the line segment from 0 to 200. Using Axiom 3′ on page 80, we can also obtain probabilities of events which are not intervals but which can be represented by the union of finitely many or countably many intervals. Thus, for two nonoverlapping intervals of length L_1 and L_2 we have a probability of

$$\frac{L_1 + L_2}{200}$$

and for an infinite sequence of nonoverlapping intervals of length L_1, L_2, L_3, . . . , we have a probability of

$$\frac{L_1 + L_2 + L_3 + \ldots}{200}$$

Note that the probability that the accident occurs at any given point is equal to zero because we can look upon a point as an interval of zero length. However, the probability that the accident occurs in a very short interval is positive; for instance, for an interval of length 1 foot the probability is $9.5(10)^{-7}$.

   Thus, in extending the concept of probability to the continuous case, we again use Axioms 1, 2, and 3′, but we shall have to restrict the meaning of the term "event." So far as practical considerations are concerned, this restriction is of no consequence; we simply do not assign probabilities to some rather abstruse point sets, which cannot be expressed as the unions or intersections of finitely many or countably many intervals.

   The way in which we assigned probabilities in the preceding example is, of course, very special; it is similar in nature to the way in which we assign equal probabilities to the six faces of a die, heads and tails, the 52 cards in a standard deck, and so forth. To treat the problem of associating probabilities with continuous random variables generally, suppose we are interested in the probability that a given random variable will take on a value on the interval from a to b, where a and b are constants with $a \leq b$. Suppose, furthermore, that we divide the interval from a to b into n equal subintervals of width Δx containing, respectively, the points $x_1, x_2, \ldots, x_n$, and that the probability that the random variable will take on a value on the subinterval containing x_i is given by $f(x_i) \cdot \Delta x$. Then, the probability that the random variable with which we are concerned will take on a value on the interval from a to b is given by

$$P(a \leq x \leq b) = \sum_{i=1}^{n} f(x_i) \cdot \Delta x$$

Now, if f is an integrable function defined for all values of the random variable with which we are concerned, we shall define the probability that the value of the random variable falls between a and b by letting $\Delta x \longrightarrow 0$, namely, as

$$P(a \leq x \leq b) = \int_a^b f(x)\, dx$$

   This definition of probability in the continuous case presupposes the existence of an appropriate function f which, integrated from any constant a to any constant b (with $a \leq b$), gives the probability that the corresponding random variable takes on a value on the interval from a to b. Note that a value $f(x)$ of f does not give the probability that the corresponding random variable takes on the value x; in the continuous case, probabilities are given by integrals and not by the values of f.

To obtain the probability that a random variable will actually take on a given value x, we might first determine the probability that it will take on a value on the interval from $x - \Delta x$ to $x + \Delta x$, and then let $\Delta x \longrightarrow 0$. However, if we did this it would become apparent that the result is always zero. The fact that the probability is always zero that a continuous random variable will take on any given value x should not be disturbing. Indeed, our definition of probability for the continuous case provides a remarkably good model for dealing with measurements or observations. Owing to the limits of our ability to measure, experimental data never seem to come from a continuous sample space. Thus, while temperatures are fruitfully thought of as points on a continuous scale, any temperature measurement actually represents an interval on this scale. If we report a temperature measurement of 74.8 degrees centigrade, we really mean that the temperature lies in the interval from 74.75 to 74.85 degrees centigrade, and not that it is exactly 74.8000. . . .

It is important to add that when we say that there is a zero probability that a random variable will take on any given value x, this does not mean that it is impossible that the random variable will take on the value x. In the continuous case, a zero probability does not imply logical impossibility, but the whole matter is largely academic since, owing to the limitations of our ability to measure and observe, we are always interested in probabilities connected with intervals and not with isolated points.

As an immediate consequence of the fact that in the continuous case probabilities associated with individual points are always zero, we find that if we speak of the probability associated with the interval from a to b, it does not matter whether either endpoint is included. Symbolically,

$$P(a \leq x \leq b) = P(a \leq x < b) = P(a < x \leq b) = P(a < x < b)$$

Drawing an analogy with the concept of a density function in physics, we call the functions f, whose existence we stipulated in extending our definition of probability to the continuous case, **probability density functions,** or simply **probability densities**. Whereas density functions are integrated to obtain weights, probability density functions are integrated to obtain probabilities.

Since a probability density, integrated between any two constants a and b, gives the probability that a random variable assumes a value between these limits, f cannot be just any real-valued integrable function. However, if we impose the conditions that

$$f(x) \geq 0 \qquad \text{for all } x \text{ within the domain of } f$$

and

$$\int_{-\infty}^{\infty} f(x)\, dx = 1$$

it can be shown that the axioms of probability (with the modification about events discussed on page 100) are satisfied. Note the similarity between these conditions and those for probability distributions given on page 55.

As in the discrete case, we shall write as $F(x)$ the probability that a random variable with the probability density f takes on a value less than or equal to x, and we shall again refer to the corresponding function F as the **distribution function** of the random variable. Thus, if a random variable with values x has the probability density f, the values of its distribution function are given by

$$F(x) = \int_{-\infty}^{x} f(t)\, dt$$

Consequently, the probability that the random variable will take on a value on the interval from a to b is $F(b) - F(a)$, and according to the fundamental theorem of integral calculus it follows that

$$\frac{dF(x)}{dx} = f(x)$$

wherever this derivative exists.

EXAMPLE   If a random variable has the probability density

$$f(x) = \begin{cases} 2e^{-2x} & \text{for } x > 0 \\ 0 & \text{for } x \leq 0 \end{cases}$$

find the probabilities that it will take on a value
(a) between 1 and 3;
(b) greater than 0.5.

   Solution   Evaluating the necessary integrals, we get

$$\int_{1}^{3} 2e^{-2x}\, dx = e^{-2} - e^{-6} = 0.135$$

for part (a) and

$$\int_{0.5}^{\infty} 2e^{-2x}\, dx = e^{-1} = 0.368$$

for part (b).

Note that in the preceding example we made the domain of f include all the real numbers even though the probability is zero that x will be negative. This is a practice we shall follow throughout this book. It is also apparent from the graph of this function in Figure 4.1 that it has a discontinuity at $x = 0$; indeed, a probability density need not be everywhere continuous, as long as it is integrable between any two limits a and b (with $a \leq b$).

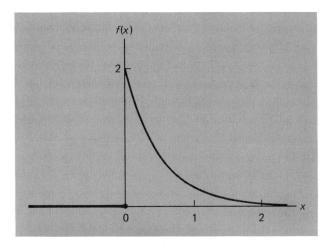

Figure 4.1 Graph of probability density.

EXAMPLE With reference to the preceding example, find the distribution function and use it to determine the probability that the random variable will take on a value less than or equal to 1.

Solution Performing the necessary integrations, we get

$$F(x) = \begin{cases} 0 & \text{for } x \leq 0 \\ \int_0^x 2e^{-2t}\,dt = 1 - e^{-2x} & \text{for } x > 0 \end{cases}$$

and substitution of $x = 1$ yields

$$F(1) = 1 - e^{-2} = 0.865$$

Note that the distribution function of this example is nondecreasing and that $F(-\infty) = 0$ and $F(\infty) = 1$. Indeed, it follows by definition that these properties are shared by all distribution functions.

EXAMPLE Find k so that the following can serve as the probability density of a random variable:

$$f(x) = \begin{cases} 0 & \text{for } x \leq 0 \\ kxe^{-4x^2} & \text{for } x > 0 \end{cases}$$

Solution To satisfy the first of the two conditions on page 101, k must be nonnegative, and to satisfy the second condition we must have

$$\int_{-\infty}^{\infty} f(x)\,dx = \int_{0}^{\infty} kxe^{-4x^2}\,dx = \int_{0}^{\infty} \frac{k}{8} \cdot e^{-u}\,du = \frac{k}{8} = 1$$

so that $k = 8$.

The statistical measures that are used to describe probability densities are very similar to the ones which we used to describe probability distributions. Replacing summations with integrals, we define the **kth moment about the origin** as

$$\mu'_k = \int_{-\infty}^{\infty} x^k \cdot f(x)\,dx$$

analogous to the definition we gave on page 73. The first moment about the origin is again referred to as the **mean**, and it is denoted by μ; as before, it is the expected, or average, value of a random variable having the probability density f. We also define the **kth moment about the mean** as

$$\mu_k = \int_{-\infty}^{\infty} (x - \mu)^k \cdot f(x)\,dx$$

In particular, the second moment about the mean is again referred to as the **variance** and it is written as σ^2; as before, it measures the spread of a probability density in the sense that it gives the expected value of the squared deviation from the mean. Also, σ is again referred to as the **standard deviation**.

EXAMPLE With reference to the example on page 102, find the mean and the variance of the given probability density.

Solution Performing the necessary integrations, we get

$$\mu = \int_{-\infty}^{\infty} xf(x)\,dx = \int_{0}^{\infty} x \cdot 2e^{-2x}\,dx = \tfrac{1}{2}$$

and

$$\sigma^2 = \int_{-\infty}^{\infty} (x - \mu)^2 f(x)\,dx = \int_{0}^{\infty} (x - \tfrac{1}{2})^2 \cdot 2e^{-2x}\,dx = \tfrac{1}{4}$$

EXERCISES

4.1 Verify that the function of the example on page 102 is, in fact, a probability density.

4.2 If the probability density of a random variable is given by

$$f(x) = \begin{cases} k(1 - x^2) & \text{for } 0 < x < 1 \\ 0 & \text{elsewhere} \end{cases}$$

find the value of k and the probabilities that a random variable having this probability density will take on a value
(a) between 0.1 and 0.2;
(b) greater than 0.5.

4.3 With reference to the preceding exercise, find the corresponding distribution function, and use it to determine the probabilities that a random variable having this distribution function will take on a value
(a) less than 0.3;
(b) between 0.4 and 0.6.

4.4 If the probability density of a random variable is given by

$$f(x) = \begin{cases} x & \text{for } 0 < x < 1 \\ 2 - x & \text{for } 1 \leq x < 2 \\ 0 & \text{elsewhere} \end{cases}$$

find the probabilities that a random variable having this probability density will take on a value
(a) between 0.2 and 0.8;
(b) between 0.6 and 1.2.

4.5 With reference to the preceding exercise, find the corresponding distribution function, and use it to determine the probabilities that a random variable having this distribution function will take on a value
(a) greater than 1.8;
(b) between 0.4 and 1.6.

4.6 Given the probability density $f(x) = \dfrac{k}{1 + x^2}$ for $-\infty < x < \infty$, find k.

4.7 If the distribution function of a random variable is given by

$$F(x) = \begin{cases} 1 - \dfrac{4}{x^2} & \text{for } x > 2 \\ 0 & \text{for } x \leq 2 \end{cases}$$

find the probabilities that this random variable will take on a value
(a) less than 3;
(b) between 4 and 5.

4.8 Find the probability density which corresponds to the distribution function of the preceding exercise. Are there any points at which it is undefined? Also sketch the graphs of the distribution function and the probability density.

4.9 In certain experiments, the error made in determining the density of a silicon compound is a random variable having the probability density

$$f(x) = \begin{cases} 25 & \text{for } -0.02 < x < 0.02 \\ 0 & \text{elsewhere} \end{cases}$$

Find the probabilities that such an error will be
(a) between -0.03 and 0.04;
(b) between -0.005 and 0.005.

4.10 The mileage (in thousands of miles) which car owners get with a certain kind of tire is a random variable having the probability density

$$f(x) = \begin{cases} \frac{1}{20}e^{-x/20} & \text{for } x > 0 \\ 0 & \text{for } x \leq 0 \end{cases}$$

Find the probabilities that one of these tires will last
(a) at most 10,000 miles;
(b) anywhere from 16,000 to 24,000 miles;
(c) at least 30,000 miles.

4.11 In a certain city, the daily consumption of electric power (in millions of kilowatt-hours) is a random variable having the probability density

$$f(x) = \begin{cases} \frac{1}{9}xe^{-x/3} & \text{for } x > 0 \\ 0 & \text{for } x \leq 0 \end{cases}$$

If the city's power plant has a daily capacity of 12 million kilowatt-hours, what is the probability that this power supply will be inadequate on any given day?

4.12 Prove that the identity $\sigma^2 = \mu'_2 - \mu^2$ holds for any probability density for which these moments exist.

4.13 Find μ and σ^2 for the probability density of Exercise 4.2.

4.14 Find μ and σ^2 for the probability density of Exercise 4.4.

4.15 Find μ and σ for the probability density obtained in Exercise 4.8.

4.16 Find μ and σ for the distribution of the errors of measurement of Exercise 4.9.

4.17 Find μ for the distribution of the mileages of Exercise 4.10.

4.18 Show that μ'_2 and, hence, σ^2 do not exist for the probability density of Exercise 4.6.

4.2 THE NORMAL DISTRIBUTION

Among the special probability densities we shall study in this chapter, the **normal probability density**, usually referred to simply as the **normal distribution**, is by far the most important.† It was studied first in the eighteenth century when scientists observed an astonishing degree of regularity in errors of measurement. They found that the patterns (distributions) they observed were closely approximated by a continuous distribution which they referred

---

†The words "density" and "distribution" are often used interchangeably in the literature of applied statistics.

to as the "normal curve of errors" and attributed to the laws of chance. The equation of the normal probability density, whose graph (shaped like the cross section of a bell) is shown in Figure 4.2, is

Normal distribution

$$f(x; \mu, \sigma^2) = \frac{1}{\sqrt{2\pi}\sigma} e^{-\frac{1}{2}\left(\frac{x-\mu}{\sigma}\right)^2} \qquad -\infty < x < \infty$$

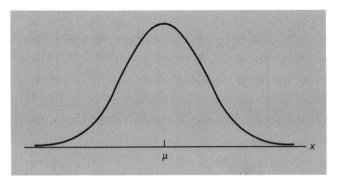

Figure 4.2 Graph of normal probability density.

and in Exercises 4.42 and 4.43 on page 114, the reader will be asked to verify that its parameters μ and σ are, indeed its mean and its standard deviation.

Since the normal probability density cannot be integrated in closed form between every pair of limits a and b, probabilities relating to normal distributions are usually obtained from special tables, such as Table 3 at the end of this book. This table pertains to the **standard normal distribution**, namely the normal distribution with $\mu = 0$ and $\sigma = 1$, and its entries are the values of

$$F(z) = \frac{1}{\sqrt{2\pi}} \int_{-\infty}^{z} e^{-\frac{1}{2}t^2} dt$$

for $z = 0.00, 0.01, 0.02, \ldots, 3.49$, and also $z = 4.00, z = 5.00$, and $z = 6.00$. To find the probability that a random variable having the standard normal distribution will take on a value between a and b, we use the equation $P(a < x < b) = F(b) - F(a)$, and if either a or b is negative, we also make use of the identity $F(-z) = 1 - F(z)$, which the reader will be asked to verify in Exercise 4.41 on page 114.

EXAMPLE Find the probabilities that a random variable having the standard normal distribution will take on a value
(a) between 0.87 and 1.28;
(b) between −0.34 and 0.62;

(c) greater than 0.85;

(d) greater than −0.65.

Solution   Looking up the necessary values in Table 3, we get

$$F(1.28) - F(0.87) = 0.8997 - 0.8078$$
$$= 0.0919$$

for part (a),

$$F(0.62) - F(-0.34) = 0.7324 - (1 - 0.6331)$$
$$= 0.3655$$

for part (b),

$$1 - F(0.85) = 1 - 0.8023$$
$$= 0.1977$$

for part (c), and

$$1 - F(-0.65) = 1 - [1 - F(0.65)]$$
$$= F(0.65)$$
$$= 0.7422$$

for part (d).

There are also problems in which we are given probabilities relating to standard normal distributions and asked to find the corresponding values of z. The results of the example which follows will be used extensively in subsequent chapters.

EXAMPLE   If z_α is such that the probability is α that it will be exceeded by a random variable having the standard normal distribution, find

(a) $z_{0.01}$;

(b) $z_{0.05}$.

Solution   (a) Since $F(z_{0.01}) = 0.99$, we look for the entry in Table 3 which is closest to 0.99 and get 0.9901 corresponding to $z = 2.33$. Thus, $z_{0.01} = 2.33$. (b) Since $F(z_{0.05}) = 0.95$, we look for the entry in Table 3 which is closest to 0.95 and get 0.9495 and 0.9505 corresponding to $z = 1.64$ and $z = 1.65$. Thus, $z_{0.05} = 1.645$.

To use Table 3 in connection with a random variable which has the values x and a normal distribution with the mean μ and the variance σ^2, we refer to the corresponding **standardized random variable** which has the values

$$z = \frac{x - \mu}{\sigma}$$

and the standard normal distribution. Thus, to find the probability that the original random variable will take on a value less than or equal to a, we look up $F\left(\dfrac{a-\mu}{\sigma}\right)$ in Table 3.

Also, if we want to find the probability that a random variable having the normal distribution with the mean μ and the variance σ^2 will take on a value between a and b, we have only to calculate the probability that a random variable having the standard normal distribution will take on a value between $\dfrac{a-\mu}{\sigma}$ and $\dfrac{b-\mu}{\sigma}$; symbolically,

$$P(a < x < b) = F\left(\frac{b-\mu}{\sigma}\right) - F\left(\frac{a-\mu}{\sigma}\right)$$

EXAMPLE If the amount of cosmic radiation to which a person is exposed while flying by jet across the United States is a random variable having the normal distribution with $\mu = 4.35$ mrem and $\sigma = 0.59$ mrem, find the probabilities that the amount of cosmic radiation to which a person will be exposed on such a flight is
(a) between 4.00 and 5.00 mrem;
(b) at least 5.50 mrem.

    Solution Looking up the necessary values in Table 3, we get

$$F\left(\frac{5.00-4.35}{0.59}\right) - F\left(\frac{4.00-4.35}{0.59}\right) = F(1.10) - F(-0.59)$$
$$= 0.8643 - (1 - 0.7224)$$
$$= 0.5867$$

for part (a) and

$$1 - F\left(\frac{5.50-4.35}{0.59}\right) = 1 - F(1.95)$$
$$= 1 - 0.9744$$
$$= 0.0256$$

for part (b).

EXAMPLE The actual amount of instant coffee which a filling machine puts into "4-ounce" jars may be looked upon as a random variable having a normal distribution with $\sigma = 0.04$ ounce. If only 2% of the jars are to contain less than 4 ounces, what is the mean fill of these jars?

    Solution To find μ such that $F\left(\dfrac{4-\mu}{0.04}\right) = 0.02$ and, hence, $F\left(-\dfrac{4-\mu}{0.04}\right) = 0.98$, we look for the entry in Table 3 closest to 0.98

and get 0.9798 corresponding to $z = 2.05$. Thus,

$$-\frac{4 - \mu}{0.04} = 2.05$$

and, solving for μ, we find that $\mu = 4.082$ ounces.

Although the normal distribution applies to continuous random variables, it is often used to approximate distributions of discrete random variables. Quite often, this yields satisfactory results, provided that we make the **continuity correction** illustrated in the following example.

EXAMPLE   In a certain city, the number of power outages per month is a random variable having a distribution with $\mu = 11.6$ and $\sigma = 3.3$. If this distribution can be approximated closely with a normal distribution, what is the probability that there will be at least eight outages in any one month?

> *Solution*   The answer is given by the area of the white region of Figure 4.3; the area to the right of 7.5, not 8. The reason for this is that the

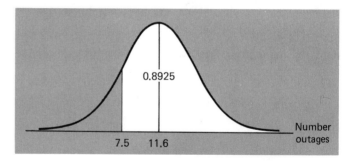

Figure 4.3 Diagram for example dealing with power outages.

number of outages is a discrete random variable, and if we want to approximate its distribution with a normal distribution, we must "spread" its values over a continuous scale. We do this by representing each integer k by the interval from $k - \frac{1}{2}$ to $k + \frac{1}{2}$. For instance, 3 is represented by the interval from 2.5 to 3.5, 10 is represented by the interval from 9.5 to 10.5, and "at least 8" is represented by the interval to the right of 7.5. Thus, the desired probability is approximated by

$$1 - F\left(\frac{7.5 - 11.6}{3.3}\right) = F(-1.24)$$
$$= F(1.24)$$
$$= 0.8925$$

4.3 THE NORMAL APPROXIMATION TO THE BINOMIAL DISTRIBUTION

In some books the normal distribution is introduced as a probability density which can be used to approximate the binomial distribution when n is large and p, the probability of a success, is close to 0.50 and, hence, not small enough to use the Poisson approximation. Thus, let us state, without proof, the following theorem:

Normal approximation to binomial distribution

Theorem 4.1 *If x is a value of a random variable having the binomial distribution with the parameters n and p, and if*

$$z = \frac{x - np}{\sqrt{np(1 - p)}}$$

then the limiting form of the distribution function of this standardized random variable as $n \longrightarrow \infty$ is given by

$$F(z) = \int_{-\infty}^{z} \frac{1}{\sqrt{2\pi}} e^{-\frac{1}{2}t^2} dt \qquad -\infty < z < \infty$$

Note that although x takes on only the values $0, 1, 2, \ldots, n$, in the limit as $n \longrightarrow \infty$ the distribution of the corresponding standardized random variable is continuous, and the corresponding probability density is the standard normal density. Note that in the following example, which is an application of Theorem 4.1, we use again the continuity correction given on page 110.

EXAMPLE If 20% of the diodes made in a certain plant are defective, what are the probabilities that in a lot of 100 randomly chosen for inspection

(a) at most 15 will be defective;

(b) exactly 15 will be defective?

Solution Since $\mu = 100(0.20) = 20$ and $\sigma = \sqrt{100(0.20)(0.80)} = 4$ for the binomial distribution with $n = 100$ and $p = 0.20$, we find that the normal approximation to the binomial distribution yields

$$F\left(\frac{15.5 - 20}{4}\right) = F(-1.13)$$

$$= 1 - F(1.13)$$

$$= 1 - 0.8708$$

$$= 0.1292$$

for part (a) and

$$F\left(\frac{15.5-20}{4}\right) - F\left(\frac{14.5-20}{4}\right) = F(-1.13) - F(-1.38)$$

$$= F(1.38) - F(1.13)$$

$$= 0.9162 - 0.8708$$

$$= 0.0454$$

for part (b).

Had we used the H. G. Romig table (referred to in the Bibliography) instead of the normal approximation in the preceding example, we would have obtained 0.1285 instead of 0.1292 for part (a) and 0.0481 instead of 0.0454 for part (b). Thus, it can be seen that both approximations are very close. A good rule of thumb is to use the normal approximation to the binomial distribution only when np and $n(1-p)$ are both greater than 5.

EXERCISES

4.19 If a random variable has the standard normal distribution, find the probability that it will take on a value
(a) less than 1.50;
(b) less than -1.20;
(c) greater than 2.16;
(d) greater than -1.75.

4.20 If a random variable has the standard normal distribution, find the probability that it will take on a value
(a) between 0 and 2.50;
(b) between 1.22 and 2.35;
(c) between -1.33 and -0.33;
(d) between -1.60 and 1.80.

4.21 Find z if the probability that a random variable having the standard normal distribution will take on a value
(a) less than z is 0.9911;
(b) greater than z is 0.1093;
(c) greater than z is 0.6443;
(d) less than z is 0.0217;
(e) between $-z$ and z is 0.9298.

4.22 If a random variable has a normal distribution, what are the probabilities that it will take on a value within
(a) 1 standard deviation of the mean;
(b) 2 standard deviations of the mean;
(c) 3 standard deviations of the mean;
(d) 4 standard deviations of the mean?

4.23 Verify that
(a) $z_{0.005} = 2.575$;
(b) $z_{0.025} = 1.96$.

4.24 Given a random variable having the normal distribution with $\mu = 16.2$ and $\sigma^2 = 1.5625$, find the probabilities that it will take on a value
(a) greater than 16.8;
(b) less than 14.9;
(c) between 13.6 and 18.8;
(d) between 16.5 and 16.7.

4.25 A random variable has a normal distribution with $\mu = 62.4$. Find its standard deviation if the probability is 0.20 that it will take on a value greater than 79.2.

4.26 A random variable has a normal distribution with $\sigma = 10$. If the probability is 0.8212 that it will take on a value less than 82.5, what is the probability that it will take on a value greater than 58.3?

4.27 The time required to assemble a piece of machinery is a random variable having approximately a normal distribution with $\mu = 12.9$ minutes and $\sigma = 2.0$ minutes. What are the probabilities that the assembly of a piece of machinery of this kind will take
(a) at least 11.5 minutes;
(b) anywhere from 11.0 to 14.8 minutes?

4.28 The burning time of an experimental rocket is a random variable having the normal distribution with $\mu = 4.76$ seconds and $\sigma = 0.04$ second. What is the probability that this kind of rocket will burn
(a) less than 4.66 seconds;
(b) more than 4.80 seconds;
(c) anywhere from 4.70 to 4.82 seconds?

4.29 In a photographic process, the developing time of prints may be looked upon as a random variable having the normal distribution with a mean of 16.28 seconds and a standard deviation of 0.12 second. Find the probability that it will take
(a) anywhere from 16.00 to 16.50 seconds to develop one of the prints;
(b) at least 16.20 seconds to develop one of the prints;
(c) at most 16.35 seconds to develop one of the prints.

4.30 With reference to the preceding example, for which value is the probability 0.95 that it will be exceeded by the time it takes to develop one of the prints?

4.31 Specifications for a certain job call for washers with an inside diameter of 0.300 ± 0.005 inch. If the inside diameters of the washers supplied by a given manufacturer may be looked upon as a random variable having the normal distribution with $\mu = 0.302$ inch and $\sigma = 0.003$ inch, what percentage of these washers will meet specifications?

4.32 With reference to the example on page 109, verify that if the variability of the filling machine is reduced to $\sigma = 0.025$ ounce, this will lower the required average amount of coffee to 4.05 ounces, yet keep 98% of the jars above 4 ounces.

4.33 A stamping machine produces can tops whose diameters are normally distributed with a standard deviation of 0.01 inch. At what "nominal" (mean) diameter should the machine be set so that no more than 5% of the can tops produced have diameters exceeding 3 inches?

4.34 Extruded plastic rods are automatically cut into nominal lengths of 6 inches. Actual lengths are normally distributed about a mean of 6 inches and their standard deviation is 0.06 inch.
(a) What proportion of the rods exceeds tolerance limits of 5.9 inches to 6.1 inches?

(b) To what value does the standard deviation need to be reduced if 99% of the rods must be within tolerance?

4.35 If a random variable has the binomial distribution with $n = 20$ and $p = 0.60$, use the normal approximation to determine the probabilities that it will take on
(a) the value 14;   (b) a value less than 12.

4.36 A manufacturer knows that on the average 2% of the electric toasters which he makes will require repairs within 90 days after they are sold. Use the normal approximation to the binomial distribution to determine the probability that among 1,200 of these toasters at least 30 will require repairs within the first 90 days after they are sold.

4.37 The probability that an electronic component will fail in less than 1,000 hours of continuous use is 0.25. Use the normal approximation to find the probability that among 200 such components fewer than 45 will fail in less than 1,000 hours of continuous use.

4.38 A safety engineer feels that 30% of all industrial accidents in his plant are caused by failure of employees to follow instructions. If this figure is correct, find, approximately, the probability that among 84 industrial accidents in this plant anywhere from 20 to 30 (inclusive) will be due to failure of employees to follow instructions.

4.39 If 62% of all clouds seeded with silver iodide show spectacular growth, what is the probability that among 40 clouds seeded with silver iodide at most 20 will show spectacular growth?

4.40 To illustrate the law of large numbers mentioned on page 77, find the probabilities that the proportion of heads will be anywhere from 0.49 to 0.51 when a balanced coin is flipped
(a) 1,000 times;   (b) 10,000 times.

4.41 Verify the identity $F(-z) = 1 - F(z)$ given on page 107.

4.42 Verify that the parameter μ in the expression for the normal density on page 107 is, in fact, its mean.

4.43 Verify that the parameter σ^2 in the expression for the normal density on page 107 is, in fact, its variance.

4.44 Show that the normal density has a relative maximum at $x = \mu$ and inflection points at $x = \mu - \sigma$ and $x = \mu + \sigma$.

4.4 OTHER PROBABILITY DENSITIES

In the application of statistics to problems in engineering and physical science, we shall encounter many probability densities other than the normal distribution. Among these are the the t, F, and chi-square distributions, the fundamental sampling distributions which we shall introduce in Chapter 6, and the exponential and Weibull distributions, which we shall apply to problems of reliability and life testing in Chapter 15.

In the remainder of this chapter we shall discuss five continuous distributions, the **uniform distribution**, the **log-normal distribution**, the **gamma distribution**, the **beta distribution**, and the **Weibull distribution**, for the twofold purpose of giving further examples of probability densities and of laying the foundation for future applications.

4.5   THE UNIFORM DISTRIBUTION

The **uniform distribution** with the parameters α and β is defined by the equation

Uniform distribution

$$f(x) = \begin{cases} \dfrac{1}{\beta - \alpha} & \text{for } \alpha < x < \beta \\ 0 & \text{elsewhere} \end{cases}$$

and its graph is shown in Figure 4.4. Note that all values of x from α to β are "equally likely" in the sense that the probability that x lies in an interval of width Δx entirely contained in the interval from α to β is equal to $\Delta x/(\beta - \alpha)$, regardless of the exact location of the interval.

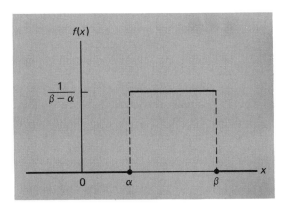

Figure 4.4 Graph of uniform distribution.

To illustrate how a physical situation might give rise to a uniform distribution, suppose that a wheel of a locomotive has the radius r and that x is the location of a point on its circumference measured along the circumference from some reference point 0. When the brakes are applied, some point will make sliding contact with the rail, and heavy wear will occur at that point. For repeated application of the brakes, it would seem reasonable to assume that x is a value of a random variable having the uniform distribution with $\alpha = 0$ and $\beta = 2\pi r$. If this assumption were incorrect, that is, if some set of points on the wheel made contact more often than others, the wheel would eventually exhibit "flat spots" or wear out of round.

To determine the mean and the variance of the uniform distribution, we first evaluate the two integrals

$$\mu_1' = \int_\alpha^\beta x \cdot \frac{1}{\beta - \alpha} dx = \frac{\alpha + \beta}{2}$$

and

$$\mu_2' = \int_\alpha^\beta x^2 \cdot \frac{1}{\beta - \alpha} dx = \frac{\alpha^2 + \alpha\beta + \beta^2}{3}$$

Thus,

Mean of uniform distribution

$$\mu = \frac{\alpha + \beta}{2}$$

and, making use of the formula $\sigma^2 = \mu_2' - \mu^2$, we find that

Variance of uniform distribution

$$\sigma^2 = \tfrac{1}{12}(\beta - \alpha)^2$$

4.6 THE LOG-NORMAL DISTRIBUTION

The **log-normal distribution** occurs in practice whenever we encounter a random variable which is such that its logarithm has a normal distribution. Its probability density is given by

Log-normal distribution

$$f(x) = \begin{cases} \dfrac{1}{\sqrt{2\pi}\,\beta} x^{-1} e^{-(\ln x - a)^2/2\beta^2} & \text{for } x > 0, \beta > 0 \\ 0 & \text{elsewhere} \end{cases}$$

where $\ln x$ is the natural logarithm of x. A graph of the log-normal distribution with $\alpha = 0$ and $\beta = 1$ is shown in Figure 4.5, and it can be seen from the figure that this distribution is positively skewed.

To find the probability that a random variable having the log-normal distribution will take on a value between a and b $(0 < a < b)$, we must

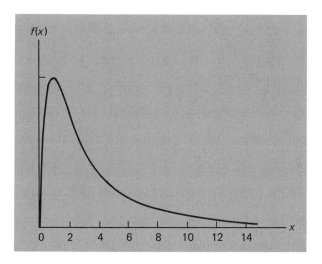

Figure 4.5 Graph of log-normal distribution.

evaluate the integral

$$\int_a^b \frac{1}{\sqrt{2\pi}\beta} x^{-1} e^{-(\ln x - \alpha)^2/2\beta^2} \, dx$$

Changing variable by letting $y = \ln x$ and identifying the integrand as the normal density with $\mu = \alpha$ and $\sigma = \beta$, we find that the desired probability is given by

$$\int_{\ln a}^{\ln b} \frac{1}{\sqrt{2\pi}\beta} e^{-(y-\alpha)^2/2\beta^2} \, dy = F\left(\frac{\ln b - \alpha}{\beta}\right) - F\left(\frac{\ln a - \alpha}{\beta}\right)$$

where **F** is the distribution function of the standard normal distribution.

EXAMPLE   The current gain of certain transistors is measured in units which make it equal to the logarithm of I_o/I_i, the ratio of the output to the input current. If it is normally distributed with $\mu = 2$ and $\sigma^2 = 0.01$, find the probability that I_o/I_i will take on a value between 6.1 and 8.2.

Solution   Since $\alpha = 2$ and $\beta = 0.1$, we get

$$F\left(\frac{\ln 8.2 - 2}{0.1}\right) - F\left(\frac{\ln 6.1 - 2}{0.1}\right) = F(1.0) - F(-2.0)$$

$$= 0.8185$$

To find a formula for the mean of the log-normal distribution, we write

$$\mu = \frac{1}{\sqrt{2\pi}\beta} \int_0^\infty x \cdot x^{-1} e^{-(\ln x - \alpha)^2 / 2\beta^2} \, dx$$

and upon letting $y = \ln x$, this becomes

$$\mu = \frac{1}{\sqrt{2\pi}\beta} \int_{-\infty}^\infty e^y e^{-(y - \alpha)^2 / 2\beta^2} \, dy$$

This integral can be evaluated by completing the square on the exponent $y - (y - \alpha)^2 / 2\beta^2$, thus obtaining an integrand which has the form of a normal density. The final result, which the reader will be asked to verify in Exercise 4.48 on page 124, is

Mean of log-normal distribution

$$\mu = e^{\alpha + \beta^2/2}$$

Similar, but more lengthy, calculations yield

Variance of log-normal distribution

$$\sigma^2 = e^{2\alpha + \beta^2}(e^{\beta^2} - 1)$$

EXAMPLE With reference to the preceding example, find the mean and the variance of the distribution of the ratio of the output to the input current.

Solution Substituting $\alpha = 2$ and $\beta = 0.1$ into the above formulas, we get

$$\mu = e^{2 + (0.1)^2/2} = 7.4$$

and

$$\sigma^2 = e^{4 + (0.1)^2}(e^{(0.1)^2} - 1) = 0.56$$

4.7 THE GAMMA DISTRIBUTION

Several important probability densities whose applications will be discussed later are special cases of the **gamma distribution**. This distribution is given by

Gamma distribution

$$f(x) = \begin{cases} \dfrac{1}{\beta^\alpha \Gamma(\alpha)} x^{\alpha - 1} e^{-x/\beta} & \text{for } x > 0, \alpha > 0, \beta > 0 \\ 0 & \text{elsewhere} \end{cases}$$

where $\Gamma(\alpha)$ is a value of the **gamma function**, defined by

$$\Gamma(\alpha) = \int_0^\infty x^{\alpha-1}e^{-x}\,dx$$

Integration by parts shows that

$$\Gamma(\alpha) = (\alpha - 1)\Gamma(\alpha - 1)$$

for any $\alpha > 1$ and, hence, that $\Gamma(\alpha) = (\alpha - 1)!$ when α is a positive integer. Graphs of several gamma distributions are shown in Figure 4.6 and they exhibit the fact that these distributions are positively skewed. In fact, the skewness decreases as α increases for any fixed value of β.

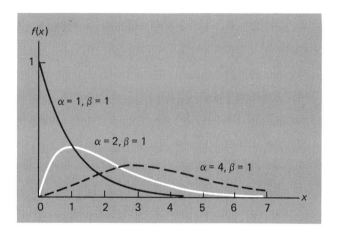

Figure 4.6 Graph of gamma distribution.

The mean and the variance of the gamma distribution may be obtained by making use of the gamma function and its special properties mentioned above. For the mean we have

$$\mu = \frac{1}{\beta^\alpha\Gamma(\alpha)} \int_0^\infty x \cdot x^{\alpha-1}e^{-x/\beta}\,dx$$

and after letting $y = x/\beta$, we get

$$\mu = \frac{\beta}{\Gamma(\alpha)} \int_0^\infty y^\alpha e^{-y}\,dy = \frac{\beta\Gamma(\alpha+1)}{\Gamma(\alpha)}$$

Then, making use of the identity $\Gamma(\alpha + 1) = \alpha \cdot \Gamma(\alpha)$, we arrive at the result that

Mean of gamma distribution

$$\mu = \alpha\beta$$

Using similar methods, it can also be shown that the variance of the gamma distribution is given by

Variance of gamma distribution

$$\sigma^2 = \alpha\beta^2$$

In the special case where $\alpha = 1$, we get the **exponential distribution**, whose probability density is thus

Exponential distribution

$$f(x) = \begin{cases} \dfrac{1}{\beta}e^{-x/\beta} & \text{for } x > 0,\ \beta > 0 \\ 0 & \text{elsewhere} \end{cases}$$

and whose mean and variance are $\mu = \beta$ and $\sigma^2 = \beta^2$. Note that the distribution of the example on page 102 is an exponential distribution with $\beta = \frac{1}{2}$.

The exponential distribution has many important applications; for instance, it can be shown that in connection with Poisson processes (see Section 3.7) the **waiting time** between successive arrivals (successes) has an exponential distribution. More specifically, it can be shown that if in a Poisson process the mean arrival rate (average number of arrivals per unit time) is α, the time until the first arrival, or the waiting time between successive arrivals, has an exponential distribution with $\beta = \dfrac{1}{\alpha}$ (see Exercise 4.62 on page 126).

EXAMPLE With reference to the example on page 84, where on the average three trucks arrived per hour to be unloaded at a warehouse, what are the probabilities that the time between the arrival of successive trucks will be
(a) less than 5 minutes;
(b) at least 45 minutes.

Solution Since $\beta = \frac{1}{3}$, we get

$$\int_0^{1/12} 3e^{-3x}\,dx = 1 - e^{-1/4} = 0.221$$

for part (a), and

$$\int_{3/4}^{\infty} 3e^{-3x}\,dx = e^{-9/4} = 0.105$$

for part (b).

4.8 THE BETA DISTRIBUTION

In Chapter 9 we shall need a probability density for a random variable which takes on values on the interval from 0 to 1, and most appropriate for this purpose is the **beta distribution**, whose probability density is

Beta distribution

$$f(x) = \begin{cases} \dfrac{\Gamma(\alpha + \beta)}{\Gamma(\alpha) \cdot \Gamma(\beta)} x^{\alpha-1}(1 - x)^{\beta-1} & \text{for } 0 < x < 1, \alpha > 0, \beta > 0 \\ 0 & \text{elsewhere} \end{cases}$$

The mean and the variance of this distribution are given by

Mean and variance of beta distribution

$$\mu = \frac{\alpha}{\alpha + \beta} \quad and \quad \sigma^2 = \frac{\alpha\beta}{(\alpha + \beta)^2(\alpha + \beta + 1)}$$

Note that for $\alpha = 1$ and $\beta = 1$ we obtain as a special case the uniform distribution of Section 4.5 defined on the interval from 0 to 1. The following example, pertaining to a proportion, illustrates a typical application of the beta distribution:

EXAMPLE In a certain county, the proportion of highway sections requiring repairs in any given year is a random variable having the beta distribution with $\alpha = 3$ and $\beta = 2$ (shown in Figure 4.7). Find

(a) on the average what percentage of the highway sections require repairs in any given year;

(b) the probability that at most half of the highway sections will require repairs in any given year.

Solution (a) $\mu = \dfrac{2}{3 + 2} = 0.60$, which means that on the average 60 % of the highway sections require repairs in any given year. (b) Substituting $\alpha = 3$ and $\beta = 2$ into the formula for the beta distribution and making use of the fact that $\Gamma(5) = 4! = 24$, $\Gamma(3) = 2! = 2$, and $\Gamma(2) = 1! = 1$, we get

$$f(x) = \begin{cases} 12x^2(1 - x) & \text{for } 0 < x < 1 \\ 0 & \text{elsewhere} \end{cases}$$

Thus, the desired probability is given by

$$\int_0^{1/2} 12x^2(1 - x)\, dx = \tfrac{5}{16}$$

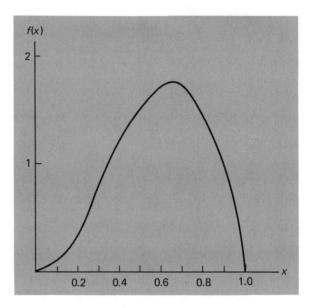

Figure 4.7 Graph of the beta distribution with $\alpha = 3$ and $\beta = 2$.

In most realistically complex situations, probabilities relating to gamma and beta distributions are obtained from special tables.

4.9 THE WEIBULL DISTRIBUTION

Weibull distribution

Closely related to the exponential distribution is the **Weibull distribution**, whose probability density is given by

$$f(x) = \begin{cases} \alpha\beta x^{\beta-1}e^{-\alpha x^{\beta}} & \text{for } x > 0, \alpha > 0, \beta > 0 \\ 0 & \text{elsewhere} \end{cases}$$

To demonstrate this relationship, we evaluate the probability that a random variable having the Weibull distribution will take on a value less than a, namely, the integral

$$\int_0^a \alpha\beta x^{\beta-1}e^{-\alpha x^{\beta}}\, dx$$

Making the change of variable $y = x^{\beta}$, we get

$$\int_0^{a^{\beta}} \alpha e^{-\alpha y}\, dy = 1 - e^{-\alpha a^{\beta}}$$

and it can be seen that y is a value of a random variable having an exponential

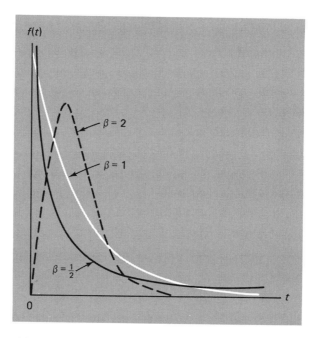

Figure 4.8 Graphs of Weibull distributions with $\alpha = 1$ and $\beta = \frac{1}{2}, 1,$ and 2.

distribution. The graphs of several Weibull distributions with $\alpha = 1$ and $\beta = \frac{1}{2}, 1,$ and 2 are shown in Figure 4.8.

The mean of the Weibull distribution having the parameters α and β may be obtained by evaluating the integral

$$\mu = \int_0^\infty x \cdot \alpha \beta x^{\beta - 1} e^{-\alpha x^\beta} \, dx$$

Making the change of variable $u = \alpha x^\beta$, we get

$$\mu = \alpha^{-1/\beta} \int_0^\infty u^{1/\beta} e^{-u} \, du$$

and recognizing the integral as $\Gamma\left(1 + \dfrac{1}{\beta}\right)$, namely, as a value of the gamma function which we defined on page 119, we find that the mean of the Weibull distribution is given by

Mean of Weibull distribution

$$\mu = \alpha^{-1/\beta} \Gamma\left(1 + \frac{1}{\beta}\right)$$

Using a similar method to determine first μ'_2, the reader will be asked to show in Exercise 4.70 on page 126 that the variance of this distribution is given by

Variance of Weibull distribution

$$\sigma^2 = \alpha^{-2/\beta}\left\{\Gamma\left(1+\frac{2}{\beta}\right) - \left[\Gamma\left(1+\frac{1}{\beta}\right)\right]^2\right\}$$

EXAMPLE Suppose that the lifetime of a certain kind of battery (in hours) is a random variable having the Weibull distribution with $\alpha = 0.1$ and $\beta = 0.5$. Find
  (a) the mean lifetime of these batteries;
  (b) the probability that such a battery will last more than 300 hours.

  Solution (a) Substitution into the formula for the mean yields

$$\mu = (0.1)^{-2}\Gamma(3) = 200 \text{ hours}$$

  (b) Performing the necessary integration, we get

$$\int_{300}^{\infty} (0.05)x^{-0.5}e^{-0.1x^{0.5}}\,dx = e^{-0.1(300)^{0.5}}$$

$$= 0.177$$

EXERCISES

4.45 Find the distribution function of a random variable which has the uniform distribution.

4.46 In certain experiments, the error made in determining the density of a substance is a random variable having the uniform density with $\alpha = -0.025$ and $\beta = 0.025$. What are the probabilities that such an error will be
  (a) between 0.010 and 0.015;
  (b) between -0.012 and 0.012?

4.47 From experience Mr. Harris has found that the low bid on a construction job can be regarded as a random variable having the uniform density

$$f(x) = \begin{cases} \dfrac{3}{4C} & \text{for } \dfrac{2C}{3} < x < 2C \\ 0 & \text{elsewhere} \end{cases}$$

where C is his own estimate of the cost of the job. What percentage should Mr. Harris add to his cost estimate when submitting bids to maximize his expected profit?

4.48 Verify the expression given on page 118 for the mean of the log-normal distribution.

4.49 With reference to the example on page 117, find the probability that I_o/I_i will take on a value between 7.0 and 7.5.

4.50 If a random variable has the log-normal distribution with $\alpha = -1$ and $\beta = 2$, find its mean and its standard deviation.

4.51 With reference to the preceding exercise, find the probabilities that the random variable will take on a value
(a) between 3.2 and 8.4;
(b) greater than 5.0.

4.52 If a random variable has the gamma distribution with $\alpha = 2$ and $\beta = 2$, find the mean and the standard deviation of this distribution.

4.53 With reference to the preceding exercise, find the probability that the random variable will take on a value less than 4.

4.54 In a certain city, the daily consumption of electric power (in millions of kilowatt-hours) can be treated as a random variable having a gamma distribution with $\alpha = 3$ and $\beta = 2$. If the power plant of this city has a daily capacity of 12 million kilowatt-hours, what is the probability that this power supply will be inadequate on any given day?

4.55 If n salesmen are employed in a door-to-door selling campaign, the gross sales volume in thousands of dollars may be regarded as a random variable having the gamma distribution with $\alpha = 100\sqrt{n}$ and $\beta = \frac{1}{2}$. If the sales costs are \$5,000 per salesman, how many salesmen should be employed to maximize the profit?

4.56 Verify the expression for the variance of the gamma distribution given on page 120.

4.57 Show that when $\alpha > 1$, the graph of the gamma density has a relative maximum at $x = \beta(\alpha - 1)$. What happens when $0 < \alpha < 1$ and when $\alpha = 1$?

4.58 The amount of time that a watch will run without having to be reset is a random variable having the exponential distribution with $\beta = 50$ days. Find the probabilities that such a watch will
(a) have to be reset in less than 20 days;
(b) not have to be reset in at least 60 days.

4.59 With reference to Exercise 3.59, find the percent of the time that the interval between breakdowns of the computer will be
(a) less than 1 week;
(b) at least 5 weeks.

4.60 With reference to Exercise 3.61, find the probabilities that the time between successive calls arriving at the switchboard of the consulting firm will be
(a) less than $\frac{1}{2}$ minute;
(b) more than 3 minutes.

4.61 Given a Poisson process with on the average α arrivals per unit time, find the probability that there will be no arrivals during a time interval of length t, namely, the probability that the waiting times between successive arrivals will be at least of length t.

4.62 Use the result of the preceding exercise to find an expression for the probability density of the waiting time between successive arrivals.

4.63 Verify for $\alpha = 3$ and $\beta = 3$ that the integral of the beta density, from 0 to 1, is equal to 1.

4.64 If the annual proportion of erroneous income tax returns filed with the IRS can be looked upon as a random variable having a beta distribution with $\alpha = 2$ and $\beta = 9$, what is the probability that in any given year there will be fewer than 10% erroneous returns?

4.65 Suppose that the proportion of defectives shipped by a vendor, which varies somewhat from shipment to shipment, may be looked upon as a random variable having the beta distribution with $\alpha = 1$ and $\beta = 4$.
(a) Find the mean of this beta distribution, namely, the average proportion of defectives in a shipment from this vendor.
(b) Find the probability that a shipment from this vendor will contain 25% or more defectives.

4.66 Show that when $\alpha > 1$ and $\beta > 1$, the beta density has a relative maximum at

$$x = \frac{\alpha - 1}{\alpha + \beta - 2}$$

4.67 With reference to the example on page 124, find the probability that such a battery will not last 100 hours.

4.68 Suppose that the time to failure (in minutes) of certain electronic components subjected to continuous vibrations may be looked upon as a random variable having the Weibull distribution with $\alpha = \frac{1}{5}$ and $\beta = \frac{1}{3}$.
(a) How long can such a component be expected to last?
(b) What is the probability that such a component will fail in less than 5 hours?

4.69 Suppose that the service life (in hours) of a semiconductor is a random variable having the Weibull distribution with $\alpha = 0.025$ and $\beta = 0.500$. What is the probability that such a semiconductor will still be in operating condition after 4,000 hours?

4.70 Verify the formula for the variance of the Weibull distribution given on page 124.

4.10 *JOINT PROBABILITY DENSITIES*

There are many situations in which we describe an outcome by giving the values of several random variables. For instance, we may measure the weight and the hardness of a rock, the volume, pressure, and temperature of a gas, or the thickness, color, compressive strength, and potassium content of a piece of glass. If $x_1, x_2, \ldots, x_k$ are the values of k random variables, we shall refer to a function **f** with values $f(x_1, x_2, \ldots, x_k)$ as the **joint probability density** of these random variables, if the probability that $a_1 \leq x_1 \leq b_1$,

$a_2 \leq x_2 \leq b_2, \ldots,$ and $a_k \leq x_k \leq b_k$ is given by the multiple integral

$$\int_{a_1}^{b_1} \int_{a_2}^{b_2} \ldots \int_{a_k}^{b_k} f(x_1, x_2, \ldots, x_k)\, dx_1\, dx_2 \ldots dx_k$$

Thus, not every function **f** with values $f(x_1, x_2, \ldots, x_k)$ can serve as a joint probability density, but if

$$f(x_1, x_2, \ldots, x_k) \geq 0$$

for all values of $x_1, x_2, \ldots, x_k$ for which the function is defined, and

$$\int_{-\infty}^{\infty} \int_{-\infty}^{\infty} \ldots \int_{-\infty}^{\infty} f(x_1, x_2, \ldots, x_k)\, dx_1\, dx_2 \ldots dx_k = 1$$

it can be shown that the axioms of probability (with the modification of the definition of "event" discussed in Section 4.1) are satisfied.

To extend the concept of a cumulative distribution function to the k-variable case, we write as $F(x_1, x_2, \ldots, x_k)$ the probability that the first random variable will take on a value less than or equal to x_1, the second random variable will take on a value less than or equal to $x_2, \ldots,$ and the kth random variable will take on a value less than or equal to x_k, and we refer to the corresponding function F as the **joint distribution function** of the k random variables.

EXAMPLE   If the joint probability density of two random variables is given by

$$f(x_1, x_2) = \begin{cases} 6e^{-2x_1 - 3x_2} & \text{for } x_1 > 0, x_2 > 0 \\ 0 & \text{elsewhere} \end{cases}$$

find the probabilities that
(a) the first random variable will take on a value between 1 and 2 and the second random variable will take on a value between 2 and 3;
(b) the first random variable will take on a value less than 2 and the second random variable will take on a value greater than 2.

Solution   Performing the necessary integrations, we get

$$\int_1^2 \int_2^3 6e^{-2x_1 - 3x_2}\, dx_1\, dx_2 = (e^{-2} - e^{-4})(e^{-6} - e^{-9})$$

$$= 0.0003$$

for part (a), and

$$\int_0^2 \int_2^{\infty} 6e^{-2x_1 - 3x_2}\, dx_1\, dx_2 = (1 - e^{-4})e^{-6}$$

$$= 0.0025$$

for part (b).

EXAMPLE   Find the joint distribution function of the two random variables of the preceding exercise, and use it to find the probability that both random variables will take on values less than 1.

Solution   By definition,

$$F(x_1, x_2) = \begin{cases} \int_0^{x_1} \int_0^{x_2} 6e^{-2u-3v} \, du \, dv & \text{for } x_1 > 0, x_2 > 0 \\ 0 & \text{elsewhere} \end{cases}$$

so that

$$F(x_1, x_2) = \begin{cases} (1 - e^{-2x_1})(1 - e^{-3x_2}) & \text{for } x_1 > 0, x_2 > 0 \\ 0 & \text{elsewhere} \end{cases}$$

and, hence,

$$F(1, 1) = (1 - e^{-2})(1 - e^{-3})$$
$$= 0.8216$$

Given the joint probability density of k random variables, the probability density of the ith random variable can be obtained by integrating out the other variables; symbolically,

Marginal density

$$f_i(x_i) = \int_{-\infty}^{\infty} \cdots \int_{-\infty}^{\infty} f(x_1, x_2, \ldots, x_k) \, dx_1 \ldots dx_{i-1} \, dx_{i+1} \ldots dx_k$$

and, in this context, the function $\mathbf{f}_i$ is called the **marginal density** of the ith random variable. Integrating out only some of the k random variables, we can similarly define **joint marginal densities** of any two, three, or more of the k random variables.

EXAMPLE   With reference to the example on page 127, find the marginal density of the first random variable.

Solution   Integrating out x_2, we get

$$f_1(x_1) = \begin{cases} \int_0^{\infty} 6e^{-2x_1-3x_2} \, dx_2 & \text{for } x_1 > 0 \\ 0 & \text{elsewhere} \end{cases}$$

or

$$f_1(x_1) = \begin{cases} 2e^{-2x_1} & \text{for } x_1 > 0 \\ 0 & \text{elsewhere} \end{cases}$$

To explain what we mean by the **independence** of continuous random variables, we could proceed as in Section 2.6 and define **conditional densities** first; however, it will be easier to say that

Independent random variables

> *k random variables are independent if and only if*
>
> $$F(x_1, x_2, \ldots, x_k) = F_1(x_1) \cdot F_2(x_2) \cdot \ldots \cdot F_k(x_k)$$
>
> *for all values of these random variables for which the respective functions are defined.*

In this notation, $F(x_1, x_2, \ldots, x_k)$ is, as before, a value of the joint distribution function of the k random variables, while $F_i(x_i)$ for $i = 1, 2, \ldots, k$ are the corresponding values of the individual distribution functions of the respective random variables.

EXAMPLE   With reference to the example on page 127, check whether the two random variables are independent.

> **Solution**   As we already saw on page 128, the joint distribution function of the two random variables is given by
>
> $$F(x_1, x_2) = \begin{cases} (1 - e^{-2x_1})(1 - e^{-3x_2}) & \text{for } x_1 > 0 \text{ and } x_2 > 0 \\ 0 & \text{elsewhere} \end{cases}$$
>
> Now, since $F_1(x_1) = F(x_1, \infty)$ and $F_2(x_2) = F(\infty, x_2)$, it follows that
>
> $$F_1(x_1) = \begin{cases} 1 - e^{-2x_1} & \text{for } x_1 > 0 \\ 0 & \text{elsewhere} \end{cases}$$
>
> and
>
> $$F_2(x_2) = \begin{cases} 1 - e^{-3x_2} & \text{for } x_2 > 0 \\ 0 & \text{elsewhere} \end{cases}$$
>
> Thus, $F(x_1, x_2) = F_1(x_1) \cdot F_2(x_2)$ and the two random variables are independent.

Analogous to the special multiplication rule for independent events and its extension on page 38, it follows from our definition of independence that if k random variables are independent, any value of their joint probability density equals the product of the corresponding values of the marginal densities of the k random variables; symbolically,

$$f(x_1, x_2, \ldots, x_k) = f_1(x_1) \cdot f_2(x_2) \cdot \ldots \cdot f_k(x_k)$$

EXAMPLE   With reference to the example on page 127, verify that $f(x_1, x_2) = f_1(x_1) \cdot f_2(x_2)$.

Solution On page 128 we showed that

$$f_1(x_1) = \begin{cases} 2e^{-2x_1} & \text{for } x_1 > 0 \\ 0 & \text{elsewhere} \end{cases}$$

and in the same way it can be shown that

$$f_2(x_2) = \begin{cases} 3e^{-3x_2} & \text{for } x_2 > 0 \\ 0 & \text{elsewhere} \end{cases}$$

Thus

$$f_1(x_1) \cdot f_2(x_2) = \begin{cases} 6e^{-2x_1 - 3x_2} & \text{for } x_1 > 0 \text{ and } x_2 > 0 \\ 0 & \text{elsewhere} \end{cases}$$

and it can be seen that $f_1(x_1) \cdot f_2(x_2) = f(x_1, x_2)$.

Given two random variables with values x_1 and x_2, we shall define the **conditional probability density** of the first given that the second takes on the value x_2 as

*Conditional
probability
density*

$$g_1(x_1 \,|\, x_2) = \frac{f(x_1, x_2)}{f_2(x_2)} \qquad f_2(x_2) \neq 0$$

where $f(x_1, x_2)$ and $f_2(x_2)$ are, as before, values of the joint density of the two random variables and the marginal density of the second. Note that this definition parallels that of conditional probability on page 36.

EXAMPLE If two random variables have the joint probability density

$$f(x_1, x_2) = \begin{cases} \frac{2}{3}(x_1 + 2x_2) & \text{for } 0 < x_1 < 1, 0 < x_2 < 1 \\ 0 & \text{elsewhere} \end{cases}$$

find the conditional density of the first given that the second takes on the value x_2.

Solution First we find the marginal density of the second random variable by integrating out x_1, and we get

$$f_2(x_2) = \int_0^1 \tfrac{2}{3}(x_1 + 2x_2)\, dx_1 = \tfrac{1}{3}(1 + 4x_2) \qquad \text{for } 0 < x_2 < 1$$

and $f_2(x_2) = 0$ elsewhere. Hence, by definition, the conditional density of the first random variable given that the second takes on the value x_2 is given by

$$g_1(x_1 \mid x_2) = \frac{\frac{2}{3}(x_1 + 2x_2)}{\frac{1}{3}(1 + 4x_2)} = \frac{2x_1 + 4x_2}{1 + 4x_2} \qquad \text{for } 0 < x_1 < 1, 0 < x_2 < 1$$

and $g_1(x_1 \mid x_2) = 0$ elsewhere.

The following is another concept which is of importance when we deal with k random variables. Suppose that a random variable takes on the value $g(x_1, x_2, \ldots, x_k)$ whenever k random variables take on the values $x_1, x_2, \ldots, x_k$. Then the expected value of this random variable (namely, the mean of its distribution) is given by

$$\int_{-\infty}^{\infty} \int_{-\infty}^{\infty} \cdots \int_{-\infty}^{\infty} g(x_1, \ldots, x_k) f(x_1, \ldots, x_k) \, dx_1, dx_2 \ldots dx_k$$

where $f(x_1, \ldots, x_k)$ is, as before, a value of the joint density of the k random variables.

EXAMPLE   With reference to the example on page 127, find the expected value of the product of the two random variables.

    Solution   Substituting into the above formula, we get

$$\int_0^{\infty} \int_0^{\infty} x_1 x_2 \cdot 6e^{-2x_1 - 3x_2} \, dx_1 \, dx_2 = \tfrac{1}{6}$$

EXERCISES

4.71 If two random variables have the joint density

$$f(x_1, x_2) = \begin{cases} x_1 x_2 & \text{for } 0 < x_1 < 1, 0 < x_2 < 2 \\ 0 & \text{elsewhere} \end{cases}$$

find the probabilities that
(a) both random variables will take on values less than 1;
(b) the sum of the values taken on by the two random variables will be less than 1.

4.72 With reference to the preceding exercise, find the marginal densities of the two random variables.

4.73 With reference to Exercise 4.71, find the joint distribution function of the two random variables, the distribution functions of the individual random variables, and check whether the two random variables are independent.

4.74 If two random variables have the joint density

$$f(x, y) = \begin{cases} \frac{6}{5}(x + y^2) & \text{for } 0 < x < 1, 0 < y < 1 \\ 0 & \text{elsewhere} \end{cases}$$

find the probability that $0.2 < x < 0.5$ and $0.4 < y < 0.6$.

4.75 With reference to the preceding exercise, find the joint distribution function of the two random variables and use it to verify the value obtained for the probability.

4.76 With reference to Exercise 4.74, find both marginal densities and use them to find the probabilities that
(a) $x > 0.8$;   (b) $y < 0.5$.

4.77 With reference to Exercise 4.74, find
(a) an expression for $g_1(x|y)$ for $0 < y < 1$;
(b) an expression for $g_1(x|\frac{1}{2})$;
(c) the mean of the conditional density of the first random variable when the second takes on the value $\frac{1}{2}$.

4.78 With reference to the example on page 130, find expressions for
(a) the conditional density of the first random variable when the second takes on the value $x_2 = 0.25$;
(b) the conditional density of the second random variable when the first takes on the value x_1.

4.79 If three random variables have the joint density

$$f(x, y, z) = \begin{cases} k(x + y)e^{-z} & \text{for } 0 < x < 1, 0 < y < 2, z > 0 \\ 0 & \text{elsewhere} \end{cases}$$

find
(a) the value of k;
(b) the probability that $x < y$ and $z > 1$.

4.80 With reference to the preceding exercise, check whether
(a) the three random variables are independent;
(b) any two of the three random variables are pairwise independent.

4.81 A pair of random variables has the **circular normal distribution** if their joint density is given by

$$f(x_1, x_2) = \frac{1}{2\pi\sigma^2} e^{-\frac{1}{2\sigma^2}[(x_1 - \mu_1)^2 + (x_2 - \mu_2)^2]}$$

for $-\infty < x_1 < \infty$ and $-\infty < x_2 < \infty$.
(a) If $\mu_1 = 2$ and $\mu_2 = -2$, and $\sigma = 10$, use Table 3 to find the probability that $-8 < x_1 < 14$ and $-9 < x_2 < 3$.
(b) If $\mu_1 = \mu_2 = 0$ and $\sigma = 3$, find the probability that (x_1, x_2) is contained in the region between the two circles $x_1^2 + x_2^2 = 9$ and $x_1^2 + x_2^2 = 36$.

4.82 A bomb aimed at a point target has a lethal radius of 500 feet. Using the target as the origin of a rectangular system of coordinates, assume that the coordinates (x, y) of the point of impact are values of a pair of random variables having the circular normal distribution (see Exercise 4.81) with $\mu_1 = \mu_2 = 0$ and $\sigma = 300$ feet. What is the probability that the target will be destroyed?

4.83 With reference to Exercise 4.71, find the expected value of the random variable whose values are given by $g(x_1, x_2) = x_1 + x_2$.

4.84 With reference to Exercise 4.74, find the expected value of the random variable whose values are given by $g(x, y) = x^2 y$.

4.85 If measurements of the length and the width of a rectangle have the joint density

$$f(x, y) = \begin{cases} \dfrac{1}{ab} & \text{for } L - \dfrac{a}{2} < x < L + \dfrac{a}{2}, \ W - \dfrac{b}{2} < y < W + \dfrac{b}{2} \\ 0 & \text{elsewhere} \end{cases}$$

find the mean and the variance of the corresponding distribution of the area of the rectangle.

4.86 Establish a relationship between $g_1(x_1 \mid x_2), g_2(x_2 \mid x_1), f_1(x_1)$, and $f_2(x_2)$.

4.11 SIMULATION

To **simulate** the observation of continuous random variables with the use of random numbers, we refer to their distribution functions, given either graphically or as tables. (The theory on which this method is based involves the so-called probability integral transformation, which is treated in all of the textbooks on mathematical statistics referred to in the Bibliography.)

To illustrate the graphical method, suppose that we want to observe the values of a random variable having a normal distribution with a given mean and standard deviation. Taking two- or three-digit random numbers (depending on the accuracy that can be attained) and treating them as decimal fractions by putting a decimal point to the left of the first digit, we refer to a graph of the distribution function of the standard normal distribution (see, for instance, Figure 4.9), mark the decimal fractions corresponding to the random digits on the $F(z)$ scale, and then read the corresponding values of a random variable having the standard normal distribution off the horizontal scale. Finally, we obtained the simulated values of the original random variable by means of the equation $x = \mu + \sigma z$, which follows from $z = \dfrac{x - \mu}{\sigma}$ (see page 108.)

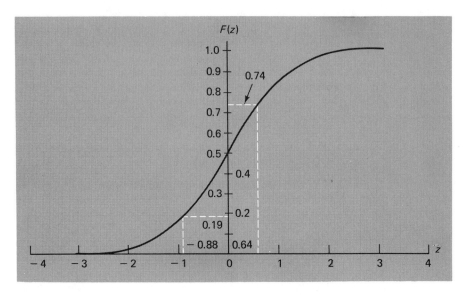

Figure 4.9 Graph of the distribution function of the standard normal distribution.

EXAMPLE   Simulate six observations of a random variable having the normal distribution with $\mu = 18.3$ and $\sigma = 20$, using the 3rd and 4th columns of the first page of Table 7, starting with the 36th row and going down the page.

> **Solution**   Proceeding as indicated, we get the two-digit random numbers 74, 19, 78, 82, 06, and 11. Treating the first of these as the decimal fraction 0.74 and marking it on the vertical scale of Figure 4.9, we find that the corresponding value of z is 0.64 and, hence, that the simulated value of the original random variable is
>
> $$x = 18.3 + 20(0.64) = 31.1$$
>
> Similarly, treating the random number 19 as a decimal fraction and marking 0.19 on the vertical scale of Figure 4.9, we find that the corresponding value of z is -0.88 and, hence, that the simulated value of the original random variable is
>
> $$x = 18.3 + 20(-0.88) = 0.7$$
>
> In Exercise 4.87 on page 135, the reader will be asked to show that the other four two-digit random numbers yield $x = 33.7$, 36.7, -12.7, and -6.3.

The graphical method we have described has the advantage that it can be used to simulate any continuous distribution, but it is generally very tedious as it requires that we carefully draw the graph of the distribution function. This can be avoided if a suitable table of cumulative probabilities is available. For instance, to simulate the observation of values of a random variable having the standard normal distribution (and, hence, any normal distribution), we refer to Table 3 and proceed as follows: Given a two- or three-digit random number, we change it, as before, into a random decimal between 0 and 1. Then, if the random decimal is greater than or equal to 0.5000, we look for the value of z for which the entry in Table 3 is closest to the random decimal; if the random decimal is less than 0.5000, we look for the value of z for which the entry in Table 3 is closest to 1 minus the random decimal and record $-z$ as the simulated observation.

EXAMPLE Use this method to rework the preceding example.

> ***Solution*** Using again the two-digit random numbers 74, 19, 78, 82, 06, and 11, we first look for the value of z for which the entry in Table 3 is closest to 0.74. Getting $z = 0.64$ corresponding to 0.7389, we proceed as before and get
>
> $$x = \mu + \sigma z = 18.3 + 20(0.64) = 31.1$$
>
> Then, checking $1 - 0.1900 = 0.8100$ in Table 3, we find that the closest entry is 0.8106 corresponding to $z = 0.88$, and we get the simulated observation
>
> $$x = 18.3 + 20(-0.88) = 0.7$$
>
> The other four values are obtained in the same way.

EXERCISES

4.87 Continue the example in the text, that is, find the simulated values of the random variable which corresponds to the random numbers 78, 82, 06, and 11, by using
(a) the diagram Figure 4.9;
(b) Table 3.

4.88 Suppose that the number of hours it takes a person to learn how to operate a certain machine is a random variable having a normal distribution with $\mu = 5.8$ and $\sigma = 1.2$. Simulate the length of time it takes eight persons to learn how to operate the machine using
(a) the diagram of Figure 4.9;
(b) Table 3.

4.89 Suppose that the increase in the pulse rate of a person performing a certain task is a random variable whose distribution can be approximated closely with a normal distribution having the mean 28.6 and the standard deviation 5.15. Simulate the increase in the pulse rate of 20 persons performing this task, using Table 3 and rounding the answers to the nearest positive integer.

4.90 Suppose that the durability of a paint (in years) is a random variable having an exponential density (see page 120) with the mean $\beta = 2$.
(a) Draw a graph of the distribution function of this random variable.
(b) Use the graph obtained in part (a) to simulate an experiment in which the paint is applied to 12 houses and it is observed in each case how long the paint lasts (that is, how long it takes until there are certain agreed-upon signs of deterioration).

4.91 Suppose that the occurrence of breakdowns of a radar tracking system is a Poisson process with $\alpha = 0.02$ per hour, and the time required to repair any breakdown is fixed at 10 hours. Simulate 1,000 hours of operation of this system and estimate the proportion of "down time." [*Hint*: Simulate the waiting times between successive breakdowns using an appropriate exponential distribution.]

4.92 Suppose that the annual number of collison-insurance claims in a given automobile insurance policy class has the log-normal distribution with $\alpha = 1.5$ and $\beta = 0.6$. Simulate 10 years of experience with this policy class.

4.93 Since many simulation experiments entail random variables having normal distributions, statisticians have constructed special tables of so-called **random normal deviates** (see the Bibliography), which are looked upon as values of a random variable having the standard normal distribution.
(a) Using the method directly based on Table 3, construct a table of 100 such random normal deviates.
(b) Use the table obtained in part (a) to simulate 20 observations of a random variable having a normal distribution with $\mu = 14.25$ and $\sigma = 2.16$.

4.12 CHECK LIST OF KEY TERMS (with page references)

5

Treatment of Data

Statistical data, obtained from surveys, experiments, or any series of measurements, are often so numerous that they are virtually useless unless they are condensed, or reduced, into a more suitable form. Thus, Sections 5.1 and 5.2 deal with problems relating to the grouping of data and the presentation of such groupings in graphical form; in Section 5.3 we discuss a relatively new way of presenting grouped data.

Sometimes it may be satisfactory to present data just as they are, and let them speak for themselves; on other occasions it may be necessary only to group the data and present the result in tabular or graphical form. However, most of the time data have to be summarized further, and in Sections 5.4 and 5.5 we introduce some of the most widely used kinds of statistical descriptions.

A **frequency distribution** is a table which divides a set of data into a suitable number of classes (categories), showing also the number of items belonging to each class. Such a table sacrifices some of the information contained in the data; instead of knowing the exact value of each item, we only know that it belongs to a certain class. On the other hand, the kind of grouping it represents often brings out important features of the data, and the gain in "legibility" usually more than compensates for the loss of information. In what follows, we shall consider mainly **numerical distributions**, that is, frequency distributions where the data are grouped according to size; if the data are grouped according to some quality, or attribute, we refer to such a distribution as a **categorical distribution**.

The first step in constructing a frequency distribution consists of deciding how many classes to use and choosing the **limits** for each class, that is, from where to where each class is to go. Generally speaking, the number of classes we use depends on the number of observations, but it is seldom profitable to use fewer than 5 or more than 15. It also depends on the range of the data, namely, the difference between the largest observation and the smallest. Then, we tally the observations and thus determine the **class frequencies**, namely, the number of observations in each class.

To illustrate the construction of a frequency distribution, let us consider the following 80 determinations of the daily emission (in tons) of sulfur oxides from an industrial plant:

15.8	26.4	17.3	11.2	23.9	24.8	18.7	13.9	9.0	13.2
22.7	9.8	6.2	14.7	17.5	26.1	12.8	28.6	17.6	23.7
26.8	22.7	18.0	20.5	11.0	20.9	15.5	19.4	16.7	10.7
19.1	15.2	22.9	26.6	20.4	21.4	19.2	21.6	16.9	19.0
18.5	23.0	24.6	20.1	16.2	18.0	7.7	13.5	23.5	14.5
14.4	29.6	19.4	17.0	20.8	24.3	22.5	24.6	18.4	18.1
8.3	21.9	12.3	22.3	13.3	11.8	19.3	20.0	25.7	31.8
25.9	10.5	15.9	27.5	18.1	17.9	9.4	24.1	20.1	28.5

Since the largest observation is 31.8, the smallest is 6.2, and the range is 25.6, we might choose the six classes having the limits 5.0–9.9, 10.0–14.9, . . . , 30.0–34.9, we might choose the seven classes 5.0–8.9, 9.0–12.9, . . . , 29.0–32.9, or we might choose the nine classes 5.0–7.9, 8.0–10.9, . . . , 29.0–31.9. Note that in each case **the classes do not overlap, they accommodate all of the data, and they are all of the same size**.

Deciding on the second of these classifications, we now tally the 80 observations and obtain the results shown in the following table:

Class limits	Tally	Frequency
5.0– 8.9	/ / /	3
9.0–12.9	┼┼┼┼ ┼┼┼┼	10
13.0–16.9	┼┼┼┼ ┼┼┼┼ / / / /	14
17.0–20.9	┼┼┼┼ ┼┼┼┼ ┼┼┼┼ ┼┼┼┼ ┼┼┼┼	25
21.0–24.9	┼┼┼┼ ┼┼┼┼ ┼┼┼┼ / /	17
25.0–28.9	┼┼┼┼ / / / /	9
29.0–32.9	/ /	2
	Total	80

Note that the class limits are given to as many decimals as the original data. Had the original data been given to two decimals we would have used the class limits 5.00–8.99, 9.00–12.99, . . . , 29.00–32.99, and if they had been rounded to the nearest ton we would have used the class limits 5–8, 9–12, . . . , 29–32.

In the preceding example, the data may be thought of as values of a continuous random variable, but if we use classes such as 5.0–9.0, 9.0–13.0, . . . , 29.0–33.0, there exists the possibility of ambiguities: 9.0 could go into the first class or into the second, 13.0 could go into the second class or into the third, and so on. To avoid this difficulty, we can let the first class go from 4.95 to 8.95, the second from 8.95 to 12.95, . . . , and the last from 28.95 to 32.95. We refer to these new limits as the **class boundaries**, and it should be noted that there will be no ambiguities even though the boundaries overlap. These class boundaries are, so to speak, "impossible" values, since the data were given to only one decimal. In practice, we use the class boundaries rather than the original class limits chiefly when we want to stress the fact that we are dealing with continuous kinds of measurements.

As we pointed out earlier, once data have been grouped, each observation has lost its identity in the sense that its exact value is no longer known. This may lead to difficulties when we want to give further descriptions of the data, but we can avoid them by representing each observation in a class by its midpoint, called the **class mark**. In general, the class marks of a frequency distribution are obtained by averaging successive class limits or successive class boundaries. If the classes of a distribution are all of equal length, as in our example, we refer to the common interval between any successive class marks as the **class interval** of the distribution. Note that the class interval may also be obtained from the difference between any successive class boundaries, but not from the difference between successive class limits.

EXAMPLE   With reference to the distribution of the sulfur oxides data, find (a) the class marks, and (b) the class interval.

> ***Solution***   (a) The class marks are $\dfrac{5.0 + 8.9}{2} = 6.95,\ \dfrac{9.0 + 12.9}{2} =$ 10.95, 14.95, 18.95, 22.95, 26.95, and 30.95. (b) The class interval is $10.95 - 6.95 = 4$.

There are several alternative forms of distributions into which data are sometimes grouped. Foremost among these are the "less than," "or less," "more than," and "or more" **cumulative distributions**. A cumulative "less than" distribution shows the total number of observations that are less than given values. These values must be class boundaries or appropriate class limits, but they may not be class marks.

EXAMPLE   Convert the distribution of the sulfur oxides emission data into a distribution showing how many of the observations are less than 4.95, less than 8.95, less than 12.95, . . . , and less than 32.95.

> ***Solution***   Since none of the values is less than 4.95, 3 are less than 8.95, $3 + 10 = 13$ are less than 12.95, $3 + 10 + 14 = 27$ are less than 16.95, . . . , and all 80 are less than 32.95, we get

Tons of sulfur oxides	*Cumulative frequency*
less than 4.95	0
less than 8.95	3
less than 12.95	13
less than 16.95	27
less than 20.95	52
less than 24.95	69
less than 28.95	78
less than 32.95	80

Note that instead of "less than 4.95" we could also have written "less than 5.0" or "4.9 or less."

Cumulative "more than" and "or more" distributions are constructed, similarly, by adding the frequencies, one by one, starting at the other end of the frequency distribution. In practice, "less than" cumulative distributions are used most widely, and it is not uncommon to refer to "less than" cumulative distributions simply as cumulative distributions.

If it is desired to compare frequency distributions, it may be necessary (or at least advantageous) to convert them into **percentage distributions**. We simply divide each class frequency by the total frequency (the total number of observations in the distribution) and multiply by 100; in this way we indicate what percentage of the data falls into each class of the distribution. The same can also be done with cumulative distributions, thus converting them to **cumulative percentage distributions.**

EXAMPLE The following are the numbers of workers absent from a factory on 50 working days:

$$
\begin{array}{cccccccccc}
13 & 5 & 13 & 37 & 10 & 16 & 2 & 11 & 6 & 12 \\
8 & 21 & 12 & 11 & 7 & 7 & 9 & 16 & 49 & 18 \\
3 & 11 & 19 & 6 & 15 & 10 & 14 & 10 & 7 & 24 \\
11 & 3 & 6 & 10 & 4 & 6 & 32 & 9 & 12 & 7 \\
29 & 12 & 9 & 19 & 8 & 20 & 15 & 5 & 17 & 10 \\
\end{array}
$$

Use the six classes 0–4, 5–9, 10–14, 15–19, 20–24, and "25 or more" to construct (a) a frequency distribution, and (b) a percentage distribution.

Solution (a) Tallying the data into the six classes, we get

Number of absences	Frequency
0–4	4
5–9	15
10–14	16
15–19	8
20–24	3
25 or more	4

and (b) dividing each class frequency by 50 and multiplying by 100, that is, multiplying each class frequency by 2, we get

Number of absences	Percentage
0–4	8
5–9	30
10–14	32
15–19	16
20–24	6
25 or more	8

Note that in these tables we replaced the five classes 25–29, 30–34, 35–39, 40–44, and 45–49, which would ordinarily have been needed, by a single class having no upper limit. This is called an **open class**, and it avoids having to use a large number of classes that are empty or have very small frequencies. Note also that when distributions are shown in their final form, the tally is usually omitted.

5.2 *GRAPHS OF FREQUENCY DISTRIBUTIONS*

Properties of frequency distributions relating to their shape are best exhibited by means of graphs, and in this section we shall introduce some of the most widely used forms of graphical presentations of frequency distributions, percentage distributions, and cumulative distributions.

The most common form of graphical presentation of a frequency distribution is the **histogram**, first mentioned on page 56 in connection with probability distributions. The histogram of a frequency distribution is constructed like that of a probability distribution; the heights of the rectangles represent the class frequencies and the bases of the rectangles extend between successive class boundaries. A histogram of the sulfur oxides emission data is shown in Figure 5.1.

In connection with histograms, it is sometimes preferable to look on the areas of the rectangles, rather than their heights, as representing the class frequencies. This applies in particular to situations where we wish to approximate histograms with smooth curves or where there are classes of unequal length (see Exercise 5.22 on page 151).

Similar to histograms are **bar charts**, such as the one shown in Figure 3.2; the heights of the rectangles, or bars, again represent the class frequencies, but there is no pretense of having a continuous horizontal scale. Another alternative way of presenting frequency distributions in graphical form is the **frequency polygon**. Here the class frequencies are plotted at the class marks, that is, we plot the points (x_i, f_i) where x_i is the class mark of the ith class

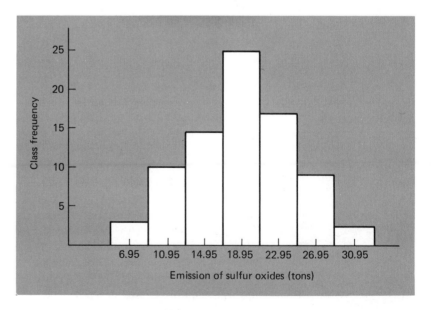

Figure 5.1 Histogram.

and f_i is the corresponding frequency, and the successive points are connected by means of straight lines after having added classes with zero frequencies at both ends of the distribution.

Inspection of the graph of a frequency distribution often brings out features that are not immediately apparent from the data themselves. Aside from the fact that such a graph presents a good overall picture of the data, it can also emphasize irregularities and unusual features. For instance, outlying observations which somehow do not fit the overall picture, that is, the overall pattern of the data, may be due to errors of measurement, equipment failure, and similar causes. Also, the fact that a histogram or a frequency polygon exhibits two or more **modes** (maxima) can provide pertinent information. The appearance of two modes may imply, for example, a shift in the process that is being measured, or it may imply that the data come from several sources. With some experience one learns to spot such irregularities or anomalies, and an experienced engineer would find it just as surprising if the histogram of a distribution of integrated-circuit failure times were symmetrical as if a distribution of American men's hat sizes were bimodal.

Cumulative distributions are usually presented graphically in the form of **ogives**, which are similar to frequency polygons, except that we plot the cumulative frequencies at the class boundaries instead of the ordinary frequencies at the class marks. The resulting points are again connected by means of straight lines, as shown in Figure 5.2, which represents the cumulative "less than" distribution of the sulfur oxides emission data.

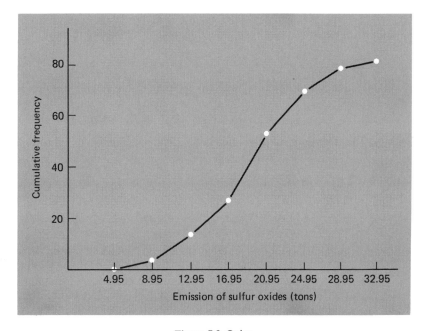

Figure 5.2 Ogive.

Ogives are nearly always S-shaped, with the relative size of the two tails of the S determined by the symmetry or lack of symmetry of the distribution. If a distribution follows closely the pattern of a normal curve, it is possible to "straighten out" the S by using a special vertical scale, called a **probability scale**, which is so designed that any cumulative normal distribution will graph as a straight line. Special graph paper with one arithmetic (ordinary) scale and one probability scale (see Figures 5.3 and 5.4), called **arithmetic probability paper**, is commercially available.

Figure 5.3 Normal probability graph (grouped data).

To illustrate, we have plotted the cumulative "less than" percentage distribution of the sulfur oxides emission data on arithmetic probability paper in Figure 5.3. Note that the cumulative percentages are plotted at the corresponding class boundaries. As can be seen from this diagram, the points all lie close to the dashed line, and we can say that the data are approximately normally distributed.

To check for "normality," probability graph paper can also be used directly with ungrouped data. In that case we first arrange the observations according to size, and if there are n observations we plot $\dfrac{100\,(i - \frac{1}{2})}{n}$ or $\dfrac{(100i - 50)}{n}$ percent on the vertical scale corresponding to the ith largest observation (see Exercise 5.23 on page 151). If n is large, it is not necessary to plot every point; every fifth or tenth point will generally suffice to check whether the distribution of the data follows closely the pattern of a normal distribution.

To illustrate, we have drawn in Figure 5.4 a probability graph for the ungrouped sulfur oxides emission data (arranged according to size), by plotting points corresponding to every tenth observation and also the smallest

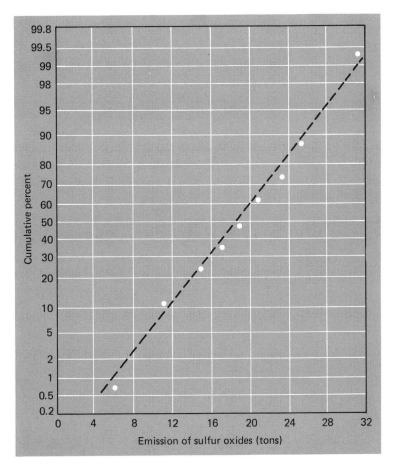

Figure 5.4 Normal probability graph (ungrouped data).

observation. Again, it is apparent that the points fall close to a straight line and, hence, that the data are approximately normally distributed.

It must be understood that the use of arithmetic probability paper, which should really be called normal probability graph paper, is only an approximate (and highly subjective) device for checking whether a distribution follows the pattern of a normal curve. Only large and obvious departures from linearity in such a graph are real evidence that the data do not follow the pattern of a normal curve.

Sometimes other special scales are used to graph cumulative distributions. For example, if it is suspected that a set of data follows the pattern of a log-normal distribution, probability graph paper with a logarithmic scale can be used. Such paper is also commercially available. Various other scales have been devised for checking whether data follow the patterns of corresponding theoretical distributions.

5.3 STEM-AND-LEAF PLOTS

In the two preceding sections we directed our attention to the grouping of relatively large sets of data, with the objective of putting such data into a manageable form. As we saw, this entailed some loss of information. In recent years, similar techniques have been proposed for the preliminary exploration of small sets of data which yield a good overall picture of the data without any loss of information.

To illustrate, consider the following humidity readings rounded to the nearest percent:

$$29 \quad 44 \quad 12 \quad 53 \quad 21 \quad 34 \quad 39 \quad 25 \quad 48 \quad 23$$
$$17 \quad 24 \quad 27 \quad 32 \quad 34 \quad 15 \quad 42 \quad 21 \quad 28 \quad 37$$

Proceeding as in Section 5.1, we might group these data into the following distribution:

Humidity readings	Tally	Frequency
10–19	/ / /	3
20–29	╫╫ / / /	8
30–39	╫╫	5
40–49	/ / /	3
50–59	/	1

where the tally pictures the overall pattern of the data like a histogram (or bar chart) lying on its side.

If we wanted to avoid the loss of information inherent in the above table, we could replace the tally marks with the last digits of the corresponding readings, getting

10–19	2 7 5
20–29	9 1 5 3 4 7 1 8
30–39	4 9 2 4 7
40–49	4 8 2
50–59	3

This can also be written as

1*	2 7 5
2*	9 1 5 3 4 7 1 8
3*	4 9 2 4 7
4*	4 8 2
5*	3

where * is a placeholder for 0, 1, 2, 3, 4, 5, 6, 7, 8, or 9, or simply as

1	2 7 5
2	9 1 5 3 4 7 1 8
3	4 9 2 4 7
4	4 8 2
5	3

In either of these final forms, the table is called a **stem-and-leaf plot** (or simply a **stem-leaf plot**)—each line is a **stem** and each digit on a stem to the right of the vertical line is a **leaf**. To the left of the vertical line are the **stem labels**, which, in our example, are 1*, 2*, . . . , and 5*, or 1, 2, . . . , and 5.

Essentially, a stem-and-leaf plot presents the same picture as the corresponding tally, yet it retains all the original information. For instance, if a stem-and-leaf plot has the stem

$$1.2* \quad | \quad 3 \quad 5 \quad 2 \quad 0 \quad 8$$

the corresponding data are 1.23, 1.25, 1.22, 1.20, and 1.28, and if a stem-and-leaf plot has the stem

$$0.3** \quad | \quad 17 \quad 03 \quad 55 \quad 89$$

with two-digit leaves, the corresponding data are 0.317, 0.303, 0.355, and 0.389.

There are various ways in which stem-and-leaf plots can be modified to meet particular needs (see Exercises 5.27 and 5.28), but we shall not go into this here in any detail as it has been our objective only to present one of the relatively new techniques which come under the general heading of **exploratory data analysis**.

EXERCISES

5.1 Measurements of the "fraction solar" of solar-assisted heat pumps in a city in the Southwest are grouped into the following classes: 0.50–0.59, 0.60–0.69, 0.65–0.79, and 0.85–0.94. If the measurements are all rounded to two decimals, none is less than 0.50 and none exceeds 0.94, explain where difficulties might arise.

5.2 Measurements of the ignition temperature of a gas vary from 1,161 to 1,319 degrees Fahrenheit. Construct a table with eight equal classes into which these data might be grouped. Give
(a) the class limits;
(b) the class marks;
(c) the class boundaries;
(d) the class interval.

5.3 The weights of certain mineral specimens, given to the nearest tenth of an ounce, are grouped into a table having the classes 10.5–11.4, 11.5–12.4, 12.5–13.4, and 13.5–14.4 ounces. Find
(a) the class marks;
(b) the class boundaries;
(c) the class interval.

5.4 With reference to the preceding exercise, is it possible to determine from the grouped data how many of the mineral specimens weigh
(a) less than 11.5 ounces;
(b) more than 11.5 ounces;
(c) at least 12.4 ounces;
(d) at most 12.4 ounces;
(e) from 11.5 to 13.5 ounces?

5.5 The following are measurements of the breaking strength (in ounces) of a sample of 60 linen threads:

32.5	15.2	35.4	21.3	28.4	26.9	34.6	29.3	24.5	31.0
21.2	28.3	27.1	25.0	32.7	29.5	30.2	23.9	23.0	26.4
27.3	33.7	29.4	21.9	29.3	17.3	29.0	36.8	29.2	23.5
20.6	29.5	21.8	37.5	33.5	29.6	26.8	28.7	34.8	18.6
25.4	34.1	27.5	29.6	22.2	22.7	31.3	33.2	37.0	28.3
36.9	24.6	28.9	24.8	28.1	25.4	34.5	23.6	38.4	24.0

Group these measurements into a distribution having the classes 15.0–19.9, 20.0–24.9, . . . , 35.0–39.9 and construct a histogram.

5.6 Convert the distribution obtained in the preceding exercise into a cumulative "less than" distribution and graph its ogive.

5.7 Convert the cumulative distribution obtained in the preceding exercise into a cumulative "less than" percentage distribution and, plotting its graph on arithmetic probability paper, decide whether it is reasonable to approximate the original distribution of the data with a normal distribution.

5.8 The class marks of a distribution of temperature readings (given to the nearest degree Celsius) are 16, 25, 34, 43, 52, and 61. Find

(a) the class boundaries;

(b) the class limits.

5.9 The following are the ignition times of certain upholstery materials exposed to a flame, given to the nearest hundredth of a second:

2.58	2.51	4.04	6.43	1.58	4.32	2.20	4.19
4.79	6.20	1.52	1.38	3.87	4.54	5.12	5.15
5.50	5.92	4.56	2.46	6.90	1.47	2.11	2.32
6.75	5.84	8.80	7.40	4.72	3.62	2.46	8.75
2.65	7.86	4.71	6.25	9.45	12.80	1.42	1.92
7.60	8.79	5.92	9.65	5.09	4.11	6.37	5.40
11.25	3.90	5.33	8.64	7.41	7.95	10.60	3.81
3.78	3.75	3.10	6.43	1.70	6.40	3.24	1.79
4.90	3.49	6.77	5.62	9.70	5.11	4.50	2.50
5.21	1.76	9.20	1.20	6.85	2.80	7.35	11.75

Group these figures into a table with a suitable number of equal classes and construct a histogram.

5.10 Convert the distribution obtained in the preceding exercise into a cumulative "less than" percentage distribution and plot its ogive.

5.11 Plot the cumulative "less than" percentages obtained in the preceding exercise on arithmetic as well as logarithmic probability paper. What do these graphs suggest about the form of the distribution obtained in Exercise 5.9? (If logarithmic probability paper is not available, plot the logarithms of the class boundaries on arithmetic probability paper.)

5.12 In a 2-week study of the productivity of workers, the following data were obtained on the total number of acceptable pieces which 100 workers produced:

65	36	49	84	79	56	28	43	67	36
43	78	37	40	68	72	55	62	22	82
88	50	60	56	57	46	39	57	73	65
59	48	76	74	70	51	40	75	56	45
35	62	52	63	32	80	64	53	74	34
76	60	48	55	51	54	45	44	35	51
21	35	61	45	33	61	77	60	85	68
45	53	34	67	42	69	52	68	52	47
62	65	55	61	73	50	53	59	41	54
41	74	82	58	26	35	47	50	38	70

Group these figures into a distribution having the classes 20–29, 30–39, 40–49, . . . , and 80–89, and plot a histogram.

5.13 Convert the distribution obtained in the preceding exercise into a cumulative "less than" percentage distribution and plot its ogive.

5.14 Plot the percentages obtained in the preceding exercise on arithmetic probability paper and interpret the result.

5.15 The following are 15 measurements of the boiling point of a silicon compound (in degrees Celsius): 166, 141, 136, 153, 170, 162, 155, 146, 183, 157, 148, 132, 160, 175, and 150. Plot the data directly on arithmetic probability paper and indicate whether they follow the pattern of a normal distribution.

5.16 The following are the number of automobile accidents which occurred at 60 major intersections in a certain city during the Fourth of July weekend:

0	2	5	0	1	4	1	0	2	1
5	0	1	3	0	0	2	1	3	1
1	4	0	2	4	1	2	4	0	4
3	5	0	1	3	6	4	2	0	2
0	2	3	0	4	2	5	1	1	2
2	1	6	5	0	3	3	0	0	4

Group these data into a frequency distribution showing how often each of the values occurs and draw a bar chart.

5.17 Convert the distribution obtained in the preceding exercise into a cumulative "or more" distribution and draw its ogive.

5.18 In view of the symmetry of the normal distribution, the mean of the normal distribution which approximates a given distribution can be obtained from its plot on arithmetic probability paper by reading off the horizontal scale the value on the line (fitted to the points) which corresponds to 50% on the vertical scale.
   (a) Use this method to estimate from Figure 5.4 the mean of the normal distribution which approximates the distribution of the sulfur oxides emission data.
   (b) Use this method to estimate from the graph obtained in Exercise 5.14 the mean of the normal distribution which approximates the distribution of the productivity data.

5.19 Since roughly 68% of the area under a normal curve falls within 1 standard deviation on either side of the mean, the standard deviation of the normal distribution which approximates a given distribution can be obtained from its plot on arithmetic probability paper by taking half the difference between the values on the horizontal scale which correspond to 84% and 16% on the vertical scale.
   (a) Use this method to estimate from Figure 5.4 the standard deviation of the normal distribution which approximates the distribution of the sulfur oxides emission data.
   (b) Use this method to estimate from the graph obtained in Exercise 5.14 the standard deviation of the normal distribution which approximates the distribution of the productivity data.

5.20 Categorical distributions are often presented graphically by means of **pie charts** in which a circle is divided into sectors proportional in size to the frequencies (or percentages) with which the data are distributed among the categories. Draw a pie chart to represent the following data, obtained in a study in which 40 drivers were asked to judge the maneuverability of a certain make car: Very good, good, good, fair, excellent, good, good, good, very good, poor, good, good, good, good, very good, good, fair, good, good, very poor, very good, fair, good, good, excellent, very good, good, good, good, fair, fair, very good, good, very good, excellent, very good, fair, good, good, and very good.

5.21 The pictogram of Figure 5.5 is intended to illustrate the fact that per capita income in the United States doubled from $4,000 in 1970 to $8,000 in 1978. Does this pictogram convey a "fair" impression of the actual change? If not, state how it might be modified.

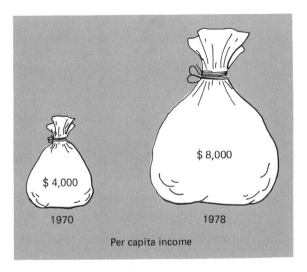

Figure 5.5 Pictogram for Exercise 5.21.

5.22 Convert the distribution of the sulfur oxides emission data on page 138 into a distribution having the classes 5.0–8.9, 9.0–20.9, 21.0–28.9, and 29.0–32.9. Draw two histograms of this distribution, one in which the class frequencies are given by the heights of the rectangles and one in which the class frequencies are given by the areas of the rectangles. Explain why the first of these histograms gives a very misleading picture.

5.23 Given a set of observations $x_1, x_2, \ldots,$ and x_n, we define their **empirical cumulative distribution** as the function whose values $F(x)$ equal the proportion of the observations less than or equal to x.
   (a) Graph the empirical cumulative distribution for the 15 measurements of Exercise 5.15.
   (b) Verify that the method of curve fitting used on page 145 (the check for "normality" for ungrouped data) amounts to plotting points corresponding to the midpoints of the steps of the empirical cumulative distribution.

5.24 The following are figures on a well's daily production of oil in barrels: 214, 203, 226, 198, 243, 225, 207, 203, 208, 200, 217, 202, 208, 212, 205, and 220. Construct a stem-and-leaf plot with the stem labels 19∗, 20∗, . . . , and 24∗.

5.25 The following are determinations of a river's annual maximum flow in m³/s: 405, 355, 419, 267, 370, 391, 612, 383, 434, 462, 288, 317, 540, 295, and 508. Construct a stem-and-left plot with two-digit leaves.

5.26 List the data which correspond to the following stems of stem-and-leaf plots:
(a) 1∗ | 3 2 5 7 1 4 8
(b) 23 | 4 0 0 1 6
(c) 2∗∗ | 35, 18, 57, 03
(d) 3.2 | 1, 7, 4, 4, 3

5.27 If we want to construct a stem-and-leaf plot with more stems than there would be otherwise, we might use ∗ as a placeholder for 0, 1, 2, 3, and 4, and · as a placeholder for 5, 6, 7, 8, and 9. For the humidity readings on page 147 we would thus get the **double-stem plot**

1∗	2
1·	7 5
2∗	1 3 4 1
2·	9 5 7 8
3∗	4 2 4
3·	9 7
4∗	4 2
4·	8
5∗	3

where we doubled the number of stems by cutting the interval covered by each stem in half. Construct a double-stem plot with one-digit leaves for the data of Exercise 5.12.

5.28 If we want to construct a stem-and-leaf plot equivalent to a distribution with a class interval of 2, we can use ∗ as a placeholder for 0 and 1, *t* for 2 and 3, *f* for 4 and 5, *s* for 6 and 7, and · for 8 and 9. The resulting stem-and-leaf plot is called a **five-stem plot**.
(a) The following are the I.Q.'s of 20 applicants to an undergraduate engineering program: 109, 111, 106, 106, 125, 108, 115, 109, 107, 109, 108, 110, 112, 104, 110, 112, 128, 106, 111, and 108. Construct a five-stem plot with one digit leaves.
(b) The following is part of a five-stem plot:

53*f*	5 4 4 4 5 4
53*s*	6 7 6 6
53·	9 8
54∗	1

List the corresponding measurements.

5.4 DESCRIPTIVE MEASURES

In Chapter 3 we introduced the mean and the variance of probability distributions as parameters which measure their center and their spread. In this section we shall define analogous measures to describe numerical data or their distribution.

Given a set of n measurements or observations, $x_1, x_2, \ldots, x_n$, there are many ways in which we can describe their center (middle, or central location). Most popular among these are the **arithmetic mean** and the **median**, although other kinds of "averages" are sometimes used for special purposes. The arithmetic mean or, more succinctly, the **mean** is defined by the formula

Mean

$$\bar{x} = \frac{\sum\limits_{i=1}^{n} x_i}{n}$$

Note that we have written the mean of the x's as $\bar{x}$ and not as μ, the symbol we used for the mean of a probability distribution (or probability density). In this connection we shall follow the general practice of using Latin letters to denote descriptions of actual data and Greek letters to denote descriptions of theoretical distributions. It is also of interest to note that the above formula would yield the mean of the distribution of a random variable which takes on the values x_i with equal probabilities of $\frac{1}{n}$; its mean would be denoted by μ.

Sometimes it is preferable to use the **median** as a descriptive measure of the center, or location, of a set of data. This is true, particularly, if it is desired to minimize the calculations or if it is desired to eliminate the effect of extreme (very large or very small) values. The median of n observations $x_1, x_2, \ldots, x_n$ can be defined loosely as the "middlemost" value once the data are arranged according to size. More precisely, if the observations are arranged according to size and n is an odd number, the median is the value of the observation numbered $\frac{n+1}{2}$; if n is an even number, the median is defined as the mean (average) of the observations numbered $\frac{n}{2}$ and $\frac{n+2}{2}$.

EXAMPLE Find the mean and the median of 15, 14, 2, 27, and 13.

Solution The mean is

$$\bar{x} = \frac{15 + 14 + 2 + 27 + 13}{5} = 14.2$$

and the median is the third largest value, namely, 14.

EXAMPLE   Find the mean and the median of 11, 9, 17, 19, 4, and 15.

> **Solution**   The mean is
>
> $$\bar{x} = \frac{11 + 9 + 17 + 19 + 4 + 15}{6} = 12.5$$
>
> and the median, the mean of the third and fourth largest values, is 13.

Although the mean and the median each provides a single number to represent an entire set of data, the mean is usually preferred in problems of estimation and other problems of statistical inference. An intuitive reason for preferring the mean is that the median does not utilize all the information contained in the observations. A related reason is that the median is generally subject to greater chance fluctuations, that is, it is apt to vary more from sample to sample. This important concept of "sampling variability" will be explored in detail in Chapter 6.

The following is an example where the median actually gives a more useful description of a set of data than the mean:

EXAMPLE   A small company employs four young engineers whom it pays $18,000 each, and the owner (also an engineer), who gets $108,000. Comment on the claim that on the average the company pays $36,000 to its engineers and, hence, is a good place to work.

> **Solution**   The mean of the five salaries is $36,000, but it hardly describes the situation. The median, on the other hand, is $18,000 and it is at least representative of what a young engineer earns with this firm. Moneywise, the company is not such a good place for young engineers.

This example illustrates that there is always an inherent danger in summarizing a set of data by means of a single number.

Like the variance of a probability distribution, the **variance** of n observations $x_1, x_2, \ldots, x_n$ measures essentially the average of their squared deviations from their mean $\bar{x}$, and it is defined by the formula

Variance

$$s^2 = \frac{\sum_{i=1}^{n} (x_i - \bar{x})^2}{n - 1}$$

Our reason for dividing by $n - 1$ instead of n is that if we look upon the x's as values of some random variable, s^2 will yield a better estimate of σ^2, the variance of its distribution. We shall say more about this in Chapters 7 and 8.

Consistent with the terminology of Chapters 3 and 4, we define the **standard deviation** of n observations $x_1, x_2, \ldots, x_n$ as the square root of their variance, namely,

Standard deviation

$$s = \sqrt{\frac{\sum_{i=1}^{n} (x_i - \bar{x})^2}{n - 1}}$$

Note that if we had divided by n instead of by $n - 1$, the resulting formula could have been used for the standard deviation of the distribution of a random variable which assumes the values x_i with equal probabilities of $\frac{1}{n}$; its standard deviation would be denoted by σ.

EXAMPLE   The delay times (handling, setting, and positioning the tools) for cutting six parts on an engine lathe are 0.6, 1.2, 0.9, 1.0, 0.6, and 0.8 minutes. Calculate s.

Solution   First we calculate the mean

$$\bar{x} = \frac{0.6 + 1.2 + 0.9 + 1.0 + 0.6 + 0.8}{6} = 0.85$$

and then we set up the work required to find $\sum (x_i - \bar{x})^2$ in the following table:

x_i	$x_i - \bar{x}$	$(x_i - \bar{x})^2$
0.6	-0.25	0.0625
1.2	0.35	0.1225
0.9	0.05	0.0025
1.0	0.15	0.0225
0.6	-0.25	0.0625
0.8	-0.05	0.0025
		0.2750

Then we divide 0.2750 by $6 - 1 = 5$, take the square root, and get

$$s = \sqrt{\frac{0.2750}{5}} = 0.23$$

The standard deviation and the variance are measures of **absolute variation**, that is, they measure the actual amount of variation present in a set of data, and they depend on the scale of measurement. To compare the variation in several sets of data, it is generally desirable to use a measure of **relative**

variation; for instance, the **coefficient of variation**, which gives the standard deviation as a percentage of the mean.

Coefficient of variation

$$V = \frac{s}{\bar{x}} \cdot 100$$

EXAMPLE Measurements made with one micrometer of the diameter of a ball bearing have a mean of 3.92 mm and a standard deviation of 0.015 mm, while measurements made with another micrometer of the unstretched length of a spring have a mean of 1.54 inches and a standard deviation of 0.008 inch. Which of these two measuring instruments is relatively more precise?

Solution For the first micrometer the coefficient of variation is

$$V = \frac{0.015}{3.92} \cdot 100 = 0.38\%$$

and for the second micrometer the coefficient of variation is

$$V = \frac{0.008}{1.54} \cdot 100 = 0.52\%$$

Thus, the measurements made with the first micrometer are relatively more precise.

In this section, we have limited the discussion to the mean, the median, the variance, and the standard deviation, but there are many other ways of describing sets of data. For instance, if we want to determine a value below which we find the lowest 25% of the data, we calculate the first **quartile** Q_1; if we want to determine a value above which we find the highest 5% of the data, we calculate the ninety-fifth **percentile** P_{95}; and so on. As we shall see, the calculation of such **fractiles** for grouped data is very similar to the calculation of the median. There are also alternative ways of describing the variability of a set of data. For instance, there is the **range** (the largest value minus the smallest), which we shall use in Section 8.1, the **mean deviation** (where we average the absolute values of the deviations from the mean), and the **interquartile range** $Q_3 - Q_1$, where Q_3 is a value below which we find the lowest 75% of the data. As the need arises, new ways and new methods of describing statistical data are constantly being developed.

5.5 THE CALCULATION OF $\bar{x}$ AND s

In this section we shall discuss methods for calculating $\bar{x}$ and s for **raw** (ungrouped) as well as grouped data. These methods are particularly well suited for small calculators, and they are both rapid and accurate.

The calculation of $\bar{x}$ for ungrouped data does not pose any problems; we have only to add the values of the observations and divide by n. On the other hand, the calculation of s^2 is usually cumbersome if we directly use the formula defining s^2 on page 154. Instead, we shall use the algebraically equivalent form

Variance (computing formula)

$$s^2 = \frac{n \cdot \sum_{i=1}^{n} x_i^2 - \left(\sum_{i=1}^{n} x_i\right)^2}{n(n-1)}$$

which has the dual advantage of requiring less labor and giving better accuracy. (In Exercise 5.48 on page 162 the reader will be asked to show that this formula is, in fact, equivalent to the one on page 154.) The decrease in labor results from the fact that we do not actually have to calculate the deviations from the mean; the increase in accuracy is due to the reduction of the number of divisions and subtractions, and their postponement until the last two steps in the calculation.

EXAMPLE Find the mean and the standard deviation of the following miles per gallon obtained in 20 test runs performed on urban roads with an intermediate-size car:

$$
\begin{array}{ccccc}
19.7 & 21.5 & 22.5 & 22.2 & 22.6 \\
21.9 & 20.5 & 19.3 & 19.9 & 21.7 \\
22.8 & 23.2 & 21.4 & 20.8 & 19.4 \\
22.0 & 23.0 & 21.1 & 20.9 & 21.3 \\
\end{array}
$$

Solution Using a calculator, we find that the sum of these figures is 427.7 and that the sum of their squares is 9,173.19. Consequently,

$$\bar{x} = \frac{427.7}{20} = 21.38$$

and

$$s^2 = \frac{20(9,173.19) - (427.7)^2}{20 \cdot 19} = 1.412$$

and it follows that $s = 1.19$. In computing the necessary sums we usually retain all decimal places, but as in this example, at the end we usually round to one more decimal than we had in the original data.

The calculations we have shown here can be made even simpler by first **coding** the observations; that is, by performing a suitable linear transformation.

EXAMPLE   Rework the preceding example by first multiplying each of the mileages by 10 (to eliminate the decimal points) and then subtracting 200 (to make the numbers easier to work with).

   Solution   The coded values, which we shall refer to as u's, are

$$
\begin{array}{rrrrr}
-3 & 15 & 25 & 22 & 26 \\
19 & 5 & -7 & -1 & 17 \\
28 & 32 & 14 & 8 & -6 \\
20 & 30 & 11 & 9 & 13
\end{array}
$$

Calculating the mean and the standard deviation of the u's by the method we used before, we get 277 for the sum of the u's and 6,519 for the sum of their squares. Hence,

$$\bar{u} = \tfrac{277}{20} = 13.85$$

and

$$s_u = \sqrt{\frac{20(6{,}519) - (277)^2}{20 \cdot 19}} = 11.88$$

In order to undo the coding, let us observe that the x's and the u's are related by the equation $u_i = 10x_i - 200$ or $x_i = \tfrac{1}{10}u_i + 20$. Continuing from here, simple algebra will show that $\bar{x} = \tfrac{1}{10}\bar{u} + 20$ and $s_x = \tfrac{1}{10}s_u$, where s_x and s_u are the standard deviations of the x's and the u's. Thus, we get

$$\bar{x} = \tfrac{1}{10}(13.85) + 20 = 21.38$$

and

$$s_x = \tfrac{1}{10}(11.88) = 1.19$$

with both results rounded to two decimals. This agrees with the values obtained on page 157.

  In Exercise 5.49 on page 162, the reader will be asked to verify that, in general, when data are coded so that $x_i = c \cdot u_i + a$, the formulas for the mean and the standard deviation become

Mean and standard deviation (with coding)

$$\bar{x} = c \cdot \bar{u} + a \quad and \quad s_x = c \cdot s_u$$

  Not too many years ago, one of the main reasons for grouping data was to simplify the calculation of descriptions such as the mean and the standard deviation. With easy access to statistical calculators and computers, this is no longer the case, but we shall nevertheless discuss here the calculation of $\bar{x}$ and s from grouped data, since some data (for instance, from government publications) may be available only in grouped from.

To calculate $\bar{x}$ and s from grouped data, we shall have to make some assumption about the distribution of the values within each class. If we represent all values within a class by the corresponding class mark, the sum of the x's and the sum of their squares can now be written

$$\sum_{i=1}^{k} x_i f_i \quad \text{and} \quad \sum_{i=1}^{k} x_i^2 f_i$$

where x_i is the classmark of the ith class, f_i is the corresponding class frequency, and k is the number of classes in the distribution. Substituting these sums into the formula for $\bar{x}$ and the computing formula for s^2, we get

Mean and variance (grouped data)

$$\bar{x} = \frac{\sum_{i=1}^{k} x_i f_i}{n}$$

$$s^2 = \frac{n \cdot \sum_{i=1}^{k} x_i^2 f_i - \left(\sum_{i=1}^{k} x_i f_i \right)^2}{n(n-1)}$$

EXAMPLE   Use the distribution obtained on page 139 to calculate the mean and the variance of the sulfur oxides emission data.

Solution   Recording the class marks and the class frequencies in the first two columns, and the products $x_i f_i$ and $x_i^2 f_i$ in the third and fourth columns, we obtain

x_i	f_i	$x_i f_i$	$x_i^2 f_i$
6.95	3	20.85	144.9075
10.95	10	109.50	1,199.0250
14.95	14	209.30	3,129.0350
18.95	25	473.75	8,977.5625
22.95	17	390.15	8,953.9425
26.95	9	242.55	6,536.7225
30.95	2	61.90	1,915.8050
	80	1,508.00	30,857.0000

Then, substitution into the formulas yields

$$\bar{x} = \frac{1,508}{80} = 18.85$$

and

$$s^2 = \frac{80(30,857) - (1,508)^2}{80 \cdot 79} = 30.77$$

These calculations were very tedious, and we went through them only to impress the reader with the simplifications that can be attained by coding the class marks. In this kind of problem, we code the class marks with successive integers, preferably with 0 near the center of the distribution or near the class with the highest frequency.

EXAMPLE   Rework the preceding example, by coding the class marks -3, -2, -1, 0, 1, 2, and 3.

Solution   Referring to the coded class marks as u's, we get

u_i	f_i	$u_i f_i$	$u_i^2 f_i$
-3	3	-9	27
-2	10	-20	40
-1	14	-14	14
0	25	0	0
1	17	17	17
2	9	18	36
3	2	6	18
	80	-2	152

and the mean and the variance of the u's are

$$\bar{u} = \frac{-2}{80} = -0.025$$

and

$$s_u^2 = \frac{80(152) - (-2)^2}{80 \cdot 79} = 1.923$$

Since the coding we used is such that $x_i = 4u_i + 18.95$, the two formulas on page 158 with $c = 4$ and $a = 18.95$ yield

$$\bar{x} = 4(-0.025) + 18.95 = 18.85$$

and

$$s_x^2 = 4^2(1.923) = 30.77$$

They both agree with the results previously obtained.

EXERCISES

5.29 The following are the numbers of twists that were required to break 12 forged alloy bars: 33, 24, 39, 48, 26, 35, 38, 54, 23, 34, 29, and 37. Find
(a) the mean;   (b) the median.

5.30 With reference to the preceding exercise, find s using
(a) the formula which defines s;
(b) the computing formula for s.

5.31 If the mean annual salary paid to the chief executives of three engineering firms is $125,000, can one of them receive $400,000?

5.32 By mistake, an instructor erased the grade which one of ten students received. If the nine other students got 43, 66, 74, 90, 40, 52, 70, 78, and 92, and the mean of all ten grades is 67, what grade did the instructor erase?

5.33 The following are the numbers of minutes which a person had to wait for the bus to work on 15 working days: 10, 1, 13, 9, 5, 9, 2, 10, 3, 8, 6, 17, 2, 10, and 15. Find
(a) the mean;   (b) the median.

5.34 With reference to the preceding exercise, find s^2 using
(a) the formula which defines s^2;
(b) the computing formula for s^2.

5.35 Records show that in Phoenix, Arizona, the normal daily maximum temperature for each month is 65, 69, 74, 84, 93, 102, 105, 102, 98, 88, 74, and 66 degrees Fahrenheit. Verify that the mean of these figures is 85 and comment on the claim that, in Phoenix, the average daily maximum temperature is a very comfortable 85 degrees.

5.36 With reference to Exercise 5.24 on page 152, find the mean and the median of the data on the well's daily production of oil.

5.37 With reference to Exercise 5.25 on page 152, find the standard deviation of the data on the river's annual maximum flow.

5.38 The following are 12 temperature readings at various locations in a large kiln (in degrees Fahrenheit): 475, 500, 460, 425, 460, 410, 470, 475, 460, 510, 450, and 415. Find
(a) the mean;   (c) the standard deviation.
(b) the median;

5.39 Rework parts (a) and (c) of the preceding exercise, using the coding
$$u = \frac{x}{5} - 90.$$

5.40 With reference to Exercise 5.15 on page 150, calculate $\bar{x}$ and s
(a) without coding;
(b) after suitably coding the 15 measurements.

5.41 Find the mean and the standard deviation of the 20 humidity readings on page 146 by using

(a) the raw (ungrouped) data;

(b) the distribution obtained on that page.

5.42 Use the distribution obtained in Exercise 5.5 on page 148 to find the mean and the variance of the breaking strengths.

5.43 Use the distribution obtained in Exercise 5.9 on page 149 to find the mean and the standard deviation of the ignition times. Also determine the coefficient of variation.

5.44 Use the distribution obtained in Exercise 5.12 on page 149, to find the coefficient of variation of the productivity data.

5.45 In three recent years, the price of copper was 69.6, 66.8, and 66.3 cents per pound, and the price of bituminous coal was 19.43, 19.82, and 22.40 dollars per short ton. Which of the two sets of prices is relatively more variable?

5.46 In five tests, one student averaged 63.2 with a standard deviation of 3.3, while another student averaged 78.8 with a standard deviation of 5.3. Which student is relatively more consistent?

5.47 Show that $\sum_{i=1}^{n} (x_i - \bar{x}) = 0$ for any set of observations $x_1, x_2, \ldots,$ and x_n.

5.48 Show that the computing formula for s^2 on page 157 is equivalent to the one used to define s^2 on page 154.

5.49 If data are coded so that $x_i = c \cdot u_i + a$, show that $\bar{x} = c \cdot \bar{u} + a$ and $s_x = c \cdot s_u$.

5.50 To find the **median** of a distribution obtained for n observations, we first determine the class into which the median must fall. Then, if there are j values in this class and k values below it, the median is located $\dfrac{(n/2) - k}{j}$ of the way into this class, and to obtain the median we multiply this fraction by the class interval and add the result to the lower class boundary of the class into which the median must fall. This method is based on the assumption that the observations in each class are "spread uniformly" throughout the class interval, and this is why we count $\dfrac{n}{2}$ of the observations instead of $\dfrac{n+1}{2}$ as on page 153.

To illustrate, let us refer to the distribution of the sulfur oxides emission data on page 138. Since $n = 80$, it can be seen that the median must fall into the class 17.0–20.9, and since $j = 25$ and $k = 27$, it follows that the median is

$$16.95 + \frac{40 - 27}{25} \cdot 4 = 19.03.$$

(a) Find the median of the distribution of the absenteeism data given on page 141.

(b) Use the distribution obtained in Exercise 5.5 on page 148 to find the median of the grouped breaking strengths.

(c) Use the distribution obtained in Exercise 5.9 on page 149 to find the median of the grouped ignition times.

5.51 For each of the following distributions decide whether it is possible to find the mean and/or the median. Explain your answers.

(a)	Grade	Frequency		(b)	I.Q.	Frequency
	40–49	5			less than 90	3
	50–59	18			90– 99	14
	60–69	27			100–109	22
	70–79	15			110–119	19
	80–89	6			more than 119	7

(c)	Weight	Frequency
	100 or less	41
	101–110	13
	111–120	8
	121–130	3
	131–140	1

5.52 To find the quartiles Q_1 and Q_3 for grouped data, we proceed as in Exercise 5.50, but count $\frac{n}{4}$ and $\frac{3n}{4}$ of the observations instead of $\frac{n}{2}$.

(a) With reference to the distribution of the sulfur oxides emission data on page 139, find Q_1, Q_3, and the interquartile range.

(b) Find Q_1 and Q_3 for the distribution of the absenteeism data given on page 141.

5.53 If k sets of data consist, respectively, of $n_1, n_2, \ldots, n_k$ observations and have the means $\bar{x}_1, \bar{x}_2, \ldots, \bar{x}_k$, then the overall mean of all the data is given by the formula

$$\bar{\bar{x}} = \frac{\sum_{i=1}^{k} n_i \bar{x}_i}{\sum_{i=1}^{k} n_i}$$

(a) The average annual salaries paid to top-level management in three companies are $84,000, $92,000, and $89,000. If the respective numbers of top-level executives in these companies are 4, 15, and 11, find the average salary paid to these 30 executives.

(b) In a nuclear engineering class there are 22 juniors, 18 seniors, and 10 graduate students. If the juniors averaged 71 in the midterm examination, the seniors averaged 78, and the graduate students averaged 89, what is the mean for the entire class?

5.54 The formula of the preceding exercise is a special case of the following formula for the **weighted mean**

$$\bar{x}_w = \frac{\sum_{i=1}^{k} w_i x_i}{\sum_{i=1}^{k} w_i}$$

where w_i is a weight indicating the relative importance of the ith observation.

(a) If an instructor counts the final examination in a course four times as much as each one-hour examination, what is the weighted average grade of a student who received grades of 69, 75, 56, and 72 in four one-hour examinations and a final examination grade of 78?

(b) From 1975 to 1980 the cost of food increased by 53% in a certain city, the cost of housing increased by 40%, and the cost of transportation increased by 34%. If the average salaried worker spent 28% of his or her income on food, 35% on housing, and 14% on transportation, what is the combined percentage increase in the cost of these items?

5.6   CHECK LIST OF KEY TERMS *(with page references)*

Absolute variation, 155
Arithmetic mean, 153
Arithmetic probability paper, 144
Bar chart, 142
Categorical distribution, 138
Class boundary, 139
Class frequency, 138
Class interval, 139
Class limit, 138
Class mark, 139
Coding, 157
Coefficient of variation, 156
Cumulative distribution, 140
Cumulative percentage distribution, 141
Double-stem plot, 152
Empirical cumulative distribution, 151
Exploratory data analysis, 148
Five-stem plot, 152
Fractile, 156
Frequency distribution, 138
Frequency polygon, 142
Histogram, 142

Interquartile range, 156
Leaf, 147
Mean, 153
Mean deviation, 156
Median, 153
Numerical distribution, 138
Ogive, 143
Open class, 141
Percentage distribution, 140
Percentile, 156
Pie chart, 150
Quartile, 156
Range, 156
Raw data, 156
Relative variation, 155
Standard deviation, 155
Stem, 147
Stem-and-leaf plot, 147
Stem label, 147
Variance, 154
Weighted mean, 163

6

Sampling Distributions

In most of the methods we shall study in this book, it will be assumed that we are dealing with a particular kind of sample called a random sample. This attention to random samples, which we shall discuss in Section 6.1, is due to their permitting valid, or logical, generalizations from sample data. Then, in Sections 6.2 through 6.4, we shall see how certain statistics (that is, certain quantities determined from samples) can be expected to vary from sample to sample. The concept of a sampling distribution, the distribution of a statistic calculated on the basis of a random sample, is basic to all of statistical inference.

Usage of the term **population** in statistics is a carryover from the days when statistics was applied mainly to sociological and economic phenomena. Nowadays, it is applied to sets or collections of objects, actual or conceptual, and mainly to sets of numbers, measurements, or observations. For example, if we are interested in determining the average number of television sets per household in the United States, the totality of these figures, one for each household, constitutes the population for this study. Similarly, the population from which inspectors draw a sample to determine some quality characteristic of a manufactured product may be the corresponding measurements for all units in a given lot; depending on the objectives of the inspection, it may also consist of the corresponding measurements for all units that may conceivably be manufactured.

In some cases, such as the above example concerning the number of television sets per household, the population is **finite**; in other cases, such as the determination of some characteristic of all units past, present, and future, that might conceivably be manufactured by a given process, it is convenient to think of the population as **infinite**. Similarly, we look upon the results obtained in a series of flips of a coin as a sample from the hypothetically finite population consisting of all conceivably possible flips of the coin.

Populations are often described by the distributions of their values, and it is common practice to refer to a population in terms of this distribution. (For finite populations, we are referring here to the actual distribution of its values; for infinite populations, we are referring to the corresponding probability distribution or probability density.) For example, we may refer to a number of flips of a coin as a sample from a "binomial population" or to certain measurements as a sample from a "normal population." Hereafter, when referring to a "population $f(x)$" we shall mean a population such that its elements have a frequency distribution, a probability distribution, or a density with values given by $f(x)$.

If a population is infinite it is impossible to observe all its values, and even if it is finite it may be impractical or uneconomical to observe it in its entirety. Thus, it is usually necessary to use a **sample**, a part of a population, and infer from it results pertaining to the entire population. Clearly, such results can be useful only if the sample is in some way "representative" of the population. It would be unreasonable, for instance, to expect useful generalizations about the population of 1984 family incomes in the United States on the basis of data pertaining to home owners only. Similarly, we can hardly expect reasonable generalizations about the performance of a tire if it is tested only on smooth roads. To assure that a sample is representative of the population from which it is obtained, and to provide a framework for the application of probability theory to problems of sampling, we shall limit our discussion to so-called **random samples**. For sampling from finite populations, they are defined as follows:

Random sample
(finite population)

> *A set of observations $x_1, x_2, \ldots, x_n$ constitutes a random sample of size n from a finite population of size N, if it is chosen so that each subset of n of the N elements of the population has the same probability of being selected.*

Note that this definition of randomness pertains essentially to the manner in which the sample values are selected. This holds also for the following definition of a random sample from an infinite population:

Random sample
(infinite
population)

> *A set of observations $x_1, x_2, \ldots, x_n$ constitutes a random sample of size n from the infinite population $f(x)$ if:*
>
> 1. *Each x_i is a value of a random variable whose distribution has the values $f(x)$.*
> 2. *These n random variables are independent.*

(In recent years, it has become the custom to apply the term "random sample" to the random variables instead of their values.)

There are several ways of assuring the selection of a sample that is at least approximately random. When dealing with a finite population, we can serially number the elements of the population and then select a sample with the aid of a table of random digits (see discussion on page 93). For instance, if a population has $N = 500$ elements and we wish to select a random sample of size $n = 10$, we can use three arbitrarily selected columns of Table 7 to obtain 10 different three-digit numbers less than or equal to 500, which will then serve as the serial numbers of the elements to be included in the sample.

When the population size is large, the use of random numbers can become very laborious and at times practically impossible. For instance, if a sample of five cartons of canned peaches is to be chosen for inspection from among the many thousands stored in a warehouse, one can hardly expect to number all the cartons, make a selection with the use of random numbers, and then pull out the ones that were chosen. In a situation like this, one really has very little choice but to make the selection relatively haphazard, hoping that this will not seriously violate the assumption of randomness which is basic to most statistical theory.

When dealing with infinite populations, the situation is somewhat different since we cannot physically number the elements of the population; but efforts should be made to approach conditions of randomness by the use of artificial devices. For example, in selecting a sample from a production line we may be able to approximate conditions of randomness by choosing one unit each half hour; when tossing a coin we can try to flip it in such a way that neither side is intentionally favored, and so forth. The proper use of artificial or mechanical devices for selecting random samples is always ✓ preferable to human judgment, as it is extremely difficult to avoid unconscious biases when making almost any kind of selection.

Even with the careful choice of artificial devices, it is all too easy to commit gross errors in the selection of a random sample. To illustrate some of these pitfalls, suppose we have the task of selecting logs being fed into a sawmill by a constant-speed conveyer belt, for the purpose of obtaining a random sample of their lengths. One sampling device, which at first sight would seem to assure randomness, consists of measuring the logs which pass a given point at the end of a certain number of ten-minute intervals. However, further thought reveals that this method of selection favors the longer logs, since they require more time to pass the given point. Thus, the sample is not random since the longer logs have a better chance of being included. Another common mistake in selecting a sample is that of sampling from the wrong population or from a poorly specified population. As we have pointed out earlier, we would hardly get a sample from which we could generalize about family incomes in the United States if we limited our sample to home owners. Similarly, if we wanted to determine the effect of vibrations on a structural member, we should be careful to delineate the frequency band of vibrations that is of relevance, and to vibrate test specimens only at frequencies selected randomly from this band.

The purpose of most statistical investigations is to generalize from information contained in random samples about the population from which the samples were obtained. In particular, we are usually concerned with the ✓ problem of making inferences about the **parameters** of populations, such as the mean μ or the standard deviation σ. In making such inferences, we use **statistics** such as $\bar{x}$ and s, namely, quantities calculated on the basis of sample observations. In practice, the term "statistic" is also applied to the corresponding random variables.

Since the selection of a random sample is controlled largely by chance, so are the values we obtain for statistics, and the remainder of this chapter will be devoted to sampling distributions, namely, to distributions which describe the chance fluctuations of statistics calculated on the basis of random samples.

6.2   THE SAMPLING DISTRIBUTION OF THE MEAN (σ KNOWN)

Suppose that a random sample of n observations has been taken from some population and that $\bar{x}$ has been computed, say, to estimate the mean of the population. It should be clear that if we took a second random sample of size n from this population, it would be quite unreasonable to expect the identical value for $\bar{x}$, and if we took several more samples, probably no two of the $\bar{x}$'s would be alike. The differences among such $\bar{x}$'s are generally attributed to chance, and this raises important questions concerning their distribution, specifically concerning the extent of their chance fluctuations.

To approach this question experimentally, suppose that 50 random samples of size $n = 10$ are to be taken from a population having the **discrete uniform distribution**

$$f(x) = \begin{cases} \frac{1}{10} & \text{for } x = 0, 1, 2, \ldots, 9 \\ 0 & \text{elsewhere} \end{cases}$$

Sampling is **with replacement**, so to speak, so that we are sampling from an infinite population. A convenient way of obtaining these samples is to use a table of random digits, like Table 7, letting each sample consist of 10 consecutive digits in arbitrarily chosen rows or columns. Actually proceeding in this way, we got 50 samples whose means are

4.4	3.2	5.0	3.5	4.1	4.4	3.6	6.5	5.3	4.4
3.1	5.3	3.8	4.3	3.3	5.0	4.9	4.8	3.1	5.3
3.0	3.0	4.6	5.8	4.6	4.0	3.7	5.2	3.7	3.8
5.3	5.5	4.8	6.4	4.9	6.5	3.5	4.5	4.9	5.3
3.6	2.7	4.0	5.0	2.6	4.2	4.4	5.6	4.7	4.3

Grouping these means into a distribution with the classes 2.0–2.9, 3.0–3.9, ..., and 6.0–6.9, we get

$\bar{x}$	Frequency
2.0–2.9	2
3.0–3.9	14
4.0–4.9	19
5.0–5.9	12
6.0–6.9	3
	50

and it is apparent from this distribution as well as its histogram shown in Figure 6.1 that the distribution of the means is fairly **bell-shaped**, even though the population itself has a uniform distribution. This raises the question whether our result is typical of what we might expect; that is, whether we would get similar distributions if we repeated the experiment again and again.

To answer this kind of question, we shall have to investigate the **theoretical sampling distribution** of the mean which, for the given example, provides us with the probabilities of getting means from 2.0 to 2.9, from 3.0 to 3.9,

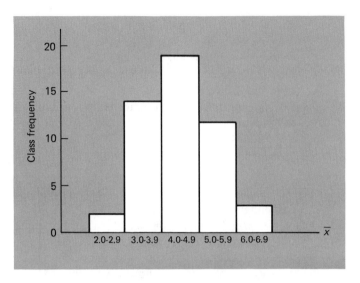

Figure 6.1 Experimental sampling distribution of the mean.

. . . , from 6.0 to 6.9, and perhaps values less than 2.0 or greater than 6.9. Although we could evaluate these probabilities for this particular example, it is usually sufficient to refer to some general theorems concerning sampling distributions. The first of these, stated as follows, gives expressions for the mean $\mu_{\bar{x}}$ and the variance $\sigma_{\bar{x}}^2$ of sampling distributions of the mean:

Formulas for
$\mu_{\bar{x}}$ *and* $\sigma_{\bar{x}}^2$

> **Theorem 6.1** *If a random sample of size n is taken from a population having the mean μ and the variance σ^2, then $\bar{x}$ is a value of a random variable whose distribution has the mean μ. For samples from* infinite *populations the variance of this distribution is σ^2/n; for samples from* finite *populations of size N the variance is $\dfrac{\sigma^2}{n} \cdot \dfrac{N-n}{N-1}$.*

To prove that $\mu_{\bar{x}} = \mu$ for the continuous case, we use the definition on page 131 and write

$$\mu_{\bar{x}} = \int_{-\infty}^{\infty} \int_{-\infty}^{\infty} \cdots \int_{-\infty}^{\infty} \sum_{i=1}^{n} \frac{x_i}{n} f(x_1, x_2, \ldots, x_n) \, dx_1 \, dx_2 \ldots dx_n$$

$$= \frac{1}{n} \sum_{i=1}^{n} \int_{-\infty}^{\infty} \int_{-\infty}^{\infty} \cdots \int_{-\infty}^{\infty} x_i f(x_1, x_2, \ldots, x_n) \, dx_1 \, dx_2 \ldots dx_n$$

where $f(x_1, x_2, \ldots, x_n)$ is a value of the joint density of the random variables whose values consitute the random sample. Using the assumption of randomness, we can write

$$f(x_1, x_2, \ldots, x_n) = f(x_1)f(x_2)\cdot \ldots \cdot f(x_n)$$

and we now have

$$\mu_{\bar{x}} = \frac{1}{n} \sum_{i=1}^{n} \int_{-\infty}^{\infty} f(x_1) \, dx_1 \cdot \ldots \cdot \int_{-\infty}^{\infty} x_i f(x_i) \, dx_i \cdot \ldots \cdot \int_{-\infty}^{\infty} f(x_n) \, dx_n$$

Since each integral except the one with the integrand $x_i f(x_i)$ equals 1, and the one with the integrand $x_i f(x_i)$ equals μ, we finally obtain

$$\mu_{\bar{x}} = \frac{1}{n} \sum_{i=1}^{n} \mu = \mu$$

and this completes the proof. (For the discrete case the proof follows the same steps, with integral signs replaced by $\sum$'s.)

To prove that $\sigma_{\bar{x}}^2 = \sigma^2/n$ for the continuous case, we shall make the simplifying assumption that $\mu = 0$, which does not involve any loss of generality, as the reader will be asked to show in Exercise 6.20 on page 177. As a matter of fact, we had a similar result on page 158, where the addition of a constant to each value did not affect the standard deviation (or the variance) of a set of data. Using the definition on page 131, we thus have

$$\sigma_{\bar{x}}^2 = \int_{-\infty}^{\infty} \int_{-\infty}^{\infty} \ldots \int_{-\infty}^{\infty} \bar{x}^2 f(x_1, x_2, \ldots, x_n) \, dx_1 \, dx_2 \ldots dx_n$$

and making use of the fact that

$$\bar{x}^2 = \frac{1}{n^2} \left(\sum_{i=1}^{n} x_i \right)^2 = \frac{1}{n^2} \left(\sum_{i=1}^{n} x_i^2 + \sum\sum_{i \neq j} x_i x_j \right)$$

we obtain

$$\sigma_{\bar{x}}^2 = \frac{1}{n^2} \sum_{i=1}^{n} \int_{-\infty}^{\infty} \int_{-\infty}^{\infty} \ldots \int_{-\infty}^{\infty} x_i^2 f(x_1, x_2, \ldots, x_n) \, dx_1 \, dx_2 \ldots dx_n$$

$$+ \frac{1}{n^2} \sum\sum_{i \neq j} \int_{-\infty}^{\infty} \int_{-\infty}^{\infty} \ldots \int_{-\infty}^{\infty} x_i x_j f(x_1, x_2, \ldots x_n) \, dx_1 \, dx_2 \ldots dx_n$$

where $\sum\sum_{i \neq j}$ extends over all i and j from 1 to n, not including the terms where $i = j$. Again using the fact that

$$f(x_1, x_2, \ldots, x_n) = f(x_1)f(x_2)\cdot \ldots \cdot f(x_n)$$

we can write each of the above multiple integrals as a product of simple integrals, with each simple integral having an integrand of the form $f(x)$ equalling 1. We thus obtain

$$\sigma_{\bar{x}}^2 = \frac{1}{n^2} \sum_{i=1}^{n} \int_{-\infty}^{\infty} x_i^2 f(x_i)\, dx_i + \frac{1}{n^2} \sum \sum_{i \neq j} \int_{-\infty}^{\infty} x_i f(x_i)\, dx_i \cdot \int_{-\infty}^{\infty} x_j f(x_j)\, dx_j$$

and since each integral in the first sum equals σ^2 while each integral in the second sum equals 0, we finally have

$$\sigma_{\bar{x}}^2 = \frac{1}{n^2} \sum_{i=1}^{n} \sigma^2 = \frac{\sigma^2}{n}$$

This completes the proof of the second part of the theorem. We shall not prove the corresponding result for random samples from finite populations, but it should be noted that in the resulting formula for $\sigma_{\bar{x}}^2$ the factor $\dfrac{N-n}{N-1}$, often called the **finite population correction factor**, is close to 1 (and can be omitted for most practical purposes) unless the sample constitutes a substantial portion of the population.

EXAMPLE Find the value of the finite population correction factor for $n = 10$ and $N = 1{,}000$.

      Solution

$$\frac{1{,}000 - 10}{1{,}000 - 1} = 0.991$$

    Although it should not come as a surprise that $\mu_{\bar{x}} = \mu$, the fact that $\sigma_{\bar{x}}^2 = \sigma^2/n$ for random samples from infinite populations is interesting and important. To point out its implications, let us apply Chebyshev's theorem to the sampling distribution of the mean, substituting $\bar{x}$ for x and $\sigma/\sqrt{n}$ for σ in the formula for the alternate form of the theorem (see page 76). We thus obtain

$$P\left(|\bar{x} - \mu| < \frac{k\sigma}{\sqrt{n}} \right) \geq 1 - \frac{1}{k^2}$$

and letting $k\sigma/\sqrt{n} = \epsilon$, we get

$$P(|\bar{x} - \mu| < \epsilon) \geq 1 - \frac{\sigma^2}{n\epsilon^2}$$

Thus, for any given $\epsilon > 0$, the probability that $\bar{x}$ differs from μ by less than ϵ can be made arbitrarily close to 1 by choosing n sufficiently large. In less rigorous language, the larger the sample size, the closer we can expect $\bar{x}$ to be to the mean of the population. In this sense we can say that the mean becomes more and more reliable as an estimate of μ as the sample size is increased.

The reliability of the mean as an estimate of μ is often measured by $\sigma_{\bar{x}} = \sigma/\sqrt{n}$, also called the **standard error of the mean**. Note that this measure of the reliability of the mean decreases in proportion to the square root of n; for instance, it is necessary to quadruple the size of the sample in order to halve the standard deviation of the sampling distribution of the mean. This also indicates what might be called a "law of diminishing returns" so far as increasing the sample size is concerned. Usually it does not pay to take excessively large samples since the extra labor and expense is not accompanied by a proportional gain in reliability. For instance, if we increase the size of a sample from 25 to 2,500, the errors to which we are exposed are reduced only by a factor of 10.

Let us now return to the experimental sampling distribution on page 169, and let us check how closely its mean and variance correspond to the values we should expect in accordance with Theorem 6.1. Since the population from which the 50 samples of size $n = 10$ were obtained has the mean

$$\mu = \sum_{x=0}^{9} x \cdot \tfrac{1}{10} = 4.5$$

and the variance

$$\sigma^2 = \sum_{x=0}^{9} (x - 4.5)^2 \tfrac{1}{10} = 8.25$$

Theorem 6.1 leads us to expect a mean of $\mu_{\bar{x}} = 4.5$ and a variance of $\sigma_{\bar{x}}^2 = 8.25/10 = 0.825$. Calculating the mean and the variance from the frequency distribution on page 169, we get $\bar{x}_{\bar{x}} = 4.45$ and $s_{\bar{x}}^2 = 0.939$, which are reasonably close to the theoretical values.

Theorem 6.1 provides only partial information about the theoretical sampling distribution of the mean. In general, it is impossible to determine such a distribution exactly without knowledge of the actual form of the population, but it is possible to find the limiting distribution as $n \longrightarrow \infty$ of a random variable whose values are closely related to $\bar{x}$, assuming only that the population has a finite variance σ^2. The random variable we are referring to here is the **standardized mean**; its values are given by

$$z = \frac{\bar{x} - \mu}{\sigma/\sqrt{n}}$$

namely, by the difference between $\bar{x}$ and μ divided by the standard error of the mean. With reference to this random variable, we can now state the following theorem, called the **central limit theorem**:

> **Theorem 6.2** *If $\bar{x}$ is the mean of a random sample of size n taken from a population having the mean μ and the finite variance σ^2, then*
>
> $$z = \frac{\bar{x} - \mu}{\sigma/\sqrt{n}}$$
>
> *is the value of a random variable whose distribution function approaches that of the standard normal distribution as $n \longrightarrow \infty$.*

Proving this theorem is beyond the scope of this text, but we can obtain experimental verification by plotting the cumulative percentage distribution corresponding to the experimental sampling distribution on page 169 on arithmetic probability paper. As can be seen from Figure 6.2, the points fall

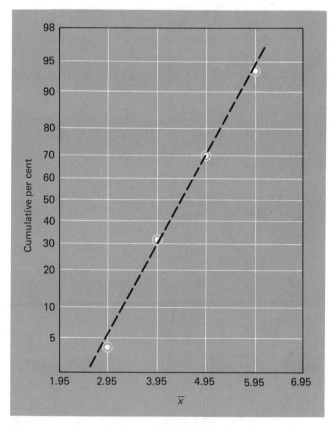

Figure 6.2 Experimental verification of central limit theorem.

close to a straight line, and it seems that even for $n = 10$ the sampling distribution of the mean for this example follows the overall pattern of a normal distribution. In practice, the normal distribution provides an excellent approximation to the sampling distribution of the mean for n as small as 25 or 30, with hardly any restrictions on the shape of the population. As we saw in our example, the sampling distribution of the mean has the general shape of a normal distribution even for samples of size $n = 10$ from a discrete uniform distribution. Note that for random samples from a normal population the sampling distribution of the mean is normal regardless of the size of the sample.

EXAMPLE If a 1-gallon can of a certain kind of paint covers on the average 513.3 square feet with a standard deviation of 31.5 square feet, what is the probability that the mean area covered by a sample of 40 of these 1-gallon cans will be anywhere from 510.0 to 520.0 square feet?

> **Solution** By Theorem 6.2 we shall have to find the normal curve area between
>
> $$z = \frac{510.0 - 513.3}{31.5/\sqrt{40}} = -0.66 \quad \text{and} \quad z = \frac{520.0 - 513.3}{31.5/\sqrt{40}} = 1.34$$
>
> and checking these values in Table 3 we obtain a probability of 0.6553. Note that if $\bar{x}$ turned out to be much less than 513.3, say, less than 500.0, this might cause serious doubt whether the sample actually came from a population having $\mu = 513.3$ and $\sigma = 31.5$; the probability of obtaining such a small value (a z-value less than -2.67) is only 0.0038.

EXERCISES

6.1 An inspector examines every twentieth piece coming off an assembly line. List some of the conditions under which this method of sampling might not yield a random sample.

6.2 In 1932 the *Literary Digest* predicted the presidential election by random sampling from telephone directories and from its list of subscribers. The prediction was grossly incorrect; explain why.

6.3 Explain why the following will not lead to random samples of the desired populations:

(a) To determine what the average person spends on a vacation, a researcher interviews passengers on a luxury cruise.

(b) To determine the average income of its graduates 10 years after graduation, the alumni office of a university sent questionnaires in 1984 to all the members of the class of 1974, and bases its estimate on the questionnaires returned.

(c) To determine public sentiment about certain import restrictions, an inter-
viewer asks voters: "Do you feel that this unfair practice should be
stopped?"

6.4 A market research organization wants to try a new product in 8 of the 50
states. Use Table 7 to make this selection.

6.5 How many different samples of size $n = 2$ can be chosen from a finite population of size
   (a) $N = 6$;   (b) $N = 25$?

6.6 With reference to the preceding exercise, what is the probability of each
sample in part (a) and the probability of each sample in part (b), if the samples
are to be random?

6.7 Take 30 slips of paper and label five each -4 and 4, four each -3 and 3,
three each -2 and 2, and two each $-1, 0$, and 1.
   (a) If each slip of paper has the same probability of being drawn, find the
       probability of getting $-4, -3, -2, -1, 0, 1, 2, 3, 4$, and find the mean
       and the variance of this distribution.
   (b) Draw 50 samples of size 10 from this population, each sample being
       drawn without replacement, and calculate their means.
   (c) Calculate the mean and the variance of the 50 means obtained in part
       (b).
   (d) Compare the results obtained in part (c) with the corresponding values
       expected according to Theorem 6.1. [Note that μ and σ^2 were obtained
       in part (a).]

6.8 Repeat the preceding exercise, but select each sample with replacement; that
is, replace each slip of paper and reshuffle before the next one is drawn.

6.9 Given the infinite population whose distribution is given by

x	$f(x)$
1	0.25
2	0.25
3	0.25
4	0.25

list the 16 possible samples of size 2 and use this list to construct the distribution of $\bar{x}$ for random samples of size 2 from the given population. Verify
that the mean and the variance of this sampling distribution are identical
with the corresponding values expected according to Theorem 6.1.

6.10 Suppose that we convert the 50 samples referred to on page 169 into 25 samples
of size $n = 20$ by combining the first two, the next two, and so on. Find the
means of these samples and calculate their mean and their standard deviation. Compare this mean and this standard deviation with the corresponding
values expected in accordance with Theorem 6.1.

6.11 When we sample from an infinite population, what happens to the standard error of the mean if the sample size is
(a) increased from 50 to 200;
(b) increased from 400 to 900;
(c) decreased from 225 to 25;
(d) decreased from 640 to 40?

6.12 What is the value of the finite population correction factor in the formula for $\sigma_{\bar{x}}^2$ when
(a) $n = 5$ and $N = 200$;   (c) $n = 100$ and $N = 5,000$?
(b) $n = 10$ and $N = 400$;

6.13 Verify that there is a fifty-fifty chance that the mean of a random sample of size n from an infinite population with the standard deviation σ will differ from μ by less than $0.6745 \cdot \dfrac{\sigma}{\sqrt{n}}$. It has been the custom to refer to this quantity as the **probable error of the mean.**

6.14 The mean of a random sample of size $n = 25$ is used to estimate the mean of an infinite population with the standard deviation $\sigma = 2.4$. What can we assert about the probability that the error will be less than 1.2, if we use
(a) Chebyshev's theorem;
(b) the central limit theorem?

6.15 A random sample of size $n = 36$ is taken from an infinite population with the mean $\mu = 63$ and the variance $\sigma^2 = 81$. What can we assert about the probability of getting a sample mean greater than 66.75, if we use
(a) Chebyshev's theorem;
(b) the central limit theorem?

6.16 A random sample of size 100 is taken from an infinite population having the mean $\mu = 76$ and the variance $\sigma^2 = 256$. What is the probability of getting an $\bar{x}$ between 75 and 78?

6.17 If measurements of the specific gravity of a metal can be looked upon as a sample from a normal population having a standard deviation of 0.04, what is the probability that the mean of a random sample of size 25 will be "off" by at most 0.02?

6.18 A wire-bonding process is said to be in control if the mean pull strength is 10 pounds. It is known that the pull-strength measurements are normally distributed with a standard deviation of 1.5 pounds. Periodic random samples of size 4 are taken from this process and the process is said to be "out of control" if a sample mean is less than 7.75 pounds. Comment.

6.19 If the distribution of the weights of all men traveling by air between Dallas and El Paso has a mean of 163 pounds and a standard deviation of 18 pounds, what is the probability that the combined gross weight of 36 men traveling on a plane between these two cities is more than 6,000 pounds?

6.20 If x is a value of a continuous random variable and $y = ax + b$, show that
(a) $\mu_y = a\mu_x + b$;   (b) $\sigma_y^2 = a^2\sigma_x^2$.

6.21 Prove that $\mu_{\bar{x}} = \mu$ for random samples from discrete (finite or countably infinite) populations.

6.3   THE SAMPLING DISTRIBUTION OF THE MEAN (σ UNKNOWN)

Application of the theory of the preceding section requires knowledge of the population standard deviation σ. If n is large, this does not pose any problems even when σ is unknown, as it is reasonable in that case to substitute for it the sample standard deviation s. However, when it comes to the random variable whose values are given by $\frac{\bar{x} - \mu}{s/\sqrt{n}}$, very little is known about its exact sampling distribution for small values of n unless we make the assumption that the sample comes from a normal population. Under this assumption, one can prove that

Value of random variable having the t distribution

> **Theorem 6.3** *If $\bar{x}$ is the mean of a random sample of size n taken from a normal population having the mean μ and the variance σ^2, then*
>
> $$t = \frac{\bar{x} - \mu}{s/\sqrt{n}}$$
>
> *is the value of a random variable having the t distribution with the parameter $v = n - 1$.*

This theorem is more general than Theorem 6.2 in the sense that it does not require knowledge of σ; on the other hand, it is less general than Theorem 6.2 in the sense that it requires the assumption of a normal population.

As can be seen from Figure 6.3, the overall shape of a **t distribution** is similar to that of a normal distribution—both are bell-shaped and symmetrical about the mean. Like the standard normal distribution, the t distribution has the mean 0, but its variance depends on the parameter v (*nu*),

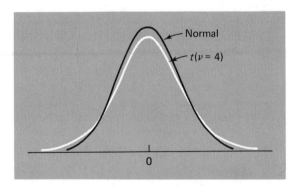

Figure 6.3 *t* distribution and standard normal distribution.

called the number of **degrees of freedom**. The variance of the t distribution exceeds 1, but it approaches 1 as $n \longrightarrow \infty$. In fact, it can be shown that the t distribution with v degrees of freedom approaches the standard normal distribution as $v \longrightarrow \infty$.

Table 4 at the end of the book contains selected values of t_α for various values of v, where t_α is such that the area under the t distribution to its right is equal to α. In this table the left-hand column contains values of v, the column headings are areas α in the right-hand tail of the t distribution, and the entries are values of t_α (see also Figure 6.4). It is not necessary to tabulate values of t_α for $\alpha > 0.50$, as it follows from the symmetry of the t distribution that $t_{1-\alpha} = -t_\alpha$; thus, the value of t that corresponds to a left-hand tail area of α is $-t_\alpha$.

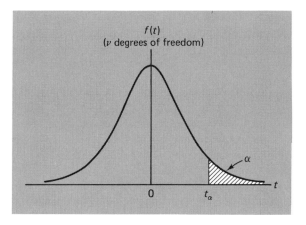

Figure 6.4 Tabulated values of t.

Note that in the bottom row of Table 4 the entries correspond to the values of z which cut off right-hand tails of area α under the standard normal curve. Using the notation z_α for such a value of z, it can be seen, for example, that $z_{.025} = 1.96 = t_{.025}$ for $v = \infty$. This result should really have been expected since the t distribution approaches the standard normal distribution as $v \longrightarrow \infty$. In fact, observing that the values of t_α for 29 or more degrees of freedom are close to the corresponding values of z_α, we conclude that **the standard normal distribution provides a good approximation to the t distribution for samples of size 30 or more**.

EXAMPLE   A manufacturer of fuses claims that with a 20% overload, his fuses will blow in 12.40 minutes on the average. To test this claim, a sample of 20 of the fuses was subjected to a 20% overload, and the times it took them to blow had a mean of 10.63 minutes and a standard deviation of 2.48 minutes. If it can be assumed that the data constitute a random sample from a normal population, do they tend to support or refute the manufacturer's claim?

Solution   First we calculate

$$t = \frac{10.63 - 12.40}{2.48/\sqrt{20}} = -3.19$$

which is a value of a random variable having the t distribution with $v = 20 - 1 = 19$ degrees of freedom. Now, from Table 4 we find that for $v = 19$ the probability that t will exceed 2.861 is 0.005, and hence that the probability that t will be less than -2.861 is also 0.005. Since $t = -3.19$ is less than -2.861 and 0.005 is a very small probability, we conclude that the data tend to refute the manufacturer's claim. In all probability, the mean blowing time of his fuses with a 20% overload is less than 12.40 minutes.

The assumption that the sample must come from a normal population is not so severe a restriction as it may seem. Studies have shown that the distribution of random variables with the values $t = \dfrac{\bar{x} - \mu}{s/\sqrt{n}}$ is fairly close to a t distribution even for samples from certain nonnormal populations. In practice, it is necessary to make sure primarily that the population from which we are sampling is approximately bell-shaped and not too skewed. A practical way of checking this assumption is to plot the observations on arithmetic probability paper as described on page 144. (If such a plot shows a distinct curve rather than a straight line, it may be possible to "straighten it out" by transforming the data—say, by taking their logarithms or their square roots.)

6.4 THE SAMPLING DISTRIBUTION OF THE VARIANCE

So far we have discussed only the sampling distribution of the mean, but if we had taken the medians or the standard deviations of the 50 samples menioned in the example on page 169, we would similarly have obtained experimental sampling distributions of these statistics. In this section we shall be concerned with the theoretical sampling distribution of the sample variance for random samples from normal populations. Since s^2 cannot be negative, we should suspect that this sampling distribution is not a normal curve; in fact, it is related to the gamma distribution (see page 118) with $\alpha = v/2$ and $\beta = 2$, called the **chi-square distribution**. Specifically, we have

Value of random variable having the chi-square distribution

Theorem 6.4 *If s^2 is the variance of a random sample of size n taken from a normal population having the variance σ^2, then*

$$\chi^2 = \frac{(n-1)s^2}{\sigma^2}$$

is a value of a random variable having the chi-square distribution with the parameter $v = n - 1$.

Table 5 at the end of the book contains selected values of χ_α^2 for various values of v, again called the **number of degrees of freedom**, where χ_α^2 is such that the area under the chi-square distribution to its right is equal to α. In this table the left-hand-column contains values of v, the column headings are areas α in the right-hand tail of the chi-square distribution, and the entries are values of χ_α^2 (see also Figure 6.5). Unlike the t distribution, it is necessary to tabulate values of χ_α^2 for $\alpha > 0.50$, because the chi-square distribution is not symmetrical.

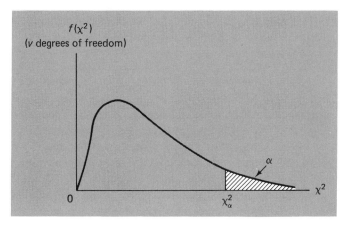

Figure 6.5 Tabulated values of chi-square.

EXAMPLE An optical firm purchases glass to be ground into lenses, and it knows from past experience that the variance of the refractive index of this kind of glass is $1.26 \cdot 10^{-4}$. As it is important that the various pieces of glass have nearly the same index of refraction, the firm rejects such a shipment if the sample variance of 20 pieces selected at random exceeds $2.00 \cdot 10^{-4}$. Assuming that the sample values may be looked upon as a random sample from a normal population, what is the probability that a shipment will be rejected even though $\sigma^2 = 1.26 \cdot 10^{-4}$?

Solution Substituting into the formula for the chi-square statistic, we get

$$\chi^2 = \frac{19(2.00 \cdot 10^{-4})}{1.26 \cdot 10^{-4}} = 30.2$$

and then we find from Table 5 that for 19 degrees of freedom $\chi_{.05}^2 = 30.1$. Thus, the probability that a good shipment will erroneously be rejected is less than 0.05.

A problem closely related to that of finding the distribution of the sample variance is that of finding the distribution of the ratio of the variances of two independent random samples. This problem is important because it arises in tests in which we want to determine whether two samples come from populations having equal variances. If they do, the two sample variances should be nearly the same; that is, their ratio should be close to 1. To determine whether the ratio of two sample variances is too small or too large, we use the theory given in the following theorem:

Value of random variable having the F distribution

> **Theorem 6.5** *If s_1^2 and s_2^2 are the variances of independent random samples of size n_1 and n_2, respectively, taken from two normal populations having the same variance, then*
>
> $$F = \frac{s_1^2}{s_2^2} \; = \; \frac{s_1^2/\sigma_1^2}{s_2^2/\sigma_2^2} \qquad \text{IF } \sigma_1^2 \neq \sigma_2^2$$
>
> *is a value of a random variable having the F distribution with the parameters $v_1 = n_1 - 1$ and $v_2 = n_2 - 1$.*

The **F distribution** is related to the beta distribution (page 121), and its two parameters, v_1 and v_2, are called the **numerator** and **denominator degrees of freedom**. As it would require too large a table to give values of F_α corresponding to many different right-hand tail probabilities α, and since $\alpha = 0.05$ and $\alpha = 0.01$ are most commonly used in practice, Table 6 contains only values of $F_{.05}$ and $F_{.01}$ for various combinations of values of v_1 and v_2 (see also Figure 6.6).

$$F_{.95 \; v_1, v_2} = \frac{1}{F_{.05 \; v_2, v_1}}$$

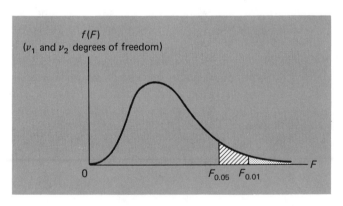

Figure 6.6 Tabulated values of *F*.

EXAMPLE   If two independent random samples of size $n_1 = 7$ and $n_2 = 13$ are taken from a normal population, what is the probability that the variance of the first sample will be at least three times as large as that of the second sample?

Solution   From Table 6 we find that $F_{.05} = 3.00$ for $v_1 = 7 - 1 = 6$ and $v_2 = 13 - 1 = 12$; thus, the desired probability is 0.05.

It is possible to use Table 6 also to find values of F corresponding to left-hand tail probabilities of 0.05 or 0.01. Writing $F_\alpha(v_1, v_2)$ for F_α with v_1 and v_2 degrees of freedom, we simply use the identity

$$F_{1-\alpha}(v_1, v_2) = \frac{1}{F_\alpha(v_2, v_1)}$$

EXAMPLE   Find the value of $F_{.95}$ (corresponding to a left-hand tail probability of 0.05) for $v_1 = 10$ and $v_2 = 20$ degrees of freedom.

Solution   Making use of the identity and Table 6, we get

$$F_{.95}(10, 20) = \frac{1}{F_{.05}(20, 10)} = \frac{1}{2.77} = 0.36$$

Note that Theorems 6.4 and 6.5 require the assumption that we are sampling from normal populations. As in the case of the t distribution, this assumption can be loosened somewhat in actual practice without materially altering the respective sampling distributions. Again, the use of arithmetic probability paper is suggested to investigate whether it is reasonable to treat the samples as coming from normal populations.

EXERCISES

6.22 A random sample of size 25 from a normal population has the mean $\bar{x} = 47.5$ and the standard deviation $s = 8.4$. Does this information tend to support or refute the claim that the mean of the population is $\mu = 42.1$?

6.23 The following are the times between six calls for an ambulance (in a certain city) and the patient's arrival at the hospital: 27, 15, 20, 32, 18, and 26 minutes. Use these figures to judge the reasonableness of the ambulance service's claim that it takes on the average 20 minutes between the call for an ambulance and the patient's arrival at the hospital.

6.24 A process for making certain bearings is under control if the diameters of the bearings have a mean of 0.5000 cm. What can we say about this process if a sample of 10 of these bearings has a mean diameter of 0.5060 cm and a standard deviation of 0.0040 cm?

6.25 The claim that the variance of a normal population is $\sigma^2 = 21.3$ is rejected if the variance of a random sample of size 15 exceeds 39.74. What is the probability that the claim will be rejected even though $\sigma^2 = 21.3$?

6.26 A random sample of 10 observations is taken from a normal population having the variance $\sigma^2 = 42.5$. Find the approximate probability of obtaining a sample standard deviation between 3.14 and 8.94.

6.27 If two independent random samples of size $n_1 = 9$ and $n_2 = 16$ are taken from a normal population, what is the probability that the variance of the first sample will be at least four times as large as the variance of the second sample?

6.28 If independent random samples of size $n_1 = n_2 = 8$ come from normal populations having the same variance, what is the probability that either sample variance will be at least seven times as large as the other?

6.29 Find the values of
(a) $F_{.95}$ for 12 and 15 degrees of freedom;
(b) $F_{.99}$ for 6 and 20 degrees of freedom.

6.30 The chi-square distribution with 4 degrees of freedom is given by

$$f(x) = \begin{cases} \frac{1}{4} \cdot x \cdot e^{-x/2} & x > 0 \\ 0 & x \le 0 \end{cases}$$

Find the probability that the variance of a random sample of size 5 from a normal population with $\sigma = 12$ will exceed 180.

6.31 The t distribution with 1 degree of freedom is given by

$$f(t) = \frac{1}{\pi}(1 + t^2)^{-1} \qquad -\infty < t < \infty$$

Verify the value given for $t_{.05}$ for $\nu = 1$ in Table 4.

6.32 The F distribution with 4 and 4 degrees of freedom is given by

$$f(F) = \begin{cases} 6F(1 + F)^{-4} & F > 0 \\ 0 & F \le 0 \end{cases}$$

If random samples of size 5 are taken from two normal populations having the same variance, find the probability that the ratio of the larger to the smaller sample variance will exceed 3.

6.5 CHECK LIST OF KEY TERMS (with page references)

Central limit theorem, 174
Chi-square distribution, 180
Degrees of freedom, 179
Discrete uniform distribution, 169
F distribution, 182
Finite population, 166
Finite population correction factor, 172
Infinite population, 166
Parameter, 168

Population, 166
Probable error of the mean, 177
Random sample, 166
Sample, 166
Sampling distribution, 169
Standard error of the mean, 173
Standardized mean, 173
Statistic, 168
t Distribution, 178

7

Inferences Concerning Means

In the beginning of Chapter 6, we stated that the purpose of most statistical investigations is to generalize from information contained in random samples about the populations from which the samples were obtained. In the classical approach the methods of statistical inference are divided into two major areas—estimation and tests of hypotheses In Sections 7.1 through 7.3 we shall present some theory and some methods which pertain to the estimation of means. Sections 7.4 and 7.5 deal with the basic concepts of hypothesis testing, and Sections 7.6 through 7.8 deal with tests of hypotheses concerning means.

7.1 POINT ESTIMATION

Basically, **point estimation** concerns the choosing of a statistic, that is, a single number calculated from sample data (and perhaps other information) for which we have some expectation, or assurance, that it is "reasonably close" to the parameter it is supposed to estimate. To explain what we mean here by "reasonably close" is not an easy task; first, the value of the parameter is unknown, and second, the value of the statistic is unknown until after the sample has been obtained. Thus, we can only ask whether, upon repeated sampling, the distribution of the statistic has certain desirable properties akin to "closeness." For instance, we know from Theorem 6.1 that the sampling distribution of the mean has the same mean as the population from which the sample is obtained; hence, we can expect that the means of repeated random samples from a given population will center on the mean of this population and not about some other value. To formulate this property more generally, let us now make the following definition:

Unbiased estimate

A statistic $\hat{\theta}$ is said to be an unbiased estimate, or the value of an unbiased estimator, if and only if the mean of the sampling distribution of the estimator equals θ.

Thus, we call a statistic unbiased if "on the average" its values will equal the parameter it is supposed to estimate. Note that we have distinguished here between an estimator, a random variable, and an estimate, which is one of its values; also, it is customary to apply the term "statistic" to both, estimates and estimators (see page 168).

Generally speaking, the property of unbiasedness is one of the more desirable properties in point estimation, although it is by no means essential and it is sometimes outweighed by other factors. One shortcoming of the criterion of unbiasedness is that it will generally not provide a unique statistic for a given problem of estimation. For instance, it can be shown that for a random sample of size $n = 2$ with the values x_1 and x_2, the mean $\frac{x_1 + x_2}{2}$ as well as the weighted mean $\frac{ax_1 + bx_2}{a + b}$, where a and b are positive constants, are unbiased estimates of the mean of the population. If we assume, furthermore, that the population is symmetric, so are the median and the midrange (the mean of the largest value and the smallest) for random samples of any size.

This suggests that we must seek a further criterion for deciding which of several unbiased estimators is "best" for estimating a given parameter. Such a criterion becomes evident when we compare the sampling distributions of the median and the mean for random samples of size n from the same normal population. Although these two sampling distributions have the same mean, the population mean μ, and although they are both symmetrical and bell-shaped, their variances differ. From Theorem 6.1, the variance of the sampling distribution of the mean for random samples from infinite populations is $\dfrac{\sigma^2}{n}$, and it can be shown that for random samples of the same size from normal populations the variance of the sampling distribution of the median is approximately $1.5708 \cdot \dfrac{\sigma^2}{n}$. Thus, it is more likely for any given sample that the mean will be closer to μ than the median. This does not imply that the sample mean is always closer than the median; in fact, in any given problem we have no way of knowing which of the two is closest.

This important property, in which we compare the variances of the sampling distributions of statistics, is formalized by the following definition:

More efficient unbiased estimator

> *A statistic $\hat{\theta}_1$ is said to be a more efficient unbiased estimate of the parameter θ than the statistic $\hat{\theta}_2$ if*
>
> 1. *$\hat{\theta}_1$ and $\hat{\theta}_2$ are both unbiased estimates of θ;*
> 2. *the variance of the sampling distribution of the first estimator is less than that of the second.*

We have thus seen that for random samples from normal populations the mean is more efficient than the median as an estimate of μ; in fact, it can be shown that in most practical situations where we estimate a population mean μ, the variance of the sampling distribution of no other statistic is less than that of the sampling distribution of the mean. In other words, in most practical situations the sample mean is an acceptable statistic for estimating a population mean μ. (There exist several other criteria for assessing the "goodness" of methods of point estimation, but we shall not discuss them in this book.)

When we use a sample mean to estimate the mean of a population, we know that although we are using a method of estimation which has certain desirable properties, the chances are slim, virtually nonexistent, that the estimate will actually equal μ. Hence, it would seem desirable to accompany such a point estimate of μ with some statement as to how close we might reasonably expect the estimate to be. The error, $\bar{x} - \mu$, is the difference between the estimate and the quantity it is supposed to estimate. To examine this error, let us make use of the fact that for large n

$$\frac{\bar{x} - \mu}{\sigma / \sqrt{n}}$$

is a value of a random variable having approximately the standard normal distribution. Consequently, we can assert with probability $1 - \alpha$ that the inequality

$$-z_{\alpha/2} \leq \frac{\bar{x} - \mu}{\sigma/\sqrt{n}} \leq z_{\alpha/2}$$

will be satisfied or that

$$\frac{|\bar{x} - \mu|}{\sigma/\sqrt{n}} \leq z_{\alpha/2}$$

where $z_{\alpha/2}$ is such that the normal curve area to its right equals $\alpha/2$. If we now let E stand for the maximum value of $|\bar{x} - \mu|$, the maximum error of estimate, we have

Maximum error of estimate

$$E = z_{\alpha/2} \cdot \frac{\sigma}{\sqrt{n}}$$

with probability $1 - \alpha$. In other words, if we intend to estimate μ with the mean of a large $(n \geq 30)$ random sample, we can assert with probability $1 - \alpha$ that the error, $|\bar{x} - \mu|$, will be at most $z_{\alpha/2} \cdot \dfrac{\sigma}{\sqrt{n}}$. The most widely

used values for $1 - \alpha$ are 0.95 and 0.99, and the corresponding values of $z_{\alpha/2}$ are $z_{0.025} = 1.96$ and $z_{0.005} = 2.575$ (see Exercise 4.23 on page 112).

EXAMPLE   An efficiency expert intends to use the mean of a random sample of size $n = 150$ to estimate the average mechanical aptitude (as measured by a certain test) of assembly line workers in a large industry. If, on the basis of experience, the efficiency expert can assume that $\sigma = 6.2$ for such data, what can he assert with probability 0.99 about the maximum size of his error?

Solution   Substituting $n = 150$, $\sigma = 6.2$, and $z_{0.005} = 2.575$ into the above formula for E, we get

$$E = 2.575 \cdot \frac{6.2}{\sqrt{150}} = 1.30$$

Thus, the efficiency expert can assert with probability 0.99 that his error will be at most 1.30.

Suppose now that the efficiency expert of this example collects his data and gets $\bar{x} = 69.5$. Can he still assert with probability 0.99 that the error is at most 1.30? After all, $\bar{x} = 69.5$ differs from the true average by at most 1.30 or it does not, and he does not know which. Well, he can, but it must be understood that the 0.99 probability applies to the method he used to determine the maximum error (getting the sample data and using the formula for E) and not directly to the parameter he is trying to estimate. To make this distinction, it has become the custom to use the word **"confidence"** here instead of "probability." **In general, we make probability statements about future values of random variables (say, the potential error of an estimate) and confidence statements once the data have been obtained**. Accordingly, we would say in our example that the efficiency expert can be 99% confident that the error of his estimate, $\bar{x} = 69.5$, is at most 1.30.

The formula for E on page 188 can also be used to determine the sample size that is needed to attain a desired degree of precision. Suppose that we want to use the mean of a large random sample to estimate the mean of a population, and we want to be able to assert with probability $1 - \alpha$ that the error will be at most some prescribed quantity E [or assert later with $(1 - \alpha)100\%$ confidence that the error is at most E]. As before, we write

$$E = z_{\alpha/2} \cdot \frac{\sigma}{\sqrt{n}}$$

and upon solving this equation for n we get

Sample size

$$n = \left[\frac{z_{\alpha/2} \cdot \sigma}{E} \right]^2$$

To be able to use this formula we must know $1 - \alpha$, E, and σ, and for the latter we often substitute an estimate based on prior data of a similar kind (or, if necessary, a good guess).

EXAMPLE A research worker wants to determine the average time it takes a mechanic to rotate the tires of a car, and she wants to be able to assert with 95% confidence that the mean of her sample is off by at most 0.50 minute. If she can presume from past experience that $\sigma = 1.6$ minutes, how large a sample will she have to take?

Solution Substituting $E = 0.50$, $\sigma = 1.6$, and $z_{0.025} = 1.96$ into the formula for n, we get

$$n = \left[\frac{1.96 \cdot 1.6}{0.50} \right]^2 = 39.3$$

or 40 rounded up to the nearest integer. Thus, the research worker will have to time 40 mechanics performing the task of rotating the tires of a car.

The methods discussed so far in this section require that σ be known or that it can be approximated with the sample standard deviation s, thus requiring that n be large. However, if it is reasonable to assume that we are sampling from a normal population, we can base our argument on Theorem 6.3 instead of Theorem 6.2, namely, on the fact that

$$t = \frac{\bar{x} - \mu}{s/\sqrt{n}}$$

is a value of a random variable having the t distribution with $n - 1$ degrees of freedom. Duplicating the steps on page 188, we thus arrive at the result that we can assert with probability $1 - \alpha$ that the error we will make in using $\bar{x}$ to estimate μ will be at most

**Maximum error
of estimate**

$$E = t_{\alpha/2} \cdot \frac{s}{\sqrt{n}}$$

or with $(1 - \alpha)100\%$ confidence that the error is less than this quantity.

EXAMPLE   In six determinations of the melting point of tin, a chemist obtained a mean of 232.26 degrees Celsius with a standard deviation of 0.14 degree. If he uses this mean as the actual melting point of tin, what can the chemist assert with 98% confidence about the maximum error?

Solution   Substituting $n = 6$, $s = 0.14$, and $t_{0.01} = 3.365$ (for $n - 1 = 5$ degrees of freedom) into the formula for E, we get

$$E = 3.365 \cdot \frac{0.14}{\sqrt{6}} = 0.19$$

Thus, the chemist can assert with 98% confidence that his figure for the melting point of tin is off by at most 0.19 degree.

7.2   INTERVAL ESTIMATION

Since point estimates cannot really be expected to coincide with the quantities they are intended to estimate, it is sometimes preferable to replace them with **interval estimates**, that is, with intervals for which we can assert with a reasonable degree of certainty that they will contain the parameter under consideration. To illustrate the construction of such an interval, suppose that we have a large ($n \geq 30$) random sample from a population with the unknown mean μ and the known variance σ^2. Referring to the inequality

$$-z_{\alpha/2} < \frac{\bar{x} - \mu}{\sigma/\sqrt{n}} < z_{\alpha/2}$$

which, as shown on page 188, will be satisfied with probability $1 - \alpha$, we now apply simple algebra and rewrite it as

Large-sample confidence interval for μ

$$\bar{x} - z_{\alpha/2} \cdot \frac{\sigma}{\sqrt{n}} < \mu < \bar{x} + z_{\alpha/2} \cdot \frac{\sigma}{\sqrt{n}}$$

Thus, when a sample has been obtained and the value of $\bar{x}$ has been calculated, we can claim with $(1 - \alpha)100\%$ confidence that the interval from $\bar{x} - z_{\alpha/2} \cdot \frac{\sigma}{\sqrt{n}}$ to $\bar{x} + z_{\alpha/2} \cdot \frac{\sigma}{\sqrt{n}}$ contains μ. It is customary to refer to an interval of this kind as a **confidence interval** for μ having the **degree of confidence** $1 - \alpha$ or $(1 - \alpha)100\%$, and to its endpoints as the **confidence limits**.

EXAMPLE A random sample of size $n = 100$ is taken from a population with $\sigma = 5.1$. Given that the sample mean is $\bar{x} = 21.6$, construct a 95% confidence interval for the population mean μ.

Solution Substituting the given values of n, $\bar{x}$, σ, and $z_{0.025} = 1.96$ into the confidence-interval formula, we get

$$21.6 - 1.96 \cdot \frac{5.1}{\sqrt{100}} < \mu < 21.6 + 1.96 \cdot \frac{5.1}{\sqrt{100}}$$

or $20.6 < \mu < 22.6$. Of course, the interval from 20.6 to 22.6 contains the population mean μ, or it does not, but we are 95% confident that it does. As was explained on page 189, this means that the method by which the interval was obtained "works" 95% of the time. In other words, in repeated applications of the confidence-interval formula, 95% of the intervals can be expected to contain the means of the respective populations.

The above confidence-interval formula is exact only for random samples from normal populations, but for large samples it will generally provide good approximations. Since σ is unknown in many practical applications, we may have to make the further approximation of substituting for σ the sample standard deviation s.

EXAMPLE   With reference to the sulfur oxides emission data, for which we had $n = 80$, $\bar{x} = 18.85$, and $s^2 = 30.77$ (hence, $s = 5.55$), construct a 99% confidence interval for the plant's true average daily emission of sulfur oxides.

> ***Solution***   Substituting into the confidence-interval formula with $s = 5.55$ in place of σ, we get
>
> $$18.85 - 2.575 \cdot \frac{5.55}{\sqrt{80}} < \mu < 18.85 + 2.575 \cdot \frac{5.55}{\sqrt{80}}$$
>
> or $17.25 < \mu < 20.45$.

For small samples ($n < 30$), we proceed as on page 190 provided it is reasonable to assume that we are sampling from a normal population. Thus, with $t_{\alpha/2}$ defined as on page 179, we get the $(1 - \alpha)100\%$ confidence-interval formula

Small-sample confidence interval for μ

$$\bar{x} - t_{\alpha/2} \cdot \frac{s}{\sqrt{n}} < \mu < \bar{x} + t_{\alpha/2} \cdot \frac{s}{\sqrt{n}}$$

This formula applies to samples from normal populations, but in accordance with the discussion on page 180, it may be used as long as the sample does not exhibit any pronounced departures from normality.

EXAMPLE   The mean weight loss of $n = 16$ grinding balls after a certain length of time in mill slurry is 3.42 grams with a standard deviation of 0.68 gram. Construct a 99% confidence interval for the true mean weight loss of such grinding balls under the stated conditions.

> ***Solution***   Substituting $n = 16$, $\bar{x} = 3.42$, $s = 0.68$, and $t_{0.005} = 2.947$ for $n - 1 = 15$ degrees of freedom into the small-sample confidence-interval formula for μ, we get
>
> $$3.42 - 2.947 \cdot \frac{0.68}{\sqrt{16}} < \mu < 3.42 + 2.947 \cdot \frac{0.68}{\sqrt{16}}$$
>
> or $2.92 < \mu < 3.92$.

7.3 BAYESIAN ESTIMATION†

In recent years there has been mounting interest in methods of inference which look upon parameters as random variables. This idea is not really new, but these **Bayesian methods**, as they are called, have received considerable impetus and much wider applicability through the concept of personal, or

---

†This section may be omitted without loss of continuity.

subjective, probability. In fact, this is why supporters of the subjective concept of probability refer to themselves as **Bayesians**, or **Bayesian statisticians**.

In this section we shall present a Bayesian method of estimating the mean of a population, looking upon μ as a random variable whose distribution is indicative of how strongly a person feels about the various values which μ can take on. He or she will feel most strongly about some particular value of μ, but this enthusiasm will diminish for values of μ which are farther and farther away from the value liked most. Like any distribution that we use in practice, this kind of subjective **prior distribution** has a mean which we shall denote by μ_0 and a standard deviation which we shall denote by σ_0.

To illustrate the concept of a prior distribution of the mean, let us refer again to the sulfur oxides emission example on page 138, and let us suppose that before the sample data were actually obtained the chief engineer of the plant said that his feelings about the true average daily emission of sulfur oxides can best be described by a normal distribution with $\mu_0 = 17.50$ tons and $\sigma_0 = 2.50$ tons. With this information, we might ask, for example, how he felt about the possibility that the plant's true average daily emission of sulfur oxides is somewhere between 18.00 and 19.00 tons. The answer to this question is given by the area of the ruled region of the upper diagram of Figure 7.1, and as can easily be verified, it represents a subjective probability of 0.15.

In **Bayesian estimation**, prior feelings about the possible values of μ are combined with direct sample evidence. Without going into any detail, let us merely point out that this leads to a **posterior distribution** of μ, which, under fairly general conditions, can be approximated by a normal distribution with

Mean and standard deviation of posterior distribution

$$\mu_1 = \frac{n\bar{x}\sigma_0^2 + \mu_0\sigma^2}{n\sigma_0^2 + \sigma^2} \quad \text{and} \quad \sigma_1 = \sqrt{\frac{\sigma^2\sigma_0^2}{n\sigma_0^2 + \sigma^2}}$$

Details about these results may be found in the texts on the theory of statistics listed in the Bibliography.

To illustrate, let us refer again to the sulfur oxides emission example and combine the prior information given above, that based on the feelings of the chief engineer, with the fact that a random sample of size $n = 80$ yielded $\bar{x} = 18.85$ and $s = 5.55$. Using s as an estimate of σ and substituting these values together with $\mu_0 = 17.50$ and $\sigma_0 = 2.5$ into the above formulas for μ_1 and σ_1, we obtain

$$\mu_1 = \frac{80(18.85)(2.5)^2 + 17.50(5.55)^2}{80(2.5)^2 + (5.55)^2} = 18.77$$

and

$$\sigma_1 = \sqrt{\frac{(5.55)^2(2.5)^2}{80(2.5)^2 + (5.55)^2}} = 0.60$$

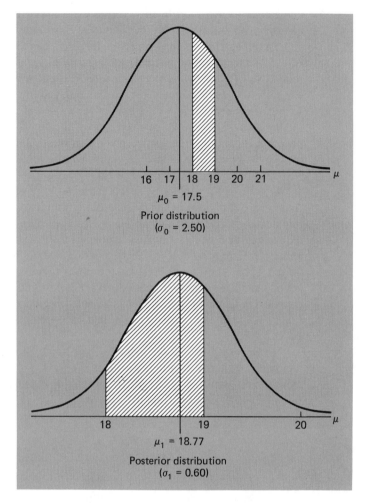

Figure 7.1 Prior and posterior distribution of μ.

Now, the posterior probability that μ lies on the interval from 18.00 to 19.00 is given by the area of the ruled region of the lower diagram of Figure 7.1, and as can easily be verified, it is 0.55. Note that the information supplied by the sample raised the probability from 0.15 to 0.55.

EXERCISES

7.1 To illustrate that the mean of a random sample is an unbiased estimate of the mean of the population, consider five slips of paper numbered 3, 6, 9, 15, and 27.

(a) List all possible samples of size 3 that can be taken *without* replacement from this finite population.

(b) Calculate the mean of each of the samples listed in (a), and, assigning each sample a probability of $\frac{1}{10}$, verify that the mean of these $\bar{x}$'s equals 12, namely, the mean of the population.

7.2 Suppose that we observe a random variable having the binomial distribution and get x successes in n trials.

(a) Show that $\frac{x}{n}$ is an unbiased estimate of the binomial parameter p.

(b) Show that $\frac{x+1}{n+2}$ is not an unbiased estimate of the binomial parameter p.

7.3 To verify the claim that the mean is generally more efficient than the median (that it is subject to smaller chance fluctuations), a student conducted an experiment consisting of 12 tosses of three dice. The following are his results: 2, 4, and 6; 5, 3, and 5; 4, 5, and 3; 5, 2, and 3; 6, 1, and 5; 2, 3, and 1; 3, 1, and 4; 5, 5, and 2; 3, 3, and 4; 1, 6, and 2; 3, 3, and 3; and 4, 5, and 3.

(a) Calculate the 12 medians and the 12 means.

(b) Group the medians and the means obtained in part (a) into separate distributions having the classes 1.5–2.5, 2.5–3.5, 3.5–4.5, and 4.5–5.5.

(c) Draw histograms of the two distributions obtained in part (b) and explain how they illustrate the claim that the mean is generally more efficient than the median.

7.4 While performing a certain task under simulated weightlessness, the pulse rate of 32 astronaut trainees increased on the average by 26.4 beats per minute with a standard deviation of 4.28 beats per minute. What can one assert with 95% confidence about the maximum error if $\bar{x} = 26.4$ is used as a point estimate of the true average increase in the pulse rate of astronaut trainees performing the given task?

7.5 With reference to the preceding exercise, construct a 95% confidence interval for the true average increase in the pulse rate of astronaut trainees performing the given task.

7.6 To estimate the average time it takes to assemble a certain computer component, the efficiency expert of an electronics firm timed 40 technicians in the performance of this task, getting a mean of 12.73 minutes and a standard deviation of 2.06 minutes.

(a) What can we say with 99% confidence about the maximum error if $\bar{x} = 12.73$ is used as a point estimate of the actual average time required to do the job?

(b) Use the given data to construct a 98% confidence interval for the true average time it takes to assemble the computer component.

7.7 With reference to the preceding exercise, with what confidence can we assert that the sample mean does not differ from the true mean by more than 30 seconds?

7.8 What is the maximum error one can expect to make with probability 0.90

when using the mean of a random sample of size $n = 64$ to estimate the mean of a population with $\sigma^2 = 2.56$?

7.9 In a study of automobile collision insurance costs, a random sample of 80 body repair costs for a particular kind of damage had a mean of $472.36 and a standard deviation of $62.35. If $\bar{x} = \$472.36$ is used as a point estimate of the true average repair cost of this kind of damage, with what confidence can one assert that the error does not exceed $10?

7.10 If we want to determine the average mechanical aptitude of a large group of workers, how large a random sample will we need to be able to assert with probability 0.95 that the sample mean will not differ from the true mean by more than 3.0 points? Assume that it is known from past experience that $\sigma = 20.0$.

7.11 The dean of a college wants to use the mean of a random sample to estimate the average amount of time students take to get from one class to the next, and she wants to be able to assert with 99% confidence that the error is at most 0.25 minute. If it can be presumed from experience that $\sigma = 1.40$ minutes, how large a sample will she have to take?

7.12 It is desired to estimate the mean number of hours of continuous use until a certain kind of computer will first require repairs. If it can be assumed that $\sigma = 48$ hours, how large a sample is needed so that one will be able to assert with 90% confidence that the sample mean is off by at most 10 hours?

7.13 Modify the formula for E on page 188 so that it applies to large samples which constitute substantial portions of finite populations, and use the resulting formula for the following problems:

(a) A sample of 50 scores on the admission test for a school of engineering is drawn at random from the scores of the 420 persons who applied to the school in 1983. If the sample mean and standard deviation are $\bar{x} = 546$ and $s = 85$, what can we assert with 95% confidence about the maximum error if $\bar{x} = 546$ is used as an estimate of the mean score of all the applicants?

(b) A random sample of 40 drums of a chemical, drawn from among 200 such drums whose weights can be expected to have the standard deviation $\sigma = 12.2$ pounds, has a mean weight of 240.8 pounds. If we estimate the mean weight of all 200 drums as 240.8 pounds, what can we assert with 99% confidence about the maximum error?

7.14 Modify the large-sample confidence-interval formula for μ on page 191 so that it applies to large samples which constitute substantial portions of finite populations, and use the resulting formula for the following problems:

(a) With reference to part (a) of the preceding exercise, construct a 99% confidence interval for the mean score of the 420 persons who applied to the school of engineering in 1983.

(b) In a random sample of 100 of 500 experimental batteries produced by a firm, the lifetimes have a mean of 148.2 hours with a standard deviation of 24.9 hours. Find a 76.60% confidence interval for the mean life of the 500 batteries.

7.15 A random sample of 100 teachers in a large metropolitan area revealed a mean weekly salary of $287 with a standard deviation of $48. With what degree of confidence can we assert that the average weekly salary of all teachers in the metropolitan area is between $272 and $302?

7.16 Instead of the large-sample confidence-interval formula for μ on page 191 we could have given the alternative formula

$$\bar{x} - z_{\alpha/3} \cdot \frac{\sigma}{\sqrt{n}} < \mu < \bar{x} + z_{2\alpha/3} \cdot \frac{\sigma}{\sqrt{n}}$$

Explain why the one on page 191 is narrower, and hence preferable, to the one given here.

7.17 In the example on page 179, 20 fuses were subjected to a 20% overload, and the times it took them to blow had a mean of 10.63 minutes and a standard deviation of 2.48 minutes. If we use $\bar{x} = 10.63$ minutes as a point estimate of the true average it takes such fuses to blow with a 20% overload, what can we assert with 95% confidence about the maximum error?

7.18 A major truck stop has kept extensive records on various transactions with its customers. If a random sample of $n = 18$ of these records shows average sales of 63.84 gallons of diesel fuel with a standard deviation of 2.75 gallons, and we use $\bar{x} = 63.84$ as an estimate of the truck stop's mean sales of diesel fuel per customer, what can we say with 99% confidence about the maximum error?

7.19 Ten bearings made by a certain process have a mean diameter of 0.5060 cm with a standard deviation of 0.0040 cm. Assuming that the data may be looked upon as a random sample from a normal population, construct a 95% confidence interval for the actual average diameter of bearings made by this process.

7.20 Inspecting ceramic tiles prior to their shipment, a quality control engineer detects 2, 3, 6, 0, 4, and 9 defectives in six cartons, each containing 144 tiles. Assuming that the data can be looked upon as a random sample from a population which can be approximated closely by a normal distribution, what can he assert with 99% confidence about the maximum error, if he uses the mean of the sample as a point estimate of the true average number of defectives per carton?

7.21 With reference to the preceding exercise, construct a 98% confidence interval for the true average number of defectives per carton.

7.22 In an air pollution study performed at an experiment station, the following amounts of suspended benzene-soluble organic matter (in micrograms per cubic meter) were obtained for eight different samples of air: 2.2, 1.8, 3.1, 2.0, 2.4, 2.0, 2.1, and 1.2. Assuming that the population sampled is normal, construct a 95% confidence interval for the corresponding true mean.

7.23 In order to test the durability of a new paint, a highway department had test strips painted across heavily traveled roads in 15 different locations. If on the average the test strips disappeared after they had been crossed by 146,692 cars with a standard deviation of 14,380 cars, construct a 99% confidence interval for the true average number of cars it takes to wear off the paint. Assume normal population.

7.24 Referring to the preceding exercise and using 14,380 as an estimate of σ, find the sample size that would have been needed to be able to assert with 95% confidence that the sample mean is off by at most 10,000. [*Hint*: First estimate n_1 by using $z = 1.96$, then use $t_{0.025}$ for $n_1 - 1$ degrees of freedom to obtain a second estimate n_2, and repeat this procedure until the last two values of n thus obtained are equal.]

7.25 A sales manager's feelings about the average monthly demand for one of his company's products may be described by means of a normal distribution with $\mu_0 = 3,800$ units and $\sigma_0 = 260$ units.
(a) What probability is he, thus, assigning to the true average monthly demand being on the interval from 3,500 to 4,000 units?
(b) If data for 9 months show an average demand for 3,702 units with a standard deviation of 390 units, how does this information affect the probability determined in part (a)? Use $s = 390$ as an estimate of σ.

7.26 A professor is making up a final examination for a course in computer-aided manufacturing which is to be given to a large group of students. His feelings about the average grade they should get is expressed subjectively by a normal distribution with $\mu_0 = 67.2$ and $\sigma_0 = 1.5$.
(a) What prior probability does he thus assign to the average grade being on the interval from 65.0 to 70.0?
(b) If the examination is tried on a random sample of 40 of the students and the mean and the standard deviation of their grades are 74.9 and 7.4, calculate the values of μ_1 and σ_1. Use $s = 7.4$ as an estimate of σ.
(c) Use the results of part (b) to calculate the posterior probability which the professor will thus assign to the average grade being on the interval from 65.0 to 70.0.

7.4 TESTS OF HYPOTHESES

There are many problems in which, rather than estimate the value of a parameter, we must decide whether a statement concerning a parameter is true or false; that is, we must test a hypothesis about a parameter. For example, in quality control work a random sample may serve to determine whether the "process mean" (for a given kind of measurement) has remained unchanged or whether it has changed to such an extent that the process has gone "out of control" and adjustments have to be made.

To illustrate the general concepts involved in this kind of decision problems, suppose that a consumer protection agency wants to test a paint manufacturer's claim that the average drying time of his new "fast-drying" paint is 20 minutes. So it instructs a member of its research staff to take 36 boards and paint them with paint from 36 different 1-gallon cans of the paint, with the intention of rejecting the claim if the mean of the drying times exceeds 20.75 minutes; otherwise, it will accept the claim and in either case it will take whatever action is called for in its plans.

This provides a clear-cut criterion for accepting or rejecting the claim, but unfortunately it is not infallible. Since the decision is based on a sample, there is the possibility that the sample mean may exceed 20.75 minutes even though the true mean drying time is $\mu = 20$ minutes, and there is also the possibility that the sample mean may be 20.75 minutes or less even though the true mean drying time is, say, $\mu = 21$ minutes. Thus, before adopting the criterion, it would seem wise to investigate the chances that the criterion may lead to a wrong decision.

Assuming that it is known from past experience that the standard deviation of such drying times can be expected to equal $\sigma = 2.4$ minutes, let us first investigate the possibility that the sample mean may exceed 20.75 minutes even though the true mean drying time is $\mu = 20$. The probability that this will happen purely due to chance is given by the area of the ruled region of Figure 7.2, and it can easily be determined by approximating the sampling distribution of the mean with a normal curve. Assuming that the population is large enough to be treated as infinite, we have $\sigma_{\bar{x}} = \dfrac{2.4}{\sqrt{36}} = 0.4$, and the

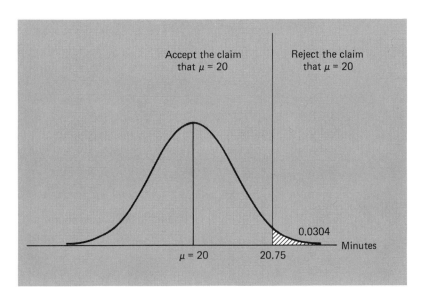

Figure 7.2 Decision criterion.

dividing line of the criterion, in standard units is

$$z = \frac{20.75 - 20}{0.4} = 1.875$$

It follows from Table 3 that the area of the ruled region of Figure 7.2 is $1 - 0.9696 = 0.0304$ (by interpolation), and hence that the probability of erroneously rejecting the hypothesis $\mu = 20$ minutes is approximately 0.03.

Let us now consider the other possibility, where the procedure fails to detect that $\mu \neq 20$ minutes. Suppose again, for the sake of argument, that the true mean drying time is $\mu = 21$ minutes, so that the probability of getting a sample mean less than or equal to 20.75 minutes (and, hence, erroneously accepting the claim that $\mu = 20$ minutes) is given by the area of the ruled region of Figure 7.3. As before, $\sigma_{\bar{x}} = 0.4$, so that the dividing line of the criterion, in standard units, is now

$$z = \frac{20.75 - 21}{0.4} = -0.625$$

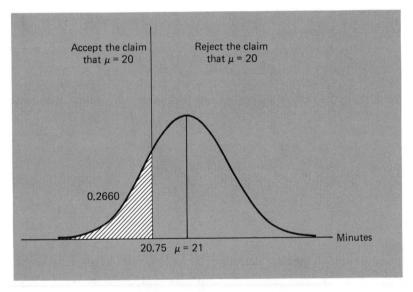

Accept the claim that $\mu = 20$

Reject the claim that $\mu = 20$

0.2660

20.75 $\mu = 21$

Minutes

Figure 7.3 Decision criterion.

It follows from Table 3 that the area of the ruled region of Figure 7.3 is $1 - 0.7340 = 0.2660$ (by interpolation), and hence that the probability of erroneously accepting the hypothesis $\mu = 20$ is approximately 0.27.

The situation described in this example is typical of testing a statistical hypothesis, and it may be summarized in the following table, where we refer to the hypothesis being tested as hypothesis H:

	Accept H	*Reject H*
H is true	Correct decision	Type I error
H is false	Type II error	Correct decision

If hypothesis H is true and accepted or false and rejected, the decision is in either case correct. If hypothesis H is true but rejected, it is rejected in error, and if hypothesis H is false but accepted, it is accepted in error. The first of these errors is called a **Type I error** and the probability of committing it is designated by the Greek letter α *(alpha)*; the second is called a **Type II error** and the probability of committing it is designated by the Greek letter β *(beta)*. Thus, in our example we showed that for the given test criterion $\alpha = 0.03$ and $\beta = 0.27$ when $\mu = 21$ minutes.

In calculating the probability of a Type II error in our example, we arbitrarily chose the alternative value $\mu = 21$ minutes. However, in this problem, as in most others, there are infinitely many other alternatives, and for each one of them there is a positive probability β of erroneously accepting the hypothesis H. What to do about this will be discussed further in Section 7.7.

7.5 NULL HYPOTHESES AND SIGNIFICANCE TESTS

In the drying-time example of the preceding section we were able to calculate the probability of a Type I error because we formulated the hypothesis H as a **simple hypothesis** about the parameter μ; that is, we formulated the hypothesis H so that μ was completely specified. Had we formulated instead a **composite hypothesis** about the parameter μ, say, $\mu \leq 20$ minutes, where μ can take on more than one possible value, we would not have been able to calculate the probability of a Type I error without specifying by how much μ is less than 20 minutes.†

To be able to calculate, or specify, the probability of a Type I error, we usually formulate hypotheses to be tested as simple hypotheses; at least, we do this whenever possible. Very often, this requires that we hypothesize the

---

†Note that we are applying the terms "simple hypothesis" and "composite hypothesis" to hypotheses about specific parameters; some statisticians use the term "simple hypothesis" only when the hypothesis completely specifies the population.

opppsite of what we hope to prove. For instance, if we want to show that one method of teaching computer programming is more efficient than another, we hypothesize that the two methods are equally effective. Similarly, if we want to show that one method of irrigating the soil is more expensive than another, we hypothesize that the two methods are equally expensive; and if we want to show that a new copper-bearing steel has a higher yield strength than ordinary steel, we hypothesize that the two yield strengths are the same. Since we hypothesize that there is no difference in the effectiveness of the two teaching methods, no difference in the cost of the two methods of irrigation, and no difference in the yield strength of the two kinds of steel, we call hypotheses like these **null hypotheses** and denote them H_0. Nowadays, the term "null hypothesis" is used for any hypothesis set up primarily to see whether it can be rejected.

The idea of setting up a null hypothesis is not an uncommon one, even in nonstatistical thinking. In fact, this is exactly what is done in an American court of law, where an accused is assumed to be innocent unless he is proven guilty "beyond a reasonable doubt." The null hypothesis states that the accused is not guilty, and the probability expressed subjectively by the phrase "beyond a reasonable doubt" reflects the probability α of risking a Type I error. Note that the "burden of proof" is always on the prosecution in the sense that the accused is found not guilty unless the null hypothesis of innocence is clearly disproved. This does not imply that the defendant has been proved innocent if found not guilty; it implies only that he has not been proved guilty. Of course, since we cannot legally "reserve judgment" if proof of guilt is not established, the accused is freed and we act as if the null hypothesis of innocence were accepted. Note that this is precisely what we may have to do in tests of statistical hypotheses, when we cannot afford the luxury of reserving judgment.

When reserving judgment is a viable alternative to the rejection of a null hypothesis, Type II errors can be sidestepped altogether. To illustrate, let us return to the drying time example of Section 7.4 and restate the criterion as follows:

Reject the hypothesis $\mu = 20$ minutes (and accept the alternative $\mu > 20$ minutes) if the mean of the 36 sample values exceeds 20.75 minutes; otherwise, reserve judgment.

If judgment is reserved as in this criterion, there is no possibility of a Type II error; no matter what happens, the null hypothesis is never accepted.

The above criterion may well be described as a test of whether $\bar{x}$ is significantly greater than $\mu = 20$ minutes, where "significantly greater" means that the discrepancy between $\bar{x}$ and $\mu = 20$ is too large to be reasonably attributed to chance. Thus, we refer to this kind of test as a **significance test**.

To approach problems of hypothesis testing systematically, it will help to proceed as outlined in the following five steps:

1. **We formulate a simple null hypothesis and an appropriate alternative hypothesis which we accept when the null hypothesis must be rejected.**†

In the drying-time example the null hypothesis is $\mu = 20$ minutes and the alternative hypothesis is $\mu > 20$ minutes (since the paint was referred to as "fast drying"). This kind of alternative, which is called a **one-sided alternative,** may also have the inequality going the other way. For instance, if we hope to be able to show that the average time required to do a certain job is less than 15 minutes, we would test the null hypothesis $\mu = 15$ against the alternative hypothesis $\mu < 15$.

The following is an example in which we would use the **two-sided alternative** $\mu \neq \mu_0$, where μ_0 is the value assumed under the null hypothesis: A food processor wants to check whether the average amount of coffee that goes into his 4-ounce jars is indeed 4 ounces. Since the food processor cannot afford to put much less than 4 ounces into each jar for fear of losing customer acceptance, nor can he afford to put much more than 4 ounces into each jar for fear of losing part of his profit, the appropriate alternative hypothesis is $\mu \neq 4$.

As in the examples of the two preceding paragraphs, alternative hypotheses usually specify that the population mean (or whatever other parameter may be of concern) is not equal to, greater than, or less than the value assumed under the null hypothesis. For any given problem, the choice of an appropriate alternative depends mostly on what we hope to be able to show, or better, where we want to put the burden of proof.

EXAMPLE An appliance manufacturer is considering the purchase of a new machine for stamping out sheet metal parts. If μ_0 is the average number of good parts stamped out per hour by his old machine and μ is the corresponding average for the new machine, the manufacturer wants to test the null hypothesis $\mu = \mu_0$ against a suitable alternative. What should the alternative be if
  (a) he does not want to buy the new machine unless it is more productive than the old one;
  (b) he wants to buy the new machine (which has some other nice features) unless it is less productive than the old one?

  Solution (a) The manufacturer should use the alternative hypothesis $\mu > \mu_0$ and purchase the new machine only if the null hypothesis can be rejected. (b) The manufacturer should use the alternative hypothesis $\mu < \mu_0$ and purchase the new machine unless the null hypothesis can be rejected.

---

†See also the discussion on page 204.

Having formulated the null hypothesis and an alternative hypothesis, we proceed with the following step:

2. We specify the probability of a Type I error; if possible, desired, or necessary, we may also specify the probabilities of Type II errors for particular alternatives.

The probability of a Type I error is also called the **level of significance**, and it is usually set at $\alpha = 0.05$ or $\alpha = 0.01$. Which value we choose in any given problem will have to depend on the risks, or consequences, of committing a Type I error. Observe, however, that we should not make the probability of a Type I error too small, because this will have the tendency to make the probabilities of serious Type II errors too large.

Step 2 cannot be performed unless the null hypothesis is a simple hypothesis, but this is not as restrictive as it may seem. To illustrate, let us investigate briefly what might be done in the drying time example if we wanted to allow for the possibility that the paint may be better (faster drying) than claimed, and hence test the null hypothesis $\mu \leq 20$ minutes against the alternative hypothesis $\mu > 20$ minutes. Now, the probability of a Type I error cannot be calculated, but observe that if μ is less than 20 minutes, the normal curve of Figure 7.2 on page 199 is shifted to the left, and the area under the curve to the right of 20.75 becomes less than 0.0304. Thus, if the null hypothesis is $\mu \leq 20$ minutes, we can say that the probability of a Type I error is at most 0.0304, and we write $\alpha \leq 0.0304$. In general, if the null hypothesis is of the form $\mu \leq \mu_0$ or $\mu \geq \mu_0$, we can only specify the maximum probability of a Type I error, and by performing the test as if the null hypothesis were $\mu = \mu_0$, we protect ourselves against the worst possibility (see the example on page 210).

After the null hypothesis, the alternative hypothesis, and the level of significance have been specified, the remaining steps are:

3. Based on the sampling distribution of an appropriate statistic, we construct a criterion for testing the null hypothesis against the given alternative.

4. We calculate from the data the value of the statistic on which the decision is to be based.

5. We decide whether to reject the null hypothesis, whether to accept it, or whether to reserve judgment.

In the drying-time example we studied the criterion using the normal-curve approximation to the sampling distribution of the mean; in general, step 3 depends not only on the statistic on which we want to base the decision and on its sampling distribution, but also on the alternative hypothesis we happen to choose. In the drying-time example we used a **one-sided criterion** (**one-sided test** or **one-tailed test**) with the one-sided alternative $\mu > 20$ minutes, rejecting the null hypothesis only for large values of the mean. In the example dealing with the food processor and his 4-ounce jars of coffee we would use a **two-sided criterion (two-sided test** or **two-tailed test)** to go with the two-sided alternative $\mu \neq 4$ ounces. In general, a test is said to be two-sided if the null hypothesis is rejected for values of the test statistic falling into either tail of its sampling distribution.

The purpose of the discussion of this and the preceding section has been to introduce some of the basic problems connected with the testing of statistical hypotheses. Although the methods we have presented are objective—that is, two experimenters analyzing the same data under the same conditions would arrive at the identical results—their use does entail some arbitrary, or subjective, considerations. For instance, in the example on page 200 it was partially a subjective decision to "draw the line" between satisfactory and unsatisfactory values of μ at 21 minutes. It was also partially a subjective decision to use a sample of 36 one-gallon cans of the paint, and to reject the manufacturer's claim for values of $\bar{x}$ exceeding 20.75 minutes. Approaching the problem differently, the government agency investigating the paint manufacturer's claim could have specified values of α and β, thus controlling the risks to which they are willing to be exposed. The choice of α, the probability of a Type I error, could have been based on the consequences of making that kind of error, namely, the manufacturer's cost of having a good product condemned, the possible cost of subsequent litigation, the manufacturer's cost of unnecessarily adjusting his machinery, the cost to the public of not having the product available when needed, and so forth. The choice of β, the probability of a Type II error, could similarly have been based on the consequences of making that kind of error, namely, the cost to the public of buying an inferior product, the manufacturer's savings in using inferior ingredients but loss in good will, again the cost of possible litigation, and so forth. It should be obvious that it would be extremely difficult to put "cash values" on all these eventualities, but they must nevertheless be considered, at least indirectly, in choosing suitable criteria for testing statistical hypotheses.

In recent years, attempts have been made to incorporate all these matters within a formal theory called **decision theory**. Although many important advances have been made, it should be recognized that such a theory does not eliminate the arbitrariness, or subjectiveness, discussed above; it merely incorporates these matters within the theory. This means that the use of decision theory requires that we actually put cash values on all possible consequences of our decisions. Although this has the advantage that it makes the experimenter (the engineer) more "cost conscious," it also has the disadvantage of requiring information which very often cannot be obtained.

In this text we shall discuss mainly the **Neyman–Pearson theory**, also called the **classical theory** of testing hypotheses. This means that we shall consider cost factors and other considerations that are partly arbitrary and partly subjective only insofar as they will affect the choice of a sample size, the choice of an alternative hypothesis, the choice of α and β, and so forth.

Before we discuss various special tests about means in the remainder of this chapter, let us point out that the concepts we have introduced here apply equally well to hypotheses concerning proportions, standard deviations, the randomness of samples, and relationships among several variables.

EXERCISES

7.27 Suppose that an engineering firm is asked to check the safety of a dam. What type of error would it commit if it erroneously rejects the null hypothesis that the dam is safe? What type of error would it commit if it erroneously accepts the null hypothesis that the dam is safe?

7.28 Suppose that we want to test the null hypothesis that an antipollution device for cars is effective. Explain under what conditions we would commit a Type I error and under what conditions we would commit a Type II error.

7.29 If the criterion on page 199 is modified so that the paint manufacturer's claim is rejected for $\bar{x} > 20.50$ minutes, find
(a) the probability of a Type I error;
(b) the probability of a Type II error when $\mu = 21$ minutes.

7.30 A process for making steel pipe is under control if the diameter of the pipe has a mean of 3.0000 inches with a standard deviation of 0.0250 inch. To check whether the process is under control, a random sample of size $n = 30$ is taken each day and the null hypothesis $\mu = 3.0000$ is rejected if $\bar{x}$ is less than 2.9960 or greater than 3.0040. Find
(a) the probability of a Type I error;
(b) The probability of a Type II error when $\mu = 3.0050$ inches.

7.31 Suppose that in the drying-time example on page 199, n is changed from 36 to 50 while everything else remains the same. Find
(a) the probability of a Type I error;
(b) the probability of a Type II error when $\mu = 21$ minutes.

7.32 It is desired to test the null hypothesis $\mu = 100$ pounds against the alternative hypothesis $\mu < 100$ pounds on the basis of a random sample of size $n = 40$ from a population with $\sigma = 12$. For what values of $\bar{x}$ must the null hypothesis be rejected if the probability of a Type I error is to be $\alpha = 0.01$?

7.33 Suppose that for a given population with $\sigma = 8.4$ in^2 we want to test the null hypothesis $\mu = 80.0$ in^2 against the alternative hypothesis $\mu < 80.0$ in^2 on the basis of a random sample of size $n = 100$.
(a) If the null hypothesis is rejected for $\bar{x} < 78.0$ in^2 and otherwise it is accepted, what is the probability of a Type I error?
(b) What is the answer to part (a) if the null hypothesis is $\mu \geq 80.0$ in^2 instead of $\mu = 80.0$ in^2?

7.34 If the null hypothesis $\mu = \mu_0$ is to be tested on the basis of a large random sample against the alternative hypothesis $\mu < \mu_0$ (or $\mu > \mu_0$), the probability of a Type I error is to be α and the probability of a Type II error is to be β for $\mu = \mu_1$, it can be shown that the required sample size is

$$n = \frac{\sigma^2 (z_\alpha + z_\beta)^2}{(\mu_1 - \mu_0)^2}$$

where σ^2 is the variance of the population.

(a) It is desired to test the null hypothesis $\mu = 40$ against the alternative hypothesis $\mu < 40$ on the basis of a large random sample from a population with $\sigma = 4$. If the probability of a Type I error is to be 0.05 and the probability of a Type II error is to be 0.12 for $\mu = 38$, find the required size of the sample.

(b) Suppose that we want to test the null hypothesis $\mu = 64$ against the alternative hypothesis $\mu < 64$ for a population whose standard deviation is $\sigma = 7.2$. How large a sample must we take if α is to be 0.05 and β is to be 0.01 for $\mu = 61$? Also, for what values of $\bar{x}$ will the null hypothesis have to be rejected?

7.35 If the alternative hypothesis is two-sided, the formula of the preceding exercise must be modified by substituting $z_{\alpha/2}$ for z_α. How large a sample must we take if it is desired to test the null hypothesis $\mu = 100$ against the alternative hypothesis $\mu \neq 100$ on the basis of a random sample from a population with $\sigma = 16$, the probability of a Type I error is to be 0.01, and the probability of a Type II error is to be 0.20 for $\mu = 92$?

7.36 A city's police department is considering replacing the tires on its cars with radial tires. If μ_1 is the average number of miles they get out of their old tires and μ_2 is the average number of miles they will get out of the new tires, the null hypothesis they shall want to test is $\mu_1 - \mu_2 = 0$.

(a) What alternative hypothesis should they use if they do not want to buy the radial tires unless they are definitely proven to give a better mileage? In other words, the burden of proof is put on the radial tires and the old tires are to be kept unless the null hypothesis can be rejected.

(b) What alternative hypothesis should they use if they are anxious to get the new tires (which have some other nice features) unless they actually give a poorer mileage than the old tires? Note that now the burden of proof is on the old tires, which will be kept only if the null hypothesis can be rejected.

7.37 A producer of extruded plastic products finds that his mean daily inventory is 1,250 pieces. A new marketing policy has been put into effect and it is desired to test the null hypothesis that the mean daily inventory is still the same. What alternative hypothesis should be used if

(a) it is desired to know whether or not the new policy changes the mean daily inventory;

(b) it is desired to demonstrate that the new policy actually reduces the mean daily inventory;

(c) the new policy will be retained so long as it cannot be shown that it actually increases the mean daily inventory?

7.6   HYPOTHESES CONCERNING ONE MEAN

Having used tests concerning means to illustrate the basic principles of hypothesis testing, let us now see how we proceed in actual practice. Suppose, for instance, that we want to test on the basis of $n = 35$ determinations and at the 0.05 level of significance whether the thermal conductivity of a certain kind of cement brick is 0.340, as has been claimed. From information gathered in similar studies, we can expect that the variability of such determinations is given by $\sigma = 0.010$.

Following the outline of the preceding section, we begin with steps 1 and 2 by writing

1. *Null hypothesis:* $\mu = 0.340$
   Alternative hypothesis: $\mu \neq 0.340$
2. *Level of significance:* $\alpha = 0.05$

The alternative hypothesis is two-sided because we shall want to reject the null hypothesis if the mean of the determinations is significantly less than or significantly greater than 0.340.

Next, in step 3, we shall depart from the procedure used in the example of the preceding section and base the test on the statistic

Statistic for test concerning mean

$$z = \frac{\bar{x} - \mu_0}{\sigma/\sqrt{n}}$$

instead of $\bar{x}$. The reason for working with standard units, or z values, is that it enables us to formulate criteria which are applicable to a great variety of problems, not just one.

If z_α is, as before, such that the area under the standard normal curve to its right equals α, the **critical regions**, namely, the sets of values of z for which we reject the null hypothesis $\mu = \mu_0$, can be expressed as in the following table:

Critical Regions for Testing $\mu = \mu_0$
(Normal population and σ known, or large sample)

Alternative hypothesis	Reject null hypothesis if:
$\mu < \mu_0$	$z < -z_\alpha$
$\mu > \mu_0$	$z > z_\alpha$
$\mu \neq \mu_0$	$z < -z_{\alpha/2}$ or $z > z_{\alpha/2}$

If $\alpha = 0.05$, the dividing lines, or **critical values**, of the criteria are -1.645 and 1.645 for the one-sided alternatives, and -1.96 and 1.96 for the two-sided alternative. If $\alpha = 0.01$, the dividing lines of the criteria are -2.33 and 2.33 for the one-sided alternatives, and -2.575 and 2.575 for the two-sided alternative. These results come from the example on page 108 and Exercise 4.23.

Returning now to the example dealing with the thermal conductivity of the cement bricks, suppose that the mean of the 35 determinations is 0.343. So we continue by writing

3. *Criterion:* Reject the null hypothesis if $z < -1.96$ or $z > 1.96$, where

$$z = \frac{\bar{x} - \mu_0}{\sigma/\sqrt{n}}$$

4. *Calculations:*

$$z = \frac{0.343 - 0.340}{0.010/\sqrt{35}} = 1.77$$

5. *Decision:* Since $z = 1.77$ falls on the interval from -1.96 to 1.96, the null hypothesis cannot be rejected; to put it another way, the difference between $\bar{x} = 0.343$ and $\mu = 0.340$ can be attributed to chance. Whether we actually accept the null hypothesis will have to depend on whether the situations call for a decision one way or the other.

In problems like this, some research workers accompany the calculated value of z with a corresponding **tail probability,** or p **value,** with the probability of getting a difference between $\bar{x}$ and μ_0 greater than or equal to that actually observed. For instance, in the above example, it is given by the total area under the standard normal curve to the left of -1.77 and to the right of 1.77, and it equals $2(1 - 0.9616) = 0.0768$. This value exceeds 0.05, which agrees with our earlier result, but observe that giving a tail probability does not relieve us of the responsibility of specifying the level of significance before the test is actually performed.

The test we have described in this section is essentially an approximate large-sample test; it is exact only when the population we are sampling is normal and σ is known. In many practical situations where σ is unknown, we must make the further approximation of substituting for it the sample standard deviation s.

EXAMPLE A trucking firm suspects the claim that the average lifetime of certain tires is at least 28,000 miles. To check the claim, the firm puts 40 of these tires on its trucks and gets a mean lifetime of 27,463 miles with a standard deviation of 1,348 miles. What can it conclude if the probability of a Type I error is to be at most 0.01?

Solution

1. *Null hypothesis*: $\mu \geq 28,000$ miles
   Alternative hypothesis: $\mu < 28,000$ miles
2. *Level of significance*: $\alpha \leq 0.01$
3. *Criterion*:   Since the probability of a Type I error is greatest when $\mu = 28,000$ miles, we proceed as if we were testing the null hypothesis $\mu = 28,000$ miles against the alternative hypothesis $\mu < 28,000$ miles at the 0.01 level of significance. Thus, the null hypothesis must be rejected if $z < -2.33$, where

$$z = \frac{\bar{x} - \mu_0}{\sigma/\sqrt{n}}$$

with σ replaced by s.

4. *Calculations*:

$$z = \frac{27,463 - 28,000}{1,348/\sqrt{40}} = -2.52$$

5. *Decision*:   Since $z = -2.52$ is less than -2.33, the null hypothesis must be rejected; in other words, the trucking firm's suspicion that $\mu < 28,000$ miles is confirmed.

If the sample size is small and σ is unknown, the tests just described cannot be used. However, if the sample comes from a normal population (to within a reasonable degree of approximation), we can make use of the theory discussed in Section 6.3 and base the test of the null hypothesis $\mu = \mu_0$ on the statistic

***Statistic for
small-sample test
concerning mean***

$$t = \frac{\bar{x} - \mu_0}{s/\sqrt{n}}$$

which is a value of a random variable having the t distribution with $n - 1$ degrees of freedom. The criteria for the **one-sample t test** based on this statistic are like those given in the table on page 208 with z replaced by t and z_α and $z_{\alpha/2}$ replaced by t_α and $t_{\alpha/2}$.

EXAMPLE   The specifications for a certain kind of ribbon call for a mean breaking strength of 180 pounds. If five pieces of the ribbon (randomly selected from different rolls) have a mean breaking strength of 169.5 pounds with a standard deviation of 5.7 pounds, test the null hypothesis $\mu = 180$ pounds against the alternative hypothesis $\mu < 180$ pounds at the 0.01 level of significance.

Solution

1. *Null hypothesis*: $\mu = 180$ pounds
   Alternative hypothesis: $\mu < 180$ pounds
2. *Level of significance*: $\alpha = 0.01$
3. *Criterion*: Reject the null hypothesis if $t < -3.747$, where 3.747 is the value of $t_{0.01}$ for $5 - 1 = 4$ degrees of freedom and

$$t = \frac{\bar{x} - \mu_0}{s/\sqrt{n}}$$

4. *Calculations*:

$$t = \frac{169.5 - 180}{5.7\sqrt{5}} = -4.12$$

5. *Decision*: Since $t = -4.12$ is less than -3.747, the null hypothesis must be rejected; in other words, the breaking strength is below specifications. The exact tail probability, or p value, cannot be determined from Table 4, but it is 0.0073.

7.7 OPERATING CHARACTERISTIC CURVES

So far we have not paid much attention to Type II errors; in the drying-time example of Section 7.4 we calculated one probability of a Type II error, and all the tests of the preceding section were looked upon as tests of significance. Since the choice of the alternative hypothesis $\mu = 21$ minutes in the drying-time example was essentially arbitrary, it may be of interest to see what happens for other values of μ. To this end, we shall investigate the probability of accepting the null hypothesis for any given value of μ, denoted by $L(\mu)$, and it should be observed that this is the probability of committing a Type II error for all values of μ which should be rejected, and the probability of not committing a Type I error for all values of μ which should be accepted.

Continuing with the drying-time example on page 199, where we had $\mu_0 = 20$, $\sigma = 2.4$, $n = 36$, and the dividing line of the criterion was $\bar{x} = 20.75$, it can easily be verified that the methods used on page 200 yield the results shown in the following table and also in Figure 7.4 (see Exercise 7.50 on page 217):

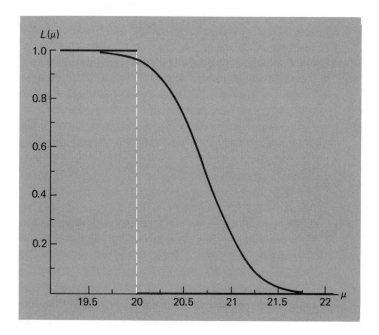

Figure 7.4 Operating characteristic curve.

Value of μ	Probability of accepting null hypothesis
19.50	0.999
19.75	0.99
20.00	0.97
20.25	0.89
20.50	0.73
20.75	0.50
21.00	0.27
21.25	0.11
21.50	0.03
21.75	0.01
22.00	0.001

Note that the probability of committing a Type II error diminishes when μ is increased, and that the probability of not committing a Type I error approaches 1 when μ becomes smaller and smaller than 20 (and the paint is even better than claimed).

 The graph of $L(\mu)$ for various values of μ, shown in Figure 7.4, is called the **operating characteristic curve**, or simply the **OC curve**, of the test criterion. Ideally, we should want to reject the null hypothesis $\mu = \mu_0$ when actually μ exceeds μ_0, and to accept it when μ is less than or equal to μ_0. Thus, the ideal OC curve for our example would be given by the heavy horizontal lines of Figure 7.4. In actual practice, OC curves can only approximate such ideal curves, with the approximation becoming better as the sample size is increased. To illustrate this point, the reader will be asked in Exercise 7.51 to recalculate the probabilities in the table on page 212 when the sample size is increased from 36 to 50.

Figure 7.4 presents the picture of a typical OC curve for the case where the alternative hypothesis is $\mu > \mu_0$. When the alternative hypothesis is $\mu < \mu_0$, the OC curve becomes the mirror image of that of Figure 7.4, reflected about the dashed vertical line through μ_0. For a two-tailed test, when the alternative hypothesis is $\mu \neq \mu_0$, the OC curve will be as shown in Figure 7.5, and it should be observed that $L(\mu)$ now is the probability of a Type II error for all values of μ except μ_0; at μ_0 it is still the probability of not committing a Type I error. Due to the symmetry of this kind of OC curve about the dashed vertical line through μ_0, all problems relating to Type II errors can be answered on the basis of the values of $L(\mu)$ for $\mu > \mu_0$.

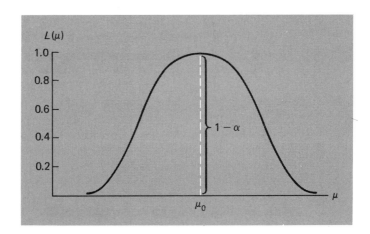

Figure 7.5 Operating characteristic curve for two-tailed test.

In many practical situations, the probabilities of Type II errors can be determined directly from charts like those given in Table 8 at the end of the book. They enable us to read off the values of β which correspond to given values of μ_0, μ_1, σ, n, and α, and they are based on the assumption that the sampling distribution of the mean is a normal distribution; hence, they can be used when n is large or the population we are sampling has roughly the shape of a normal distribution. Charts (a) and (b) apply to one-tailed tests with $\alpha = 0.05$ or $\alpha = 0.01$, and they apply regardless of whether the alternative hypothesis is $\mu > \mu_0$ or $\mu < \mu_0$, since the quantity which we plot on the horizontal scale is not μ but

Quantity needed for use of Table 8

$$d = \frac{|\mu - \mu_0|}{\sigma}$$

namely, the absolute value of the difference between μ and μ_0 divided by σ. The following are some examples which illustrate this technique:

EXAMPLE Suppose we want to investigate the claim that the sound intensity of certain vacuum cleaners is a random variable having a normal distribution with a mean of 75.20 dB and a standard deviation of 3.6 dB. Specifically, we shall want to test the null hypothesis $\mu = 75.20$ against the alternative hypothesis $\mu > 75.20$ on the basis of measurements of the sound intensity of $n = 15$ of these machines. If the probability of a Type I error is to be $\alpha = 0.05$, what is the probability of a Type II error for $\mu = 77.00$ dB?

Solution Since $d = \dfrac{|77.00 - 75.20|}{3.6} = 0.50$, we mark this point on the horizontal scale of chart (a) of Table 8, and, going up vertically until we come to the point where the line is crossed by the *OC* curve for $n = 15$, we find that the corresponding probability of a Type II error is about 0.38.

EXAMPLE Suppose we want to check the claim that in city traffic the mileage a person will get per tankful with a certain kind of car may be looked upon as a random variable having a normal distribution with a mean of 25.0 miles per gallon and a standard deviation of 0.8 mile per gallon. Specifically, we shall want to test the null hypothesis $\mu = 25.0$ against the alternative hypothesis $\mu < 25.0$ on the basis of the mileage obtained in city traffic with a tankful of gas by 10 of these cars. If the probability of a Type I error is to be $\alpha = 0.01$, what is the probability of a Type II error for $\mu = 24.0$ miles per gallon?

Solution Since $d = \dfrac{|24.0 - 25.0|}{0.8} = 1.25$, we mark this point on the horizontal scale of chart (b) of Table 8, and, going up vertically until we come to the point where the line is crossed by the *OC* curve for $n = 10$, we find that the corresponding probability of a Type II error is about 0.06.

For two-tailed tests, that is, for tests of the null hypothesis $\mu = \mu_0$ against the alternative hypothesis $\mu \neq \mu_0$, the procedure is the same as before, with the exception that we must now refer to charts (c) and (d) of Table 8. Note that these charts contain only the right-hand halves of the respective *OC* curves (as shown in Figure 7.5 on page 213), and that is all we need since these *OC* curves are symmetrical about μ_0.

EXAMPLE Suppose that the length of certain machine parts may be looked upon as a random variable having a normal distribution with a mean of 2.000 cm and a standard deviation of 0.050 cm. Specifically, we shall want to test the null hypothesis $\mu = 2.000$ against the alternative hypothesis $\mu \neq 2.000$ on the basis of the mean of a random sample of size $n = 30$. If the probability of a Type I error is to be $\alpha = 0.05$, what is the probability of a Type II error for $\mu = 2.010$?

Solution Since $d = \dfrac{|2.010 - 2.000|}{0.050} = 0.20$, we mark this point on the horizontal scale of chart (c) of Table 8, and, going up vertically until we come to the point where the line is crossed by the *OC* curve for $n = 30$, we find that the corresponding probability of a Type II error is about 0.80.

Table 8 can also be used to solve problems similar to the one of Exercise 7.34 on page 206, in which we are asked to determine the sample size that is required by given values of μ_0, μ_1, α, and β.

EXAMPLE With reference to the drying time example on page 199, where we had $\mu_0 = 20$, $\sigma = 2.4$, and $\alpha = 0.05$, how large a sample would we need so that $\beta = 0.10$ for $\mu = 21$?

Solution Since $d = \dfrac{|21 - 20|}{2.4} = 0.42$, we locate the point corresponding to $d = 0.42$ and $\beta = 0.10$ on chart (c) of Table 8, and we find that it lies barely above the *OC* curve for $n = 50$. Thus, a sample of size $n = 50$ will serve the stated purpose.

EXERCISES

7.38 According to the norms established for a mechanical aptitude test, persons who are 18 years old should average 73.2 with a standard deviation of 8.6. If 45 randomly selected persons of that age averaged 76.7, test the null hypothesis $\mu = 73.2$ against the alternative hypothesis $\mu > 73.2$ at the 0.01 level of significance.

7.39 Tests performed with a random sample of 40 diesel engines produced by a large manufacturer show that they have a mean thermal efficiency of 31.4% with a standard deviation of 1.6%. At the 0.01 level of significance, test the null hypothesis $\mu = 32.3\%$ against the alternative hypothesis $\mu \neq 32.3\%$.

7.40 In 64 randomly selected hours of production, the mean and the standard deviation of the number of acceptable pieces produced by an automatic stamping machine are $\bar{x} = 1,038$ and $s = 146$. At the 0.05 level of significance does this enable us to reject the null hypothesis $\mu = 1,000$ against the alternative hypothesis $\mu > 1,000$?

7.41 An oceanographer wants to check whether the average depth of the ocean in a certain region is 67.4 fathoms, as has previously been recorded. What can he conclude at the 0.01 level of significance if soundings taken at 40 random locations in the given region yielded a mean of 69.3 fathoms with a standard deviation of 5.4 fathoms?

7.42 In a labor-management discussion it was brought up that workers at a certain large plant take on the average 32.6 minutes to get to work. If a random sample of 60 workers took on the average 33.8 minutes with a standard deviation of 6.1 minutes, can we reject the null hypothesis $\mu = 32.6$ against the alternative hypothesis $\mu > 32.6$ at the 0.05 level of significance?

7.43 A random sample of six steel beams has a mean compressive strength of 58,392 psi (pounds per square inch) with a standard deviation of 648 psi. Use this information and the level of significance $\alpha = 0.05$ to test whether the true average compressive strength of the steel from which this sample came is 58,000 psi.

7.44 Given a random sample of 5 pints from different production lots, we want to test whether the fat content of a certain kind of ice cream exceeds 14%. What can we conclude at the 0.01 level of significance about the null hypothesis $\mu = 14\%$ if the sample has the mean $\bar{x} = 14.9\%$ and the standard deviation $s = 0.42\%$?

7.45 A laboratory technician claims that on the average it takes no more than 7.5 minutes to perform a certain task. If he is timed 20 times in the performance of this task, getting $\bar{x} = 7.9$ and $s = 1.2$, and the probability of a Type I error is to be at most 0.05, does this constitute evidence against the technician's claim?

7.46 Test runs with six models of an experimental engine showed that they operated for 24, 28, 21, 23, 32, and 22 minutes with a gallon of a certain kind of fuel. If the probability of a type I error is to be at most 0.01, is this evidence against the claim that on the average this kind of engine will operate for at least 29 minutes per gallon with this kind of fuel?

7.47 A random sample from a company's very extensive files shows that orders for a certain piece of machinery were filled, respectively, in 10, 12, 19, 14, 15, 18, 11, and 13 days. Use the level of significance $\alpha = 0.01$ to test the claim that on the average such orders are filled in 10.5 days. Choose the alternative hypothesis so that rejection of the null hypothesis $\mu = 10.5$ implies that it takes longer than indicated.

7.48 Five measurements of the tar content of a certain kind of cigarette yielded 14.5, 14.2, 14.4, 14.3, and 14.6 mg per cigarette. Show that the difference between the mean of this sample, $\bar{x} = 14.4$, and the average tar content claimed by the manufacturer, $\mu = 14.0$, is significant at $\alpha = 0.05$.

7.49 Suppose that in the preceding exercise the first measurement is recorded incorrectly as 16.0 instead of 14.5. Show that now the difference between the mean of the sample, $\bar{x} = 14.7$, and the average tar content claimed by the cigarette manufacturer, $\mu = 14.0$, is not significant at $\alpha = 0.05$. Explain the apparent paradox that even though the difference between $\bar{x}$ and μ has increased, it is no longer significant.

7.50 Use the method by which we calculated the probability of a Type II error on page 200, to verify the probabilities given in the table on page 212.

7.51 Suppose that in the drying-time example on page 199, n is changed from 36 to 50, while the other quantities remain $\mu_0 = 20$, $\sigma = 2.4$, and $\alpha = 0.03$. Find
(a) the new dividing line of the test criterion;
(b) the probability of Type II errors for the same values of μ as shown in the table on page 212.
Also plot the *OC* curve and compare it with the one shown in Figure 7.4.

7.52 With reference to the vacuum cleaner example on page 214, use Table 8 to find the probabilities of Type II errors for
(a) $\mu = 76.00$;   (b) $\mu = 78.00$.

7.53 With reference to the car mileage example on page 214, use Table 8 to find the probability of a Type II error for $\mu = 24.5$.

7.54 With reference to the machine parts example on page 215, use Table 8 to find the probabilities of Type II errors for
(a) $\mu = 2.020$;   (c) $\mu = 2.040$.
(b) $\mu = 2.030$;

7.55 If a random sample of size $n = 6$ is used to test the hypothesis that the mean of a normal population with $\sigma = 4$ is $\mu_0 = 50$ against the alternative hypothesis that $\mu > 50$, and $\alpha = 0.01$, use Table 8 to determine the probability of committing a Type II error when
(a) $\mu = 52$;   (d) $\mu = 55$;
(b) $\mu = 53$;   (e) $\mu = 56$;
(c) $\mu = 54$;   (f) $\mu = 57$.

7.56 Rework the preceding exercise, using the alternative hypothesis $\mu \neq 50$ instead of the alternative hypothesis $\mu > 50$.

7.57 If a random sample of size $n = 8$ is used to test the hypothesis that the mean of a normal population with $\sigma = 20$ is $\mu_0 = 200$ against the alternative hypothesis $\mu \neq 200$, and $\alpha = 0.05$, use Table 8 to determine the probability of committing a Type II error when
(a) $\mu = 190$;   (d) $\mu = 175$;
(b) $\mu = 185$;   (e) $\mu = 170$.
(c) $\mu = 180$;

7.58 Rework the preceding exercise, using the alternative hypothesis $\mu < 200$ instead of the alternative hypothesis $\mu \neq 200$.

7.59 Use Table 8 to rework part (b) of Exercise 7.31 on page 206.

7.60 Use Table 8 to rework part (a) of Exercise 7.34 on page 206.

7.61 Use Table 8 to rework part (b) of Exercise 7.34 on page 206.

7.62 Use Table 8 to rework Exercise 7.35 on page 207.

7.8 HYPOTHESES CONCERNING TWO MEANS

There are many statistical problems in which we are faced with decisions about the relative size of the means of two or more populations. Leaving the general problem until Chapter 12, we shall devote this section to tests concerning the difference between two means. For example, if two methods of welding are being considered for use with railroad rails, we may take samples and decide which is better by comparing their mean strengths; also, if a licensing examination is given to engineers who graduated from two different colleges, we may want to decide whether any observed difference between the means of the scores of the students from the two colleges is significant or whether it may be attributed to chance.

Formulating the problem more generally, we shall consider two populations having the means μ_1 and μ_2 and the variances σ_1^2 and σ_2^2, and we shall want to test the null hypothesis $\mu_1 - \mu_2 = \delta$, where δ is a specified constant, on the basis of independent random samples of size n_1 and n_2. Analogous to the tests concerning one mean, we shall consider tests of this null hypothesis against each of the alternatives $\mu_1 - \mu_2 < \delta$, $\mu_1 - \mu_2 > \delta$, and $\mu_1 - \mu_2 \neq \delta$. The test, itself, will depend on the difference between the sample means, $\bar{x}_1 - \bar{x}_2$, and if both samples come from normal populations with known variances, it can be based on the statistic

$$z = \frac{(\bar{x}_1 - \bar{x}_2) - \delta}{\sigma_{\bar{x}_1 - \bar{x}_2}}$$

which is a value of a random variable having the standard normal distribution. Here $\sigma_{\bar{x}_1 - \bar{x}_2}$ is the standard deviation of the sampling distribution of the difference between the sample means, and its value for random samples from infinite populations may be obtained with the use of the following theorem, which we shall state without proof:

Mean and variance of sum (or difference) of two random variables

> **Theorem 7.1** *If the distributions of two independent random variables have the means μ_1 and μ_2 and the variances σ_1^2 and σ_2^2, then the distribution of their* sum (or difference) *has the mean $\mu_1 + \mu_2$ (or $\mu_1 - \mu_2$) and the variance $\sigma_1^2 + \sigma_2^2$.*

To find the variance of the difference between the means of two independent random samples of size n_1 and n_2 from infinite populations, note first that the variances of the two means, themselves, are

$$\sigma_{\bar{x}_1}^2 = \frac{\sigma_1^2}{n_1} \quad \text{and} \quad \sigma_{\bar{x}_2}^2 = \frac{\sigma_2^2}{n_2}$$

where σ_1^2 and σ_2^2 are the variances of the respective populations. Thus, by Theorem 7.1 we have

$$\sigma_{\bar{x}_1 - \bar{x}_2}^2 = \frac{\sigma_1^2}{n_1} + \frac{\sigma_2^2}{n_2}$$

and the test statistic can be written as

Statistic for test concerning difference between two means

$$z = \frac{(\bar{x}_1 - \bar{x}_2) - \delta}{\sqrt{\dfrac{\sigma_1^2}{n_1} + \dfrac{\sigma_2^2}{n_2}}}$$

which is a value of a random variable having the standard normal distribution. In deriving this result we assumed that we are sampling normal populations. However, the above statistic can also be used when our samples are large enough so that we can apply the central limit theorem and approximate σ_1 and σ_2 with s_1 and s_2, namely, when n_1 and n_2 are both greater than or equal to 30.

Analogous to the table on page 208, the critical regions for testing the null hypothesis $\mu_1 - \mu_2 = \delta$ are as follows:

Critical Regions for Testing $\mu_1 - \mu_2 = \delta$
(Normal populations with σ_1 and σ_2 known, or large samples)

Alternative hypothesis	*Reject null hypothesis if:*
$\mu_1 - \mu_2 < \delta$	$z < -z_\alpha$
$\mu_1 - \mu_2 > \delta$	$z > z_\alpha$
$\mu_1 - \mu_2 \neq \delta$	$z < -z_{\alpha/2}$ or $z > z_{\alpha/2}$

Although δ can be any constant, it is worth noting that in the great majority of problems its value is zero and we test the null hypothesis of "no difference," namely, the null hypothesis $\mu_1 = \mu_2$.

EXAMPLE To test the claim that the resistance of electric wire can be reduced by more than 0.050 ohm by alloying, 32 values obtained for standard wire yielded $\bar{x}_1 = 0.136$ ohm and $s_1 = 0.004$ ohm, and 32 values obtained for alloyed wire yielded $\bar{x}_2 = 0.083$ ohm and $s_2 = 0.005$ ohm. At the 0.05 level of significance, does this support the claim?

Solution

1. *Null hypothesis*: $\mu_1 - \mu_2 = 0.050$
   Alternative hypothesis: $\mu_1 - \mu_2 > 0.050$
2. *Level of significance*: $\alpha = 0.05$
3. *Criterion*:   Reject the null hypothesis if $z > 1.645$, where z is given by the formula on page 219.
4. *Calculations*:

$$z = \frac{0.136 - 0.083 - 0.050}{\sqrt{\frac{(0.004)^2}{32} + \frac{(0.005)^2}{32}}} = 2.65$$

5. *Decision*:   Since $z = 2.65$ exceeds 1.645, the null hypothesis must be rejected; that is, the data substantiate the claim.

EXAMPLE   A company claims that its light bulbs are superior to those of its main competitor. If a study showed that a sample of $n_1 = 40$ of its bulbs had a mean lifetime of 647 hours of continuous use with a standard deviation of 27 hours, while a sample of $n_2 = 40$ bulbs made by its main competitor had a mean lifetime of 638 hours of continuous use with a standard deviation of 31 hours, does this substantiate the claim at the 0.05 level of significance?

Solution

1. *Null hypothesis*: $\mu_1 - \mu_2 = 0$
   Alternative hypothesis: $\mu_1 - \mu_2 > 0$
2. *Level of significance*: $\alpha = 0.05$
3. *Criterion*:   Reject the null hypothesis if $z > 1.645$, where z is given by the formula on page 219.
4. *Calculations*:

$$z = \frac{647 - 638}{\sqrt{\frac{27^2}{40} + \frac{31^2}{40}}} = 1.38$$

5. *Decision*:   Since $z = 1.38$ does not exceed 1.645, the null hypothesis cannot be rejected; that is, the observed difference between the two sample means is not significant.

If, in some problems, we actually want to accept the null hypothesis, we would risk Type II errors, for which the probabilities depend on the actual alternative differences $\delta' = \mu_1 - \mu_2$. Fortunately, these can be determined with the use of Table 8 (so long as we are sampling from normal populations with known standard deviations or both samples are large). The quantity which we mark on the horizontal scale is

***Quantity needed
for use of Table 8***

$$d = \frac{|\delta - \delta'|}{\sqrt{\sigma_1^2 + \sigma_2^2}}$$

where δ is the value of $\mu_1 - \mu_2$ assumed under the null hypothesis and δ' is the alternative value of $\mu_1 - \mu_2$ with which we are concerned. If $n_1 = n_2 = n$, the probability of a Type II error is read off the OC curve corresponding to this value of n; if $n_1 \neq n_2$, it is read off the OC curve corresponding to

$$n = \frac{\sigma_1^2 + \sigma_2^2}{\dfrac{\sigma_1^2}{n_1} + \dfrac{\sigma_2^2}{n_2}}$$

EXAMPLE With reference to the preceding example, what is the probability of a Type II error for $\delta' = 16$ hours?

Solution Since $d = \dfrac{|0 - 16|}{\sqrt{27^2 + 31^2}} = 0.39$, we mark this point on the horizontal scale of chart (a) of Table 8, and, going up vertically until we come to the OC curve for $n = 40$, we find that the probability of a Type II error is approximately 0.20.

When n_1, n_2, or both are small and the population variances are unknown, we can base tests of the null hypothesis $\mu_1 - \mu_2 = \delta$ on a suitable t statistic, provided we can assume that both populations are normal with $\sigma_1 = \sigma_2 (=\sigma)$. Under these conditions, it can be shown that

$$t = \frac{(\bar{x}_1 - \bar{x}_2) - \delta}{\hat{\sigma}_{\bar{x}_1 - \bar{x}_2}}$$

is a value of a random variable having the t distribution with $n_1 + n_2 - 2$ degrees of freedom, where $\hat{\sigma}_{\bar{x}_1 - \bar{x}_2}$ is the square root of an estimate of

$$\sigma_{\bar{x}_1 - \bar{x}_2}^2 = \frac{\sigma_1^2}{n_1} + \frac{\sigma_2^2}{n_2} = \sigma^2 \left(\frac{1}{n_1} + \frac{1}{n_2} \right)$$

with σ^2 estimated by **pooling** the sums of the squared deviations from the respective sample means. That is, we estimate σ^2 as

$$\frac{\sum (x_1 - \bar{x}_1)^2 + \sum (x_2 - \bar{x}_2)^2}{n_1 + n_2 - 2} = \frac{(n_1 - 1)s_1^2 + (n_2 - 1)s_2^2}{n_1 + n_2 - 2}$$

where $\sum (x_1 - \bar{x}_1)^2$ is the sum of the squared deviations from the mean for the first sample, while $\sum (x_2 - \bar{x}_2)^2$ is the sum of the squared deviations from the mean for the second sample. We divide by $n_1 + n_2 - 2$, since there are $n_1 - 1$ independent deviations from the mean in the first sample, $n_2 - 1$ in the second, and we thus have $n_1 + n_2 - 2$ independent deviations from the mean to estimate the population variance. Substituting this estimate of σ^2 into the above expression for $\sigma^2_{\bar{x}_1 - \bar{x}_2}$ and then substituting the square root of the result into the denominator of the formula for t above, we finally obtain

Statistic for small-sample test concerning difference between two means

$$t = \frac{(\bar{x}_1 - \bar{x}_2) - \delta}{\sqrt{(n_1 - 1)s_1^2 + (n_2 - 1)s_2^2}} \sqrt{\frac{n_1 n_2 (n_1 + n_2 - 2)}{n_1 + n_2}}$$

which is a value of a random variable having the t distribution with $n_1 + n_2 - 2$ degrees of freedom. The criteria for the **two-sample t test** based on this statistic are like those given in the table on page 219 with z replaced by t and z_α and $z_{\alpha/2}$ replaced by t_α and $t_{\alpha/2}$. In the application of this test, n_1 and n_2 may be small, yet $n_1 + n_2 - 2$ may be 30 or more; in that case we use the critical values given in the bottom line of Table 4.

EXAMPLE   The following random samples are measurements of the heat-producing capacity (in millions of calories per ton) of specimens of coal from two mines:

> *Mine 1:*   8,260,   8,130,   8,350,   8,070,   8,340
> *Mine 2:*   7,950,   7,890,   7,900,   8,140,   7,920,   7,840

Use the 0.01 level of significance to test whether the difference between the means of these two samples is significant.

Solution

1. *Null hypothesis:* $\mu_1 - \mu_2 = 0$
   Alternative hypothesis: $\mu_1 - \mu_2 \neq 0$
2. *Level of significance:* $\alpha = 0.01$
3. *Criterion:*   Reject the null hypothesis if $t < -3.250$ or $t > 3.250$, where 3.250 is the value of $t_{0.005}$ for $5 + 6 - 2 = 9$ degrees of freedom and t is given by the formula above.
4. *Calculations:*   The means and the variances of the two samples are $\bar{x}_1 = 8,230$, $\bar{x}_2 = 7,940$, $s_1^2 = \dfrac{63,000}{4} = 15,750$, and $s_2^2 = \dfrac{54,600}{5} = 10,920$, so that

$$t = \frac{(8,230 - 7,940)}{\sqrt{63,000 + 54,600}} \sqrt{\frac{5 \cdot 6 \cdot 9}{11}} = 4.19$$

5. *Decision*: Since $t = 4.19$ exceeds 3.250, the null hypothesis must be rejected; we conclude that the average heat-producing capacity of the coal from the two mines is not the same.

In the preceding example we went ahead and performed the two-sample

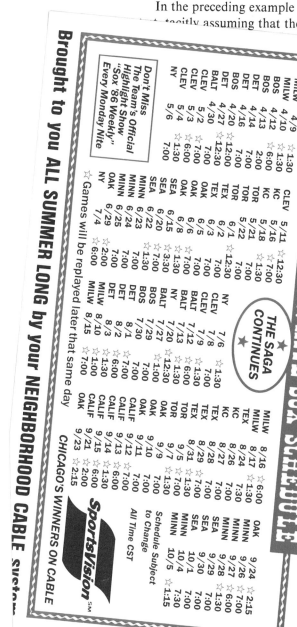

tacitly assuming that the population variances are equal. Fortunately,
differences between the population
instance is justifiable. To be on safer
tested whether the difference be-
tributed to chance; a procedure for
Chapter 8. If the difference between
ial, the two-sample t test cannot be
ests, among them the Smith–Satter-

le t test we shall also have to watch
istance, the test cannot be used when
of data, the I.Q.'s of husbands and
uations where the data are naturally
signed) differences of the paired data
be looked upon as a random sample
ually $\mu = 0$. If the sample is small, we
; otherwise, we use the corresponding

losses of man-hours due to accidents
a certain safety program was put into

d 44, 124 and 119, 33 and 35,
d 29, 26 and 24, and 17 and 11

to test whether the safety program is

$= 0$ (where μ is the mean of the population
)
: $\mu > 0$
$\alpha = 0.05$
he null hypothesis if $t > 1.833$, the value
9 degrees of freedom, where

$$t = \frac{\bar{x} - \mu_0}{s/\sqrt{n}}$$

and $\bar{x}$ and s are the mean and the standard deviation of the differences.

4. *Calculations*: The differences are 9, 13, 2, 5, −2, 6, 6, 5, 2, and 6, their mean is $\bar{x} = 5.2$, their standard deviation is $s = 4.08$, so that

$$t = \frac{5.2 - 0}{4.08/\sqrt{10}} = 4.03$$

5. *Decision*: Since $t = 4.03$ exceeds 1.833, the null hypothesis must be rejected; we conclude that the industrial safety program is effective.

In connection with this kind of problem, the one-sample t test is referred to as the **paired-sample t test**.

EXERCISES

7.63 The diameters of rotor shafts in a lot have a mean of 0.249 inch and a standard deviation of 0.003 inch. The inner diameters of bearings in another lot have a mean of 0.255 inch and a standard deviation of 0.002 inch.
(a) What are the mean and the standard deviation of the clearances between shafts and bearings selected from these lots?
(b) If a shaft and a bearing are selected at random, what is the probability that the shaft will not fit inside the bearing? (Assume that both dimensions are normally distributed.)

7.64 An investigation of two kinds of photocopying equipment showed that 75 failures of the first kind of equipment took on the average 83.2 minutes to repair with a standard deviation of 19.3 minutes, while 75 failures of the second kind of equipment took on the average 90.8 minutes to repair with a standard deviation of 21.4 minutes.
(a) Test the null hypothesis $\mu_1 - \mu_2 = 0$ (namely, the hypothesis that on the average it takes an equal amount of time to repair either kind of equipment) against the alternative hypothesis $\mu_1 - \mu_2 \neq 0$ at the level of significance $\alpha = 0.05$.
(b) Using 19.3 and 21.4 as estimates of σ_1 and σ_2 and referring to Table 8, find the probability of accepting the null hypothesis $\mu_1 - \mu_2 = 0$ with the criterion of part (a) when actually $\mu_1 - \mu_2 = -12$.

7.65 Suppose that we want to investigate whether on the average men earn more than $20 per week more than women in a certain industry. If sample data show that 60 men earn on the average $\bar{x}_1 = \$292.50$ per week with a standard deviation of $s_1 = \$15.60$, while 60 women earn on the average $\bar{x}_2 = \$266.10$ per week with a standard deviation of $s_2 = \$18.20$, what can we conclude at the 0.01 level of significance?

7.66 Random samples are taken from two normal populations with $\sigma_1 = 10.8$ and $\sigma_2 = 14.4$ to test the null hypothesis $\mu_1 - \mu_2 = 53.2$ against the alternative hypothesis $\mu_1 - \mu_2 > 53.2$ at the level of significance $\alpha = 0.01$. Use Table 8 to determine the common sample size $n = n_1 = n_2$ that is required if the probability of accepting the null hypothesis is to be 0.09 when $\mu_1 - \mu_2 = 66.7$.

7.67 Studying the flow of traffic at two busy intersections between 4 P.M. and 6 P.M. (to determine the possible need for turn signals), it was found that on 40 weekdays there were on the average 247.3 cars approaching the first intersection from the south which made left turns, while on 30 weekdays there were on the average 254.1 cars approaching the second intersection from the south which made left turns. The corresponding sample standard deviations are $s_1 = 15.2$ and $s_2 = 18.7$.

   (a) Test the null hypothesis $\mu_1 - \mu_2 = 0$ against the alternative hypothesis $\mu_1 - \mu_2 \neq 0$ at the level of significance $\alpha = 0.01$.

   (b) Using 15.2 and 18.7 as estimates of σ_1 and σ_2 and referring to Table 8, find the probability of accepting the null hypothesis $\mu_1 - \mu_2 = 0$ when actually $|\mu_1 - \mu_2| = 15.6$.

7.68 Measuring specimens of nylon yarn taken from two spinning machines, it was found that eight specimens from the first machine had a mean denier of 9.67 with a standard deviation of 1.81 while 10 specimens from the second machine had a mean denier of 7.43 with a standard deviation of 1.48. Assuming that the populations sampled are normal and have the same variance, test the null hypothesis $\mu_1 - \mu_2 = 1.5$ against the alternative hypothesis $\mu_1 - \mu_2 > 1.5$ at the 0.05 level of significance.

7.69 As part of an industrial training program, some trainees are instructed by Method *A* which is straight teaching-machine instruction, and some are instructed by Method *B* which also involves the personal attention of an instructor. If random samples of size 10 are taken from large groups of trainees instructed by each of these two methods, and the scores which they obtained in an appropriate achievement test are

Method A: 71, 75, 65, 69, 73, 66, 68, 71, 74, 68
Method B: 72, 77, 84, 78, 69, 70, 77, 73, 65, 75

use the 0.05 level of significance to test the claim that Method *B* is more effective. Assume that the populations sampled can be approximated closely with normal distributions having the same variance.

7.70 The following are the number of sales which a sample of nine salesmen of industrial chemicals in California and a sample of six salesmen of industrial chemicals in Oregon made over a certain fixed period of time:

> *California:*   59,   68,   44,   71,   63,   46,   69,   54,   48
> *Oregon:*   50,   36,   62,   52,   70,   41

Assuming that the populations sampled can be approximated closely with normal distributions having the same variance, test the null hypothesis $\mu_1 - \mu_2 = 0$ against the alternative hypothesis $\mu_1 - \mu_2 \neq 0$ at the 0.01 level of significance.

7.71 When we deal with two independent random samples from normal populations whose variances seem to be unequal, the following test, called the **Smith–Satterthwaite test**, can be used instead of the two-sample t test. The test statistic

$$t' = \frac{(\bar{x}_1 - \bar{x}_2) - \delta}{\sqrt{\dfrac{s_1^2}{n_1} + \dfrac{s_2^2}{n_2}}}$$

is a value of a random variable having approximately the t distribution with

$$\frac{\left(\dfrac{s_1^2}{n_1} + \dfrac{s_2^2}{n_2}\right)^2}{\dfrac{(s_1^2/n_1)^2}{n_1 - 1} + \dfrac{(s_2^2/n_2)^2}{n_2 - 1}}$$

degrees of freedom. Use this test in the following problems:
(a) The following are the Brinell hardness values obtained for samples of two magnesium alloys:

> *Alloy 1:*   66.3,   63.5,   64.9,   61.8,   64.3,   64.7,   65.1,
>   64.5,   68.4,   63.2
> *Alloy 2:*   71.3,   60.4,   62.6,   63.9,   68.8,   70.1,   64.8,
>   68.9,   65.8,   66.2

Use the 0.05 level of significance to test the null hypothesis $\mu_1 - \mu_2 = 0$ against the alternative hypothesis $\mu_1 - \mu_2 < 0$.
(b) To compare two kinds of bumper guards, six of each kind were mounted on a certain kind of compact car. Then each car was run into a concrete wall at 5 miles per hour, and the following are the costs of the repairs (in dollars):

> *Bumper guard 1:*   107,   148,   123,   165,   102,   119
> *Bumper guard 2:*   134,   115,   112,   151,   133,   129

Use the 0.01 level of significance to test whether the difference between the two sample means is significant.

7.72 The following data were obtained in an experiment designed to check whether there is a systematic difference in the weights obtained with two different scales:

	Weight in grams	
	Scale I	Scale II
Rock specimen 1	11.23	11.27
Rock specimen 2	14.36	14.41
Rock specimen 3	8.33	8.35
Rock specimen 4	10.50	10.52
Rock specimen 5	23.42	23.41
Rock specimen 6	9.15	9.17
Rock specimen 7	13.47	13.52
Rock specimen 8	6.47	6.46
Rock specimen 9	12.40	12.45
Rock specimen 10	19.38	19.35

Use the paired-sample t test at the 0.05 level of significance to test whether the difference of the means of the weights obtained with the two scales is significant.

7.73 In a study of the effectiveness of physical exercise in weight reduction, a group of 16 persons engaged in a prescribed program of physical exercise for one month showed the following results:

Weight before (pounds)	Weight after (pounds)	Weight before (pounds)	Weight after (pounds)
209	196	170	164
178	171	153	152
169	170	183	179
212	207	165	162
180	177	201	199
192	190	179	173
158	159	243	231
180	180	144	140

Use the 0.01 level of significance to test whether the prescribed program of exercise is effective.

7.74 Use the formula for t on page 222 to construct a $(1 - \alpha)100\%$ confidence interval for the difference, δ, between the two population means.

7.75 With reference to the preceding exercise and the example on page 219, construct a 95% confidence interval for the true difference between the average resistance of the two kinds of wire.

7.9 *CHECK LIST OF KEY TERMS* (*with page references*)

Alternative hypothesis, 203
Bayesian estimation, 193
Classical theory, 205
Composite hypothesis, 201
Confidence, 189
Confidence interval, 191
Confidence limits, 191
Critical regions, 208
Critical values, 209
Decision theory, 205
Degree of confidence, 191
Interval estimate, 190
Level of significance, 204
Neyman-Pearson theory, 205
Null hypothesis, 202
One-sample t *test, 210*
One-sided alternative, 203
One-sided criterion, 204
One-sided test, 204
One-tailed test, 204

Operating characteristic curve, 212
Paired-sample t *test, 224*
Point estimatation, 186
Pooling, 221
Posterior distribution, 193
Prior distribution, 193
p *value, 209*
Significance test, 202
Simple hypothesis, 201
Smith-Satterthwaite test, 226
Tail probability, 209
Two-sample t *test, 222*
Two-sided alternative, 203
Two-sided criterion, 205
Two-sided test, 205
Two-tailed test, 205
Type I error, 201
Type II error, 201
Unbiased estimate, 186

8

Inferences Concerning Variances

In Chapter 7 we learned how to judge the size of the error in estimating a population mean, how to construct confidence intervals for means, and how to perform tests of hypotheses about the means of one and of two populations. As we shall see in this and in subsequent chapters, very similar methods apply to inferences about other population parameters.

In this chapter we shall concentrate on population variances, or standard deviations, which are not only important in their own right, but which must sometimes be estimated before inferences about other parameters can be made. Section 8.1 is devoted to the estimation of σ^2 and σ, and Sections 8.2 and 8.3 deal with tests of hypotheses about these parameters.

8.1 THE ESTIMATION OF VARIANCES

In the preceding chapter, there were several instances where we estimated a population standard deviation by means of a sample standard deviation—we substituted s for σ in the large-sample confidence interval for μ on page 192, in the large-sample test concerning μ on page 210, and in the large-sample test concerning the difference between two means on page 220. Since there are many statistical procedures in which s is thus substituted for σ, or s^2 for σ^2, let us show first that the sample variance is, in fact, an unbiased estimate of the population variance.

If $f(x_1, x_2, \ldots, x_n)$ is the joint density of the sample values $x_1, x_2, \ldots,$ and x_n, it follows from the discussion on page 131 that the mean of the sampling distribution of s^2 is given by

$$\int_{-\infty}^{\infty} \int_{-\infty}^{\infty} \cdots \int_{-\infty}^{\infty} s^2 f(x_1, x_2, \ldots, x_n)\, dx_1\, dx_2 \ldots dx_n$$

$$= \int_{-\infty}^{\infty} \int_{-\infty}^{\infty} \cdots \int_{-\infty}^{\infty} \sum_{i=1}^{n} \frac{(x_i - \bar{x})^2}{n-1} f(x_1, x_2, \ldots, x_n)\, dx_1\, dx_2 \ldots dx_n$$

If we now write

$$\sum_{i=1}^{n} (x_i - \bar{x})^2 = \sum_{i=1}^{n} x_i^2 - n\bar{x}^2$$

and interchange the operations of summation and integration, the expression for the mean of the distribution of s^2 becomes

$$\frac{1}{n-1} \sum_{i=1}^{n} \int_{-\infty}^{\infty} \int_{-\infty}^{\infty} \cdots \int_{-\infty}^{\infty} x_i^2 f(x_1, x_2, \ldots, x_n)\, dx_1\, dx_2 \ldots dx_n$$

$$- \frac{n}{n-1} \int_{-\infty}^{\infty} \int_{-\infty}^{\infty} \cdots \int_{-\infty}^{\infty} \bar{x}^2 f(x_1, x_2, \ldots x_n)\, dx_1\, dx_2 \ldots dx_n$$

Assuming without loss of generality that the population mean μ is equal to zero, we find that these last two integrals have already been evaluated on pages 171 and 172, where it was shown that their respective values are σ^2 and σ^2/n. Thus, the mean of the sampling distribution of s^2 is given by

$$\frac{1}{n-1} \sum_{i=1}^{n} \sigma^2 - \frac{n}{n-1} \cdot \frac{\sigma^2}{n} = \frac{n\sigma^2}{n-1} - \frac{\sigma^2}{n-1} = \sigma^2$$

and this completes the proof of the unbiasedness of s^2 as an estimate of σ^2. (Note that, had we divided by n instead of $n-1$ in defining s^2, the resulting estimator would have been biased; the mean of its sampling distribution would have been $\frac{n-1}{n} \cdot \sigma^2$.)

Although the sample variance is an unbiased estimator of σ^2, it does not follow that the sample standard deviation is also an unbiased estimator of σ; in fact, it is not. However, for large samples the bias is small and it is common practice to estimate σ with s.

Besides s, population standard deviations are sometimes estimated in terms of the **sample range** R, which we defined in Section 5.4 as the largest value of a sample minus the smallest. Given a random sample of size n from a normal population, it can be shown that the sampling distribution of the range has the mean $d_2\sigma$ and the standard deviation $d_3\sigma$, where d_2 and d_3 are constants which depend on the size of the sample. For $n = 1, 2, \ldots,$ and 10, their values are as shown in the following table:

n	2	3	4	5	6	7	8	9	10
d_2	1.128	1.693	2.059	2.326	2.534	2.704	2.847	2.970	3.078
d_3	0.853	0.888	0.880	0.864	0.848	0.833	0.820	0.808	0.797

Thus, R/d_2 is an unbiased estimate of σ, and for very small samples, $n \leq 5$, it provides nearly as good an estimate of σ as does s; as the sample size increases, it becomes more efficient to use s instead of R/d_2. Nowadays, the range is used to estimate σ primarily in problems of industrial quality control, where sample sizes are usually small and computational ease is of prime concern. This application will be discussed in Chapter 14, where we shall need the above values of the constant d_3.

EXAMPLE With reference to the example on page 222, use the range of the first sample to estimate σ for the heat-producing capacity of coal from the first mine.

Solution Since the smallest value is 8,070, the largest value is 8,350, and $n = 5$ so that $d_2 = 2.326$, we get

$$\frac{R}{d_2} = \frac{8,350 - 8,070}{2.326} = 120.4$$

Note that this is fairly close to the sample standard deviation $s = 125.5$.

In most practical applications, interval estimates of σ or σ^2 are based on the sample standard deviation or the sample variance. For random samples from normal populations, we make use of Theorem 6.4, according to which

$$\frac{(n-1)s^2}{\sigma^2}$$

is a value of a random variable having the chi-square distribution with $n - 1$ degrees of freedom. Thus, with χ_α^2 defined as on page 181 for a chi-square distribution with $n - 1$ degrees of freedom, we can assert with probability $1 - \alpha$ that the inequality

$$\chi_{1-\alpha/2}^2 < \frac{(n-1)s^2}{\sigma^2} < \chi_{\alpha/2}^2$$

will be satisfied; once the data have been obtained, we make the same assertion with $(1 - \alpha)100\%$ confidence. Solving this inequality for σ^2, we obtain the following result:

Confidence interval for σ^2

$$\frac{(n-1)s^2}{\chi_{\alpha/2}^2} < \sigma^2 < \frac{(n-1)s^2}{\chi_{1-\alpha/2}^2}$$

If we take the square root of each member of this inequality, we obtain a corresponding $(1 - \alpha)100\%$ confidence interval for σ.

Note that confidence intervals for σ or σ^2 obtained by taking "equal tails," as in the above formula, do not actually give the narrowest confidence intervals, because the chi-square distribution is not symmetrical (see Exercise 7.16). Nevertheless, they are used in most applications in order to avoid fairly complicated calculations.

EXAMPLE   Returning to the example on page 181, suppose that the refractive indices of 20 pieces of glass (randomly selected from a large shipment purchased by the optical firm) have a variance of $1.20 \cdot 10^{-4}$. Construct a 95% confidence interval for σ, the standard deviation of the population sampled.

   Solution   For $20 - 1 = 19$ degrees of freedom, $\chi_{0.975}^2 = 8.907$ and $\chi_{0.025}^2 = 32.852$ according to Table 5, so that substitution into the formula yields

$$\frac{(19)(1.20 \cdot 10^{-4})}{32.852} < \sigma^2 < \frac{(19)(1.20 \cdot 10^{-4})}{8.907}$$

$$0.000069 < \sigma^2 < 0.000256$$

and, hence,

$$0.0083 < \sigma < 0.0160$$

The method which we have discussed applies only to random samples from normal populations (or at least to random samples from populations which can be approximated closely with normal distributions). If the sample size is large, it can be shown under fairly general conditions that the sampling distribution of the sample standard deviation can be approximated closely with a normal distribution having the mean σ and the standard deviation $\dfrac{\sigma}{\sqrt{2n}}$. Hence,

$$z = \frac{s - \sigma}{\sigma/\sqrt{2n}}$$

is a value of a random variable having approximately the standard normal distribution, and solving the inequality

$$-z_{\alpha/2} < \frac{s - \sigma}{\sigma/\sqrt{2n}} < z_{\alpha/2}$$

for σ, we get

Large-sample confidence interval for σ }

$$\frac{s}{1 + \dfrac{z_{\alpha/2}}{\sqrt{2n}}} < \sigma < \frac{s}{1 - \dfrac{z_{\alpha/2}}{\sqrt{2n}}}$$

The use of this $(1 - \alpha)100\%$ large-sample confidence-interval formula is illustrated by the following example:

EXAMPLE With reference to the example on page 159, where we showed that $s^2 = 30.77$ tons squared, and hence $s = 5.55$ tons, for a large industrial plant's emission of sulfur oxides on 80 days, construct a 95% confidence interval for the standard deviation of the population sampled.

> **Solution** Substituting $n = 80$, $s = 5.55$, and $z_{0.025} = 1.96$ into the above confidence interval formula, we get
>
> $$\frac{5.55}{1 + \dfrac{1.96}{\sqrt{160}}} < \sigma < \frac{5.55}{1 - \dfrac{1.96}{\sqrt{160}}}$$
>
> and
>
> $$4.80 < \sigma < 6.57$$

This means that we are 95% confident that the interval from 4.80 tons to 6.57 tons contains σ, the true standard deviation of the plant's daily emission of sulfur oxides.

EXERCISES

8.1 Use the data of Exercise 7.46 on page 216 to estimate σ for the length of time the experimental engine will operate with the given fuel in terms of

(a) the sample standard deviation;

(b) the sample range.

Compare the two estimates by expressing their difference as a percentage of the first.

8.2 With reference to the example on page 222, use the range of the second sample to estimate σ for the heat-producing capacity of coal from the second mine, and compare the result with the standard deviation of the second sample.

8.3 Use the data of part (a) of Exercise 7.71 to estimate σ for the Brinell hardness of Alloy 1 in terms of

(a) the sample standard deviation;

(b) the sample range.

Compare the two estimates by expressing their difference as a percentage of the first.

8.4 With reference to the example on page 210, construct a 95% confidence interval for the true standard deviation of the breaking strength of the given kind of ribbon.

8.5 With reference to Exercise 7.47 on page 216, construct a 99% confidence interval for the variance of the amount of time it takes the company to fill an order for a piece of the given kind of machinery.

8.6 While performing a strenuous task, the pulse rate of 25 workers increased on the average by 18.4 beats per minute with a standard deviation of 4.9 beats per minute. Find a 95% confidence interval for the corresponding population standard deviation, using

(a) the assumption that the sample comes from a normal population;

(b) the large-sample approximation.

Compare the two confidence intervals.

8.7 With reference to Exercise 7.48, construct a 99% confidence interval for the variance of the population sampled.

8.8 Use the value of s obtained in Exercise 8.3 to construct a 98% confidence interval for σ, measuring the actual variability in the hardness of Alloy 1.

8.9 With reference to Exercise 7.39 on page 215, construct a 95% large-sample confidence interval for σ, measuring the true variability of the thermal efficiency of such diesel engines.

8.10 With reference to Exercise 7.41 on page 216, construct a 99% confidence interval for σ, measuring the true variability of the depth of the ocean in the region sampled.

8.11 If 32 measurements of the boiling point of sulfur have a standard deviation of 0.83 degree Celsius, construct a 98% confidence interval for the true standard deviation of such measurements.

8.12 In the example on page 232 we indicated that the refractive indices of 20 pieces of glass have a variance of $1.20 \cdot 10^{-4}$. Using the square root of this quantity, namely, 0.011, as a preliminary estimate of σ and the large-sample theory on page 233, find the minimum sample size needed to estimate σ with a maximum error of 0.0025 and 95% confidence.

8.2 HYPOTHESES CONCERNING ONE VARIANCE

In this section we shall consider the problem of testing the null hypothesis that a population variance equals a specified constant against a suitable one-sided or two-sided alternative; that is, we shall test the null hypothesis $\sigma^2 = \sigma_0^2$ against one of the alternatives $\sigma^2 < \sigma_0^2, \sigma^2 > \sigma_0^2$, or $\sigma^2 \neq \sigma_0^2$. Tests like these are important whenever it is desired to control the uniformity of a product or an operation. For example, suppose that a silicon disc, or "wafer," is to be cut into small squares, or "dice," to be used in the manufacture of a semiconductor device. Since certain electrical characteristics of the finished device may depend on the thickness of the die, it is important that all dice cut from a wafer have approximately the same thickness. Thus, not only must the mean thickness of a wafer be kept within specifications, but also the variation in thickness from location to location on the wafer.

Using the same sampling theory as on page 232, we base such tests on the fact that for random samples from a normal population with the variance σ_0^2

Statistic for test concerning variance

$$\chi^2 = \frac{(n-1)s^2}{\sigma_0^2}$$

is a value of a random variable having the chi-square distribution with $n - 1$ degrees of freedom. The critical regions for such tests are as shown in the following table:

Critical Regions for Testing $\sigma^2 = \sigma_0^2$
(Normal population)

Alternative hypothesis	Reject null hypothesis if:
$\sigma^2 < \sigma_0^2$	$\chi^2 < \chi^2_{1-\alpha}$
$\sigma^2 > \sigma_0^2$	$\chi^2 > \chi^2_{\alpha}$
$\sigma^2 \neq \sigma_0^2$	$\chi^2 < \chi^2_{1-\alpha/2}$ or $\chi^2 > \chi^2_{\alpha/2}$

In this table χ^2_{α} is as defined on page 181. Note that "equal tails" are used for the two-sided alternative, and this is actually not the best procedure since the chi-square distribution is not symmetrical.

EXAMPLE The lapping process which is used to grind certain silicon wafers to the proper thickness is acceptable only if σ, the population standard deviation of the thickness of dice cut from the wafers, is at most 0.50 mil. Use the 0.05 level of significance to test the null hypothesis $\sigma = 0.50$ against the alternative hypothesis $\sigma > 0.50$, if the thicknesses of 15 dice cut from such wafers have a standard deviation of 0.64 mil.

Solution

1. *Null hypothesis*: $\sigma = 0.50$
   Alternative hypothesis: $\sigma > 0.50$
2. *Level of significance*: $\alpha = 0.05$
3. *Criterion*: Reject the null hypothesis if $\chi^2 > 23.685$, the value of $\chi^2_{0.05}$ for 14 degrees of freedom, where

$$\chi^2 = \frac{(n-1)s^2}{\sigma^2}$$

4. *Calculations*:

$$\chi^2 = \frac{(15-1)(0.64)^2}{(0.50)^2} = 22.94$$

5. *Decision*: Since $\chi^2 = 22.94$ does not exceed 23.685, the null hypothesis cannot be rejected; even though the sample standard deviation exceeds 0.50, there is not sufficient evidence to conclude that the lapping process is unsatisfactory.

There exist tables, like Table 8, which enable us to read off the probabilities of Type II errors connected with this kind of test. As given in the *National Bureau of Standards Handbook 91* (see the Bibliography), they contain the *OC* curves for the different one-sided and two-sided alternatives, for $\alpha = 0.05$ and $\alpha = 0.01$, and for various values of n. The quantity which we mark on the horizontal scale is the ratio $\frac{\sigma_1}{\sigma_0}$, where σ_1 is the alternative value of σ for which we want to determine the probability of erroneously accepting the null hypothesis $\sigma = \sigma_0$.

If the population from which we are sampling is not normal but the sample size is large, $n \geq 30$, the null hypothesis $\sigma = \sigma_0$ can be tested with the use of the statistic

Statistic for large-sample test concerning standard deviation

$$z = \frac{s - \sigma_0}{\sigma_0/\sqrt{2n}}$$

which is a value of a random variable having approximately the standard normal distribution. The critical regions are like those shown in the table on page 235 with z and z_α replacing χ^2 and χ^2_α, and $-z_\alpha$ replacing $\chi^2_{1-\alpha}$.

EXAMPLE The specifications for the mass production of certain bearings require, among other things, that the standard deviation of their diameters should not exceed 0.0040 cm. Given that a random sample of 35 of the bearings yielded $s = 0.0053$, test the null hypothesis $\sigma = 0.0040$ against the alternative hypothesis $\sigma > 0.0040$ at the 0.01 level of significance.

> **Solution**
>
> 1. *Null hypothesis*: $\sigma = 0.0040$
>   *Alternative hypothesis*: $\sigma > 0.0040$
> 2. *Level of significance*: $\alpha = 0.01$
> 3. *Criterion*: Reject the null hypothesis if $z > 2.33$, where
>
> $$z = \frac{s - \sigma_0}{\sigma_0 / \sqrt{2n}}$$
>
> 4. *Calculations*:
>
> $$z = \frac{0.0053 - 0.0040}{0.0040 / \sqrt{70}} = 2.72$$
>
> 5. *Decision*: Since $z = 2.72$ exceeds 2.33, the null hypothesis must be rejected; in other words, the bearings do not meet specifications.

8.3 HYPOTHESES CONCERNING TWO VARIANCES

The two-sample t test, described in Section 7.8, requires that the variances of the two populations sampled are equal. Therefore, before proceeding with this test, it is desirable to put this requirement to a test. In this section we shall describe a test of the null hypothesis $\sigma_1^2 = \sigma_2^2$, which applies to independent random samples from two normal populations; it must be used with some discretion as it is very sensitive to departures from this assumption.

If independent random samples of size n_1 and n_2 are taken from normal populations having the same variance, it follows from Theorem 6.5 that

Statistic for test of equality of two variances

$$F = \frac{s_1^2}{s_2^2}$$

is a value of a random variable having the F distribution with $n_1 - 1$ and $n_2 - 1$ degrees of freedom. Thus, if the null hypothesis $\sigma_1^2 = \sigma_2^2$ is true, the ratio of the sample variances s_1^2 and s_2^2 provides a statistic on which tests of the null hypothesis can be based.

The critical region for testing the null hypothesis $\sigma_1^2 = \sigma_2^2$ against the alternative hypothesis $\sigma_1^2 > \sigma_2^2$ is $F > F_\alpha$, where F_α is as defined on page 182. Similarly, the critical region for testing the null hypothesis against the alternative hypothesis $\sigma_1^2 < \sigma_2^2$ is $F < F_{1-\alpha}$, and this causes some difficulties since Table 6 only contains values corresponding to right-hand tails of $\alpha = 0.05$ and $\alpha = 0.01$. As a result, we use the reciprocal of the original test statistic and make use of the relation

$$F_{1-\alpha}(\nu_1, \nu_2) = \frac{1}{F_\alpha(\nu_2, \nu_1)}$$

first given on page 183. Thus, we base the test on the statistic $F = s_2^2/s_1^2$ and the critical region for testing the null hypothesis $\sigma_1^2 = \sigma_2^2$ against the alternative hypothesis $\sigma_1^2 < \sigma_2^2$ becomes $F > F_\alpha$, where F_α is the appropriate critical value of F for $n_2 - 1$ and $n_1 - 1$ degrees of freedom.

For the two-sided alternative $\sigma_1^2 \neq \sigma_2^2$ the critical region is $F < F_{1-\alpha/2}$ or $F > F_{\alpha/2}$, where $F = s_1^2/s_2^2$ and the degrees of freedom are $n_1 - 1$ and $n_2 - 1$. In practice, we modify this test as in the preceding paragraph, so that we can again use the table of F values corresponding to right-hand tails of $\alpha = 0.05$ and $\alpha = 0.01$. To this end we let s_M^2 represent the larger of the two sample variances, s_m^2 the smaller, and we write the corresponding sample sizes as n_M and n_m. Thus, the test statistic becomes $F = s_M^2/s_m^2$ and the critical region is as shown in the following table:

Critical Regions for Testing $\sigma_1^2 = \sigma_2^2$
(Normal populations)

Alternative hypothesis	Test statistic	Reject null hypothesis if:
$\sigma_1^2 < \sigma_2^2$	$F = \dfrac{s_2^2}{s_1^2}$	$F > F_\alpha(n_2 - 1, n_1 - 1)$
$\sigma_1^2 > \sigma_2^2$	$F = \dfrac{s_1^2}{s_2^2}$	$F > F_\alpha(n_1 - 1, n_2 - 1)$
$\sigma_1^2 \neq \sigma_2^2$	$F = \dfrac{s_M^2}{s_m^2}$	$F > F_{\alpha/2}(n_M - 1, n_m - 1)$

The level of significance of these tests is α and the figures indicated in parentheses are the respective degrees of freedom. Note that, as in the chi-square test, "equal tails" are used in the two-tailed test as a matter of mathematical convenience, even though the F distribution is not symmetrical.

EXAMPLE It is desired to determine whether there is less variability in the silver plating done by Company 1 than in that done by Company 2. If independent random samples of the work done by the two companies yield $s_1 = 0.035$ mil and

$s_2 = 0.062$ mil, test the null hypothesis $\sigma_1^2 = \sigma_2^2$ against the alternative hypothesis $\sigma_1^2 < \sigma_2^2$ at the 0.05 level of significance.

Solution

1. *Null hypothesis*: $\sigma_1^2 = \sigma_2^2$
   Alternative hypothesis: $\sigma_1^2 < \sigma_2^2$
2. *Level of significance*: $\alpha = 0.05$
3. *Criterion*: Reject the null hypothesis if $F > 2.82$, the value of $F_{0.05}$ for 11 and 11 degrees of freedom, where

$$F = \frac{s_2^2}{s_1^2}$$

4. *Calculations*:

$$F = \frac{(0.062)^2}{(0.035)^2} = 3.14$$

5. *Decision*: Since $F = 3.14$ exceeds 2.82, the null hypothesis must be rejected; in other words, the data support the contention that the plating done by Company 1 is less variable than that done by Company 2.

EXAMPLE With reference to the example dealing with the heat-producing capacity of coal from two mines on page 222, use the 0.02 level of significance to test whether it is reasonable to assume that the variances of the two populations sampled are equal.

Solution

1. *Null hypothesis*: $\sigma_1^2 = \sigma_2^2$
   Alternative hypothesis: $\sigma_1^2 \neq \sigma_2^2$
2. *Level of significance*: $\alpha = 0.02$
3. *Criterion*: Reject the null hypothesis if $F > 11.4$, the value of $F_{0.01}$ for 4 and 5 degrees of freedom, where

$$F = \frac{s_1^2}{s_2^2}$$

since $s_1^2 = 15,750$ is greater than $s_2^2 = 10,920$.

4. *Calculations*:

$$F = \frac{15,750}{10,920} = 1.44$$

5. *Decision*: Since $F = 1.44$ does not exceed 11.4, the null hypothesis cannot be rejected; there is no real reason to doubt the equality of the variances of the two populations.

Had we wanted to use the level of significance $\alpha = 0.05$ or $\alpha = 0.01$ in this example, we would have required tables of the values of $F_{0.025}(v_1, v_2)$ or $F_{0.005}(v_1, v_2)$; such tables may be found in the *Biometrika Tables for Statisticians* listed in the Bibliography under Pearson and Hartley. Also, *OC* curves for the various F tests discussed in this section may be found in the book by A. H. Bowker and G. J. Lieberman listed in the Bibliography, while those for the one-tailed tests only may be found in the *National Bureau of Standards Handbook 91*.

EXERCISES

8.13 With reference to Exercise 7.43 on page 216, test the null hypothesis $\sigma = 600$ psi for the compressive strength of the given kind of steel against the alternative hypothesis $\sigma > 600$ psi. Use the 0.05 level of significance.

8.14 If 12 determinations of the specific heat of iron have a standard deviation of 0.0086, test the null hypothesis that $\sigma = 0.010$ for such determinations. Use the alternative hypothesis $\sigma \neq 0.010$ and the level of significance $\alpha = 0.01$.

8.15 With reference to Exercise 7.64, test the null hypothesis that $\sigma = 15.0$ minutes for the time that is required for repairs of the first kind of photocopying equipment against the alternative hypothesis that $\sigma > 15.0$ minutes. Use the 0.05 level of significance.

8.16 Use the 0.01 level of significance to test the null hypothesis that $\sigma = 0.015$ inch for the diameters of certain bolts against the alternative hypothesis that $\sigma \neq 0.015$ inch, given that a random sample of size 15 yielded $s^2 = 0.00011$.

8.17 With reference to Exercise 8.6 on page 234, use the 0.05 level of significance to test the null hypothesis that $\sigma^2 = 30.0$ for such increases in the pulse rate (while performing the given task) against the alternative hypothesis that $\sigma^2 < 30.0$.

8.18 Playing 10 rounds of golf on his home course, a golf professional averaged 71.3 with a standard deviation of 1.32. Test the null hypothesis that the consistency of his game on his home course is actually measured by $\sigma = 1.20$, against the alternative hypothesis that he is less consistent. Use the level of significance $\alpha = 0.05$.

8.19 Past data indicate that the variance of measurements made on sheet metal stampings by experienced quality control inspectors is 0.18 square inch. Such measurements made by an inexperienced inspector could have too large a variance (perhaps because of inability to read instruments properly) or too small a variance (perhaps because unusually high or low measurements are discarded). If a new inspector measures 100 stampings with a variance of 0.13 square inch, test at the 0.05 level of significance whether the inspector is making satisfactory measurements.

8.20 The security department of   arge office building wants to test the null hypothesis that $\sigma = 2.0$ minutes for the time it takes a guard to walk his round against the alternative hypothesis that $\sigma \neq 2.0$ minutes. What can it conclude at the 0.01 level of significance if a random sample of size $n = 32$ yields $s = 1.8$ minutes?

8.21 Justify the use of the two-sample t test in Exercise 7.68 on page 225 by testing the null hypothesis that the two populations have equal variance. Use the 0.02 level of significance.

8.22 With reference to Exercise 7.69 on page 225, use the 0.02 level of significance to test the assumption that the two populations have equal variances.

8.23 With reference to Exercise 7.70 on page 225, use the 0.02 level of significance to test the assumption that the two populations have equal variances.

8.24 Two different lighting techniques are compared by measuring the intensity of light at selected locations in areas lighted by the two methods. If 15 measurements in the first area had a standard deviation of 2.7 foot-candles and 21 measurements in the second area had a standard deviation of 4.2 foot-candles, can it be concluded that the lighting in the second area is less uniform? Use a 0.01 level of significance. What assumptions must be made as to how the two samples are obtained?

8.25 With reference to Exercise 7.67, where we had $n_1 = 40$, $n_2 = 30$, $s_1 = 15.2$, and $s_2 = 18.7$, use the 0.05 level of significance to test the claim that there is a greater variability in the number of cars which make left turns approaching from the south between 4 P.M. and 6 P.M. at the second intersection.

8.26 Pull-strength tests on 10 soldered leads for a semiconductor device yield the following results in pounds force required to rupture the bond:

$$15.8, \quad 12.7, \quad 13.2, \quad 16.9, \quad 10.6, \quad 18.8, \quad 11.1, \quad 14.3, \quad 17.0, \quad 12.5$$

Another set of eight leads was tested after encapsulation to determine whether the pull strength has been increased by encapsulation of the device, with the following results:

$$24.9, \quad 23.6, \quad 19.8, \quad 22.1, \quad 20.4, \quad 21.6, \quad 21.8, \quad 22.5$$

As a preliminary to the two-sample t test, use the 0.02 level of significance to test whether it is reasonable to assume that the two samples come from populations with equal variances.

8.27 Random samples of size n_1 and n_2, respectively, are taken from two log-normal populations, and the resulting sample means are $\bar{x}_1 = 3.74$ and $\bar{x}_2 = 13.91$. You wish to test whether the second population has a mean value four times as large as the first.
(a) Can you directly use a two-sample test? Why?
(b) Is there a transformation that can be made on the data that could conceivably allow the use of a two-sample test?

8.4 CHECK LIST OF KEY TERMS (with page reference)

9

Inferences Concerning Proportions

Many engineering problems deal with proportions, percentages, or probabilities. In acceptance sampling we are concerned with the proportion of defectives in a lot, and in life testing we are concerned with the percentage of certain components which will perform satisfactorily during a stated period of time, or the probability that a given component will last at least a given number of hours. It should be clear from these examples that problems concerning proportions, percentages, or probabilities are really equivalent; a percentage is merely a proportion multiplied by 100, and a probability may be interpreted as a proportion in the long run.

Sections 9.1 and 9.2 deal with the estimation of proportions; Section 9.3 deals with tests concerning proportions; Section 9.4 deals with tests concerning two or more proportions; in Section 9.5 we shall learn how to analyze data tallied into a two-way classification; and in Section 9.6 we shall learn how to judge whether differences between an observed frequency distribution and corresponding expectations can be attributed to chance.

9.1 ESTIMATION OF PROPORTIONS

The information that is usually available for the estimation of a proportion is the number of times, x, that an appropriate event has occurred in n trials, occasions, or observations. The point estimate, itself, is usually the **sample proportion** $\frac{x}{n}$, namely, the proportion of the time that the event has actually occurred. If the n trials satisfy the assumptions underlying the binomial distribution listed on page 57, we know that the mean and the standard deviation of the number of successes are given by np and $\sqrt{np(1-p)}$. If we divide both of these quantities by n, we find that the mean and the standard deviation of the proportion of successes (namely, of the sample proportion) are given by

$$\frac{np}{n} = p \quad \text{and} \quad \frac{\sqrt{np(1-p)}}{n} = \sqrt{\frac{p(1-p)}{n}}$$

The first of these results shows that the sample proportion is an unbiased estimator of the binomial parameter p, namely, of the true proportion we are trying to estimate on the basis of a sample.

In the construction of confidence intervals for the binomial parameter p, we meet several obstacles. First, since x and $\frac{x}{n}$ are values of discrete random variables, it may be impossible to get an interval for which the degree of confidence is exactly $(1-\alpha)100\%$. Second, the standard deviation of the sampling distribution of the number of successes, as well as that of the proportion of successes, involves the parameter p that we are trying to estimate.

To construct a confidence interval for p having approximately the degree of confidence $(1-\alpha)100\%$, we first determine for a given set of values of p the corresponding quantities x_0 and x_1, where x_0 is the largest integer for which

$$\sum_{k=0}^{x_0} b(k; n, p) \leq \frac{\alpha}{2}$$

while x_1 is the smallest integer for which

$$\sum_{k=x_1}^{n} b(k; n, p) \leq \frac{\alpha}{2}$$

To emphasize the point that x_0 and x_1 depend on the value of p, we shall write these quantities as $x_0(p)$ and $x_1(p)$. Thus, we can assert with a probability of approximately $1 - \alpha$, and at least $1 - \alpha$, that the inequality

$$x_0(p) < x < x_1(p)$$

will be satisfied; here x is a value of a random variable and p is a fixed constant. To change inequalities like these into confidence intervals for p, we can use a simple graphical method which is illustrated by the following example: Suppose, for instance, that we want to find approximate 95 % confidence intervals for p for samples of size $n = 20$. Using Table 1 at the end of the book, we first determine x_0 and x_1 for selected values of p such that x_0 is the largest integer for which

$$B(x_0; 20, p) \leq 0.025$$

while x_1 is the smallest integer for which

$$1 - B(x_1 - 1; 20, p) \leq 0.025$$

Letting p equal $0.1, 0.2, \ldots$, and 0.9, we thus obtain the values shown in the following table:

p	0.1	0.2	0.3	0.4	0.5	0.6	0.7	0.8	0.9
x_0	—	0	1	3	5	7	9	11	14
x_1	6	9	11	13	15	17	19	20	—

Plotting the points with coordinates p and $x(p)$ as in Figure 9.1, and drawing smooth curves, one through the x_0 points and one through the x_1 points, we can now "solve" for p. For any given value of x we can obtain approximate 95 % confidence limits for p by going horizontally to the two curves and marking off the corresponding values of p (see Figure 9.1). Thus, for $x = 4$ we obtain the approximate 95 % confidence interval

$$0.06 < p < 0.45$$

Graphs similar to the one shown in Figure 9.1 are given in Tables 9(a) and 9(b) at the end of the book for various values of n and for the 95 % and 99 % degrees of confidence. These tables differ from the one of Figure 9.1 in that the sample proportion $\frac{x}{n}$ is used instead of x, thus making it possible to graph curves corresponding to various values of n on the same diagram. Also, for increased accuracy, Tables 9(a) and 9(b) are arranged so that values of $\frac{x}{n}$ from 0.00 to 0.50 are marked on the bottom scale while those from 0.50 to 1.00 are marked on the top scale of the diagram. For values of $\frac{x}{n}$ from 0.00

to 0.50 the confidence limits for p are read off the left-hand scale of the diagram, while for values of $\frac{x}{n}$ from 0.50 to 1.00 they are read off the right-hand scale. Note that for $n = 20$ and $x = 4$, Table 9(a) yields the 95% confidence interval $0.06 < p < 0.44$, which is very close, indeed, to the results obtained with Figure 9.1.

Figure 9.1 95% confidence intervals for proportions ($n = 20$).

On page 112 we gave the general rule of thumb that the normal distribution provides a good approximation to the binomial distribution when np and $n(1 - p)$ are both greater than 5. Thus, for $n = 50$ the normal curve approximation may be used if it can be assumed that p lies between 0.10 and 0.90; for $n = 100$ it may be used if it can be assumed that p lies between 0.05 and 0.95; for $n = 200$ it may be used if it can be assumed that p lies between 0.025 and 0.975; and so forth. This is what we shall mean here, and later in this chapter, by "n being large."

When n is large, we can construct approximate confidence intervals for the binomial parameter p by using the normal approximation to the binomial distribution. Accordingly, we can assert with probability $1 - \alpha$ that the inequality

$$-z_{\alpha/2} < \frac{x - np}{\sqrt{np(1 - p)}} < z_{\alpha/2}$$

will be satisfied. Solving this quadratic inequality for p, we can obtain a corresponding set of approximate confidence limits for p (see Exercise 9.14 on page 252), but since the necessary calculations are involved, we shall make the further approximation of substituting $\frac{x}{n}$ for p in $\sqrt{np(1-p)}$. This yields

Large-sample confidence interval for p

$$\frac{x}{n} - z_{\alpha/2}\sqrt{\frac{\frac{x}{n}\left(1-\frac{x}{n}\right)}{n}} < p < \frac{x}{n} + z_{\alpha/2}\sqrt{\frac{\frac{x}{n}\left(1-\frac{x}{n}\right)}{n}}$$

where the degree of confidence is $(1-\alpha)100\%$.

EXAMPLE   If $x = 36$ of $n = 100$ persons interviewed are familiar with the tax incentives for installing certain energy-saving devices, construct a 95% confidence interval for the corresponding true proportion.

Solution   Substituting $\frac{x}{n} = \frac{36}{100} = 0.36$ and $z_{\alpha/2} = 1.96$ into the above formula, we get

$$0.36 - 1.96\sqrt{\frac{(0.36)(0.64)}{100}} < p < 0.36 + 1.96\sqrt{\frac{(0.36)(0.64)}{100}}$$

or

$$0.266 < p < 0.454$$

Note that if we had used Table 9(a), we would have obtained

$$0.27 < p < 0.46$$

The magnitude of the error we make when we use $\frac{x}{n}$ as an estimate of p is given by $\left|\frac{x}{n} - p\right|$. Again using the normal approximation, we can thus assert with probability $1 - \alpha$ that the inequality

$$\left|\frac{x}{n} - p\right| \leq z_{\alpha/2}\sqrt{\frac{p(1-p)}{n}}$$

will be satisfied, namely, that the error will be at most $z_{\alpha/2}\sqrt{\frac{p(1-p)}{n}}$. With $\frac{x}{n}$ substituted for p this yields

Maximum error of estimate

$$E = z_{\alpha/2} \sqrt{\frac{\frac{x}{n}\left(1 - \frac{x}{n}\right)}{n}}$$

EXAMPLE In a sample survey conducted in a large city, 136 of 400 persons answered "Yes" to the question whether their city's public transportation is adequate. With 99 % confidence, what can we say about the maximum error, if $\dfrac{x}{n} = \dfrac{136}{400}$ = 0.34 is used as an estimate of the corresponding true proportion?

Solution Substituting $\dfrac{x}{n} = 0.34$ and $z_{\alpha/2} = 2.575$ into the above formula, we find that the error is at most

$$E = 2.575 \sqrt{\frac{(0.34)(0.66)}{400}} = 0.061$$

The above formula for E can also be used to determine the sample size that is needed to attain a desired degree of precision. Solving for n, we get

Sample size

$$n = p(1 - p)\left[\frac{z_{\alpha/2}}{E}\right]^2$$

but this formula cannot be used as it stands unless we have some information about the possible size of p (on the basis of collateral data, say, a pilot sample). If no such information is available, we can make use of the fact that $p(1 - p)$ is at most $\frac{1}{4}$, corresponding to $p = \frac{1}{2}$, as can be shown by the methods of elementary calculus. Thus, if

Sample size

$$n = \frac{1}{4}\left[\frac{z_{\alpha/2}}{E}\right]^2$$

we can assert with a probability of at least $1 - \alpha$ that the error in using $\dfrac{x}{n}$ as an estimate of p will not exceed E; once the data have been obtained, we will be able to assert with at least $(1 - \alpha)100\%$ confidence that the error does not exceed E.

EXAMPLE   Suppose that we want to estimate the true proportion of defectives in a very large shipment of adobe bricks, and that we want to be at least 95% confident that the error is at most 0.04. How large a sample will we need if
(a) we have no idea what the true proportion might be;
(b) we know that the true proportion does not exceed 0.12?

Solution   (a) Using the second of the two formulas for the sample size, we get

$$n = \frac{1}{4}\left[\frac{1.96}{0.04}\right]^2 = 600.25$$

or $n = 601$ rounded up to the nearest integer. (b) Using the first of the two formulas for the sample size with $p = 0.12$ (the possible value closest to $p = \frac{1}{2}$), we get

$$n = (0.12)(0.88)\left[\frac{1.96}{0.04}\right]^2 = 253.55$$

or $n = 254$ rounded up to the nearest integer. This serves to illustrate how some collateral information about the possible size of p can substantially reduce the size of the required sample.

When p is very close to 0, as is the case in problems of high reliability and p is the probability of failure, none of the confidence intervals we have discussed provides a satisfactory solution. What we really need here are **one-sided confidence intervals** of the form $p < C$, where C is a constant depending on the degree of confidence and the size of the sample. As we already pointed out on page 79, the binomial distribution is best approximated with a Poisson distribution with $\lambda = np$ when p is small and n is large. Based on this approximation, it can be shown that

One-sided confidence interval for p

$$p < \frac{1}{2n}\cdot\chi_\alpha^2$$

is a one-sided confidence interval for p, where χ_α^2 is as defined on page 181 and the number of degrees of freedom equals $2(x + 1)$. A discussion of this result may be found in the book by A. Hald mentioned in the Bibliography.

EXAMPLE   If there are $x = 4$ failures among $n = 2,000$ parts used continuously for a month, construct a one-sided 99% confidence interval for the probability that one such part will fail under the stated conditions.

Solution   Since $\chi_{0.01}^2 = 23.209$ for $2(4 + 1) = 10$ degrees of freedom, substitution into the formula yields

$$p < \frac{1}{2(2,000)} \cdot 23.209$$

and, hence

$$p < 0.0058$$

9.2 BAYESIAN ESTIMATION†

In the preceding section we looked upon the true proportions we tried to estimate as unknown constants; in Bayesian estimation these parameters are looked upon as random variables having prior distributions which reflect the strength of one's belief about the possible values they can take on, or other indirect information. As in Section 7.3, we are thus faced with the problem of combining prior information with direct sample evidence.

To illustrate how this might be done, suppose that a manufacturer, who regularly receives large shipments of electronic components from a vendor, knows that about 25% of the time 0.005 (half of 1%) of the components are defective, about 25% of the time 0.01 of the components are defective, and about 50% of the time 0.02 of the components are defective. Thus, before a shipment from this vendor is inspected, we have the following prior distribution for the proportion of defectives:

Value of p	Prior probability
0.005	0.25
0.01	0.25
0.02	0.50

Now suppose that 200 of these components, randomly selected from the shipment, are inspected, and only one of them is found to be defective. The probability of this happening when $p = 0.005, p = 0.01$, or $p = 0.02$ are, respectively,

$$\binom{200}{1}(0.005)^1(0.995)^{199} = 0.37$$

$$\binom{200}{1}(0.01)^1(0.99)^{199} = 0.27, \quad \text{and} \quad \binom{200}{1}(0.02)^1(0.98)^{199} = 0.07$$

where we used the formula for the binomial distribution on page 58 and logarithms to simplify the calculations. Combining these probabilities by means of the formula for Bayes' theorem (Theorem 2.11 on page 41), we

---

†This section may be omitted without loss of continuity.

find that the posterior probability for $p = 0.005$ is

$$\frac{(0.25)(0.37)}{(0.25)(0.37) + (0.25)(0.27) + (0.50)(0.07)} = 0.47$$

and that the corresponding posterior probabilities for $p = 0.01$ and $p = 0.02$ are 0.35 and 0.18. We have thus arrived at the following posterior distribution for the proportion of defective components:

Value of p	Posterior probability
0.005	0.47
0.01	0.35
0.02	0.18

Note that whereas the odds were originally 3 to 1 against $p = 0.005$, it is now almost an even bet; of course, this shift is accounted for by the fact that in the sample only $\frac{1}{200} = 0.005$ of the components inspected were defective.

In the preceding example we assumed that p had to be $0.005, 0.01$, or 0.02, and this restriction was imposed mainly to simplify the calculations; the method would have been the same if we had considered 10 different values of p, or even 100. It would be more logical, perhaps, to let p take on any value on the continuous interval from 0 to 1, and in that case it is customary to use as the prior distribution the beta distribution of Section 4.8. The parameters of this distribution are α and β, and its mean and variance can be expressed in terms of α and β in accordance with the formulas on page 121. It can then be shown that the posterior distribution of p, namely, the conditional distribution of p for a given (observed) value of x, is also a beta distribution, and that its parameters are $x + \alpha$ and $n - x + \beta$ instead of α and β.[†] Thus, the mean and the variance of the posterior distribution may be obtained by substituting $x + \alpha$ for α and $n - x + \beta$ for β in the formulas on page 121.

EXAMPLE A person doing research for a large oil company feels that the proportion of persons requiring oil as well as gasoline at one of the oil company's service stations is a random variable having the beta distribution with $\alpha = 10$ and $\beta = 400$. In a random sample of size $n = 800$, she finds that only $x = 3$ persons required oil as well as gasoline. Find the mean and the variance of
(a) the prior distribution of p;
(b) the posterior distribution of p.

[†]Proofs of these results may be found in the book by John E. Freund and Ronald E. Walpole listed in the Bibliography.

Solution (a) For the prior distribution we get

$$\mu_0 = \frac{10}{10 + 400} = 0.024$$

and

$$\sigma_0^2 = \frac{10 \cdot 400}{410^2 \cdot 411} = 0.000058$$

(b) For the posterior distribution we get

$$\mu_1 = \frac{3 + 10}{10 + 400 + 800} = 0.011$$

and

$$\sigma_1^2 = \frac{(3 + 10)(800 - 3 + 400)}{(10 + 400 + 800)^2(10 + 400 + 800 + 1)}$$
$$= 0.0000088$$

If we have mathematical tables giving the values of beta integrals, we can continue with an example like this and calculate prior as well as posterior probabilities associated with various intervals of the values of *p*.

EXERCISES

9.1 In a random sample of 200 claims filed against an insurance company writing collision insurance on cars, 84 exceeded $1,200. Construct a 95% confidence interval for the true proportion of claims filed against this insurance company that exceed $1,200, using
(a) Table 9;
(b) the large-sample confidence-interval formula.

9.2 With reference to the preceding exercise, what can we say with 99% confidence about the maximum error, if we use the sample proportion as an estimate of the true proportion of claims filed against this insurance company that exceed $1,200?

9.3 In a random sample of 400 industrial accidents, it was found that 231 were due at least partially to unsafe working conditions. Construct a 99% confidence interval for the corresponding true proportion using
(a) Table 9;
(b) the large-sample confidence-interval formula.

9.4 With reference to the preceding exercise, what can we say with 95% confidence about the maximum error if we use the sample proportion to estimate the corresponding true proportion?

9.5 In a sample survey of the "safety explosives" used in certain mining operations, explosives containing potassium nitrate were found to be used in 95 of 250 cases.

   (a) Use Table 9 to construct a 95% confidence interval for the corresponding true proportion.

   (b) If $\frac{95}{250} = 0.38$ is used as an estimate of the corresponding true proportion, what can we say with 95% confidence about the maximum error?

9.6 In a random sample of 160 workers exposed to a certain amount of radiation, 24 experienced some ill effects. Construct a 99% confidence interval for the corresponding true percentage using

   (a) Table 9;

   (b) the large-sample confidence-interval formula.

9.7 In a recent study, 69 of 120 meteorites were observed to enter the earth's atmosphere with a velocity of less than 26 miles per second. If we estimate the corresponding true proportion as $\frac{69}{120} = 0.575$, what can we say with 95% confidence about the maximum error?

9.8 Among 100 fish caught in a large lake, 18 were inedible due to the pollution of the environment. If we use $\frac{18}{100} = 0.18$ as an estimate of the corresponding true proportion, with what confidence can we assert that the error of this estimate is at most 0.065?

9.9 A random sample of 300 shoppers at a supermarket includes 204 who regularly use cents-off coupons. Construct a 98% confidence interval for the probability that any one shopper at the supermarket, selected at random, will regularly use cents-off coupons.

9.10 What is the size of the smallest sample required to estimate an unknown proportion to within a maximum error of 0.06 with at least 95% confidence?

9.11 With reference to the preceding exercise, how would the required sample size be affected if it is known that the proportion to be estimated is at least 0.75?

9.12 Suppose that we want to estimate what percentage of all drivers exceed the 55-mile per hour speed limit on a certain stretch of road. How large a sample will we need to be at least 99% confident that the error of our estimate, the sample percentage, is at most 3.5%?

9.13 With reference to the preceding exercise, how would the required sample size be affected if it is known that the percentage to be estimated is at most 40%?

9.14 Show that the inequality on page 245 leads to the following $(1 - \alpha)100\%$ confidence limits:

$$\frac{x + \frac{1}{2}z_{\alpha/2}^2 \pm z_{\alpha/2}\sqrt{\dfrac{x(n - x)}{n} + \dfrac{1}{4}z_{\alpha/2}^2}}{n + z_{\alpha/2}^2}$$

9.15 Use the formula of the preceding exercise to rework Exercise 9.3.

9.16 Use the formula of Exercise 9.14 to rework Exercise 9.6.

9.17 In 4,000 firings of a certain kind of rocket there were 10 instances in which a rocket exploded upon ignition. Construct an upper 95% confidence limit for the probability that such a rocket will explode upon ignition.

9.18 Observing the amount of pollutants in the air in a western city on 500 days, it was found that it exceeded 200 micrograms per cubic meter only four times. Construct an upper 99% confidence limit for the probability that the air pollution in this city will exceed 200 micrograms per cubic meter on any one day.

9.19 The head of a highway department feels that four out of five road building jobs stay within cost estimates, while his assistant feels that it should be only three out of five.

(a) If the head of the highway department is regarded to be "three times as good" as his assistant in determining figures like these, what prior probabilities should we assign to their claims?

(b) What posterior probabilities should we assign to their claims if it is found that among 12 road building jobs (randomly selected from the department's files) only two stayed within cost estimates?

9.20 The purchasing agent of a firm feels that the probability is 0.80 that any one of several shipments of steel recently received will meet specifications. The head of the firm's quality control department feels that this probability is 0.90, and the chief engineer feels (somewhat more pessimistically) that it is 0.60.

(a) If the managing director of the firm feels that in this matter the purchasing agent is 10 times as reliable as the chief engineer while the head of the quality control department is 14 times as reliable as the chief engineer, what prior probabilities would he assign to their claims?

(b) If five of the shipments are inspected and only two meet specifications, what posterior probabilities should the managing director of the firm assign to the respective claims?

9.21 The output of a certain transistor production line is checked daily by inspecting a sample of 200 units. Over a long period of time, the process has maintained a yield of 80%, that is, a proportion defective of 0.20, and the variation of the proportion defective (from day to day) is measured by a standard deviation of 0.0125. If on a certain day the sample contains 86 defectives, find the mean of the posterior distribution of the proportion defective as an estimate of that day's proportion defective. Assume that the prior distribution of the proportion defective can be approximated closely with a beta distribution.

9.22 Records of the dean of an engineering school (collected over many years) show that on the average 75% of all applicants have an I.Q. of at least 115. Of course, the percentage varies somewhat from year to year and this variation is measured by a standard deviation of 2.15%.

(a) Verify that if the prior distribution of the proportion of applicants with an I.Q. of at least 115 can be approximated closely with a beta distribution, we can use the beta distribution with $\alpha = 300$ and $\beta = 100$.

(b) If a sample check of 25 of this year's applicants shows that only 16 of them have an I.Q. of at least 115, use the results and the assumptions of part (a) to find the mean and the standard deviation of the posterior distribution of the proportion of this year's applicants who have an I.Q. of at least 115.

9.3 HYPOTHESES CONCERNING ONE PROPORTION

Many of the methods used in sampling inspection, quality control, and reliability verification are based on tests of the null hypothesis that a proportion (percentage, or probability) equals some specified constant. The details of the application of such tests to quality control will be discussed in Chapter 14, where we shall also go into some problems of sampling inspection; applications to reliability and life testing will be taken up in Chapter 15.

Although there are exact tests based on the binomial distribution which can be performed with the use of Table 1, we shall consider here only approximate large-sample tests based on the normal approximation to the binomial distribution. In other words, we shall test the null hypothesis $p = p_0$ against one of the alternatives $p < p_0$, $p > p_0$, or $p \neq p_0$ with the use of the statistic

Statistic for large-sample test concerning p

$$z = \frac{x - np_0}{\sqrt{np_0(1 - p_0)}}$$

which is a value of a random variable having approximately the standard normal distribution.† The critical regions are like those shown in the table on page 208 with p and p_0 substituted for μ and μ_0.

EXAMPLE In a study designed to investigate whether certain detonators used with explosives in coal mining meet the requirement that at least 90% will ignite the explosive when charged, it is found that 174 of 200 detonators function properly. Test the null hypothesis $p = 0.90$ against the alternative hypothesis $p < 0.90$ at the 0.05 level of significance.

Solution

1. *Null hypothesis*: $p = 0.90$
   Alternative hypothesis: $p < 0.90$
2. *Level of significance*: $\alpha = 0.05$

---

†Some authors write the numerator of this formula for z as $x \pm \frac{1}{2} - np_0$, whichever is numerically smaller, but there is generally no need for this continuity correction so long as n is large.

3. *Criterion*: Reject the null hypothesis if $z < -1.645$, where

$$z = \frac{x - np_0}{\sqrt{np_0(1 - p_0)}}$$

4. *Calculations*: Substituting $x = 174, n = 200$, and $p_0 = 0.90$ into the formula for z, we get

$$z = \frac{174 - 200(0.90)}{\sqrt{200(0.90)(0.10)}} = -1.41$$

5. *Decision*: Since $z = -1.41$ is not less than -1.645, the null hypothesis cannot be rejected; in other words, there is not sufficient evidence to say that the given kind of detonator fails to meet the required standard.

9.4 HYPOTHESES CONCERNING SEVERAL PROPORTIONS

When we compare the consumer response (percentage favorable and percentage unfavorable) to two different products, when we decide whether the proportion of defectives of a given process remains constant from day to day, when we judge whether there is a difference in political persuasion among several nationality groups, and in many similar situations, we are interested in testing whether two or more binomial populations have the same parameter p. Referring to these parameters as $p_1, p_2, \ldots$, and p_k, we are, in fact, interested in testing the null hypothesis

$$p_1 = p_2 = \ldots p_k = p$$

against the alternative hypothesis that these population proportions are not all equal. To perform a suitable large-sample test of this hypothesis, we require independent random samples of size $n_1, n_2, \ldots$, and n_k from the k populations; then, if the corresponding numbers of "successes" are $x_1, x_2, \ldots$, and x_k, the test we shall use is based on the fact that (1) for large samples the sampling distribution of

$$z_i = \frac{x_i - n_i p_i}{\sqrt{n_i p_i(1 - p_i)}}$$

is approximately the standard normal distribution, (2) the square of a random variable having the standard normal distribution is a random variable having the chi-square distribution with 1 degree of freedom, and (3) the sum of k independent random variables having chi-square distributions with 1 degree of freedom is a random variable having the chi-square distribution with k degrees of freedom. (Proofs of these last two results may be found in the book by John E. Freund and Ronald E. Walpole mentioned in the Bibliography.) Thus,

$$\chi^2 = \sum_{i=1}^{k} \frac{(x_i - n_i p_i)^2}{n_i p_i (1 - p_i)}$$

is a value of a random variable having approximately the chi-square distribution with k degrees of freedom, and in practice we substitute for the p_i, which under the null hypothesis are all equal, the pooled estimate

$$\hat{p} = \frac{x_1 + x_2 + \ldots + x_k}{n_1 + n_2 + \ldots + n_k}$$

Since the null hypothesis should be rejected if the differences between the x_i and the $n_i \hat{p}$ are large, the critical region is $\chi^2 > \chi_\alpha^2$, where χ_α^2 is as defined on page 181 and the number of degrees of freedom is $k - 1$. The loss of one degree of freedom results from substituting for p the estimate $\hat{p}$.

In actual practice, when we compare two or more sample proportions it is convenient to determine the value of the χ^2 statistic by looking at the data as arranged in the following way:

	Sample 1	Sample 2	...	Sample k	Total
Successes	x_1	x_2	...	x_k	x
Failures	$n_1 - x_1$	$n_2 - x_2$	...	$n_k - x_k$	$n - x$
Total	n_1	n_2	...	n_k	n

The notation is the same as before, except for x and n, which represent, respectively, the total number of successes and the total number of trials for all samples combined. With reference to this table, the entry in the cell belonging to the ith row and jth column is called the **observed cell frequency** o_{ij} with $i = 1, 2$ and $j = 1, 2, \ldots, k$.

Under the null hypothesis $p_1 = p_2 = \ldots = p_k = p$, we estimate p, as before, as the total number of successes divided by the total number of trials, which we now write as $\hat{p} = \frac{x}{n}$. Hence, the expected number of successes and failures for the jth sample are estimated by

$$e_{1j} = n_j \cdot \hat{p} = \frac{n_j \cdot x}{n}$$

and

$$e_{2j} = n_j (1 - \hat{p}) = \frac{n_j \cdot (n - x)}{n}$$

The quantities e_{1j} and e_{2j} are called the **expected cell frequencies** for $j = 1, 2,$ ..., k. Note that **the expected frequency for any given cell may be obtained by multiplying the totals of the column and the row to which it belongs, and then dividing by the grand total n.**

In this notation, the χ^2 statistic on page 256, with $\hat{p}$ substituted for the p_i, can be written in the form

Statistic for test concerning difference among proportions

$$\chi^2 = \sum_{i=1}^{2} \sum_{j=1}^{k} \frac{(o_{ij} - e_{ij})^2}{e_{ij}}$$

as the reader will be asked to verify in Exercise 9.40 on page 262. This formula has the advantage that it can easily be extended to the more general case, to be treated in Section 9.5, where each trial permits more than two possible outcomes, and there are, thus, more than two rows in the tabular presentation of the various frequencies.

EXAMPLE Samples of three kinds of materials, subjected to extreme temperature changes, produced the results shown in the following table:

	Material A	Material B	Material C	Total
Crumbled	41	27	22	90
Remained intact	79	53	78	210
Total	120	80	100	300

Use the 0.05 level of significance to test whether, under the stated conditions, the probability of crumbling is the same for the three kinds of materials.

Solution

1. *Null hypothesis*: $p_1 = p_2 = p_3$
   Alternative hypothesis: $p_1, p_2,$ and p_3 are not all equal.
2. *Level of significance*: $\alpha = 0.05$
3. *Criterion*: Reject the null hypothesis if $\chi^2 > 5.991$, the value of $\chi^2_{0.05}$ for $3 - 1 = 2$ degrees of freedom, where χ^2 is given by the formula above.
4. *Calculations*: The expected frequencies for the first two cells of the first row are

$$e_{11} = \frac{90 \cdot 120}{300} = 36 \quad \text{and} \quad e_{12} = \frac{90 \cdot 80}{300} = 24$$

and, as it can be shown that **the sum of the expected frequencies for any row or column equals that of the corresponding observed**

frequencies (see Exercise 9.41 on page 262), we find by subtraction that $e_{13} = 90 - (36 + 24) = 30$, and that the expected frequencies for the second row are $e_{21} = 120 - 36 = 84$, $e_{22} = 80 - 24 = 56$, and $e_{23} = 100 - 30 = 70$. Then, substituting these values together with the observed frequencies into the formula for χ^2, we get

$$\chi^2 = \frac{(41 - 36)^2}{36} + \frac{(27 - 24)^2}{24} + \frac{(22 - 30)^2}{30}$$

$$+ \frac{(79 - 84)^2}{84} + \frac{(53 - 56)^2}{56} + \frac{(78 - 70)^2}{70}$$

$$= 4.575$$

5. *Decision*: Since $\chi^2 = 4.575$ does not exceed 5.991, the null hypothesis cannot be rejected; in other words, the data do not refute the hypothesis that, under the stated conditions, the probability of crumbling is the same for the three kinds of material.

It is customary in problems of this kind to round the expected cell frequencies to the nearest integer or to one decimal. Most of the entries of Table 5 are given to three decimals, but, since rounding errors tend to average out, there is seldom any need to carry more than two decimals when calculating the value of the χ^2 statistic. Also, the test we have been discussing here is only an approximate test since the sampling distribution of the χ^2 statistic is only approximately the chi-square distribution, and it should not be used when one or more of the expected frequencies is less than 5. If this is the case, we can sometimes combine two or more of the samples in such a way that none of the e's is less than 5.

So far, the alternative hypothesis has been that $p_1, p_2, \ldots,$ and p_k are not all equal, and for $k = 2$ this reduces to the alternative hypothesis $p_1 \neq p_2$. In problems where the alternative hypothesis may also be $p_1 < p_2$ or $p_1 > p_2$, we can base the test on the statistic

Statistic for test concerning difference between two proportions

$$z = \frac{\dfrac{x_1}{n_1} - \dfrac{x_2}{n_2}}{\sqrt{\hat{p}(1 - \hat{p})\left(\dfrac{1}{n_1} + \dfrac{1}{n_2}\right)}} \quad \text{with} \quad \hat{p} = \frac{x_1 + x_2}{n_1 + n_2}$$

which, for large samples, is a value of a random variable having approximately the standard normal distribution. The test based on this statistic is equivalent to the one based on the χ^2 statistic on page 257 with $k = 2$, in the sense that the square of this z statistic actually equals the χ^2 statistic (see Exercise 9.42 on page 262). The critical regions for this alternative test of the null hypothesis $p_1 = p_2$ are like those shown in the table on page 208 with p_1 and p_2 substituted for μ and μ_0.

EXAMPLE A study shows that 16 of 200 tractors produced on one assembly line required extensive adjustments before they could be shipped, while the same was true for 14 of 400 tractors produced on another assembly line. At the 0.01 level of significance, does this support the claim that the second production line does superior work?

Solution

1. *Null hypothesis*: $p_1 = p_2$
   Alternative hypothesis: $p_1 > p_2$
2. *Level of significance*: $\alpha = 0.01$
3. *Criterion*: Reject the null hypothesis if $z > 2.33$, where z is given by the above formula.
4. *Calculations*: Substituting $x_1 = 16, n_1 = 20, x_2 = 14, n_2 = 400$, and

$$\hat{p} = \frac{16 + 14}{200 + 400} = 0.05$$

into the formula for z, we get

$$z = \frac{\frac{16}{200} - \frac{14}{400}}{\sqrt{(0.05)(0.95)(\frac{1}{200} + \frac{1}{400})}} = 2.38$$

5. *Decision*: Since $z = 2.38$ exceeds 2.33, the null hypothesis must be rejected; we conclude that the true proportion of tractors requiring extensive adjustments is greater for the first assembly line than for the second.

The test we have described here applies to the null hypothesis $p_1 = p_2$, but it can easily be modified (see Exercise 9.39 on page 261) so that it applies also to the null hypothesis $p_1 - p_2 = \delta$.

EXERCISES

9.23 A manufacturer of submersible pumps claims that at most 30% of his pumps require repairs within the first 5 years of operation. If a random sample of 120 of these pumps includes 47 which required repairs within the first 5 years, test the null hypothesis $p = 0.30$ against the alternative hypothesis $p > 0.30$ at the 0.05 level of significance.

9.24 The performance of a computer is observed over a period of 2 years to check the claim that the probability is 0.20 that its down time will exceed 5 hours in any given week. Testing the null hypothesis $p = 0.20$ against the alternative hypothesis $p \neq 0.20$, what can we conclude at the level of significance $\alpha = 0.05$, if there were only 11 weeks in which the downtime of the computer exceeded 5 hours?

9.25 To check on an ambulance service's claim that at least 40% of its calls are life-threatening emergencies, a random sample was taken from its files, and it was found that only 49 of 150 calls were life-threatening emergencies. Can the null hypothesis $p \geq 0.40$ be rejected against the alternative hypothesis $p < 0.40$ if the probability of a Type I error is to be at most 0.01?

9.26 In a random sample of 600 cars making a right turn at a certain intersection, 157 pulled into the wrong lane. Test the null hypothesis that actually 30% of all drivers make this mistake at the given intersection, using the alternative hypothesis $p \neq 0.30$ and the level of significance
(a) $\alpha = 0.05$;   (b) $\alpha = 0.01$.

9.27 An airline claims that only 6% of all lost luggage is never found. If, in a random sample, 17 of 200 pieces of lost luggage are not found, test the null hypothesis $p = 0.06$ against the alternative hypothesis $p > 0.06$ at the 0.05 level of significance.

9.28 Suppose that 4 of 13 undergraduate engineering students state that they will go on to graduate school. Test the dean's claim that 60% of the undergraduate students will go on to graduate school, using the alternative hypothesis $p < 0.60$ and the level of significance $\alpha = 0.05$. [*Hint*: Use Table 1 to determine the probability of getting "at most 4 successes in 13 trials" when $p = 0.60$.]

9.29 Suppose that we want to test the "honesty" of a coin on the basis of the number of heads we will get in 15 flips. Using Table 1, determine how few or how many heads we would have to get so that we could reject the null hypothesis $p = 0.50$ against the alternative hypothesis $p \neq 0.50$ at the level of significance $\alpha = 0.05$. What is the *actual* level of significance we would be using with this criterion?

9.30 It costs more to test a certain type of ammunition than to manufacture it, and, hence, only three rounds are tested from each large lot. If the lot is rejected unless all three rounds function according to specifications,
(a) sketch the *OC* curve for this test;
(b) find the actual proportion of defectives for which the test procedure will cause a lot to be rejected with a probability of 0.10.

9.31 If 26 of 200 Brand A tires fail to last 20,000 miles, while the corresponding figures for 200 tires each of Brands B, C, and D are 23, 15, and 32, use the 0.05 level of significance to test the null hypothesis that there is no difference in the quality of the four kinds of tires with regard to their durability.

9.32 Tests are made on the proportion of defective castings produced by five different molds. If there were 14 defectives among 100 castings made with Mold I, 33 defectives among 200 castings made with Mold II, 21 defectives among 180 castings made with Mold III, 17 defectives among 120 castings made with Mold IV, and 25 defectives among 150 castings made with Mold V, use the 0.01 level of significance to test whether the true proportion of defectives is the same for each mold.

9.33 A study showed that 64 of 180 persons who saw a photocopying machine advertized during the telecast of a baseball game and 75 of 180 other persons who saw it advertized on a variety show remembered 2 hours later the brand name. Use the χ^2 statistic to test at the 0.05 level of significance whether the difference between the corresponding sample proportions is significant.

9.34 The following data come from a study in which random samples of the employees of three government agencies were asked questions about their pension plan:

	Agency 1	*Agency 2*	*Agency 3*
For the pension plan	67	84	109
Against the pension plan	33	66	41

Use the 0.01 level of significance to test the null hypothesis that the actual proportions of employees favoring the pension plan are the same.

9.35 The owner of a machine shop must decide which of two cigarette-vending machines to install in his shop. If each machine is tested 250 times, the first machine fails to work (neither delivers the cigarettes nor returns the money) 13 times, and the second machine fails to work 7 times, test at the 0.05 level of significance whether the difference between the corresponding sample proportions is significant, using
(a) the χ^2 statistic on page 257;
(b) the z statistic on page 258.

9.36 With reference to the preceding exercise, verify that the square of the value obtained for z in part (b) equals the value obtained for χ^2 in part (a).

9.37 One method of seeding clouds was successful in 57 of 150 attempts while another method was successful in 33 of 100 attempts. At the 0.05 level of significance, can we conclude that the first method is better than the second?

9.38 Two groups of 80 patients each took part in an experiment in which one group received pills containing an antiallergy drug, while the other group received a placebo, a pill containing no drug. If in the group given the drug 23 exhibited allergic symptoms while in the group given the placebo 41 exhibited such symptoms, is this sufficient evidence to conclude at the 0.01 level of significance that the drug is effective in reducing these symptoms?

9.39 To test the null hypothesis that the difference between two population proportions equals some constant δ, not necessarily 0, we can use the statistic

$$z = \frac{\dfrac{x_1}{n_1} - \dfrac{x_2}{n_2} - \delta}{\sqrt{\dfrac{\dfrac{x_1}{n_1}\left(1 - \dfrac{x_1}{n_1}\right)}{n_1} + \dfrac{\dfrac{x_2}{n_2}\left(1 - \dfrac{x_2}{n_2}\right)}{n_2}}}$$

which, for large samples, is a value of a random variable having the standard normal distribution.

(a) With reference to the preceding exercise, use this statistic to test at the 0.05 level of significance whether the true percentage of patients exhibiting allergic symptoms is at least 8% less for those who actually receive the drug.

(b) In a true-false test, a test item is considered to be good if it discriminates between well-prepared students and poorly-prepared students. If 205 of 250 well-prepared students and 137 of 250 poorly-prepared students answer a certain item correctly, test at the 0.01 level of significance whether for the given item the proportion of correct answers can be expected to be at least 15% higher among well-prepared students than among poorly-prepared students.

9.40 Verify that the formulas for the χ^2 statistic on page 257 (with $\hat{p}$ substituted for the p_i) and on page 256 are equivalent.

9.41 Verify that if the expected frequencies are determined in accordance with the rule on page 256, the sum of the expected frequencies for each row and column equals the sum of the corresponding observed frequencies.

9.42 Verify that the square of the z statistic on page 258 equals the χ^2 statistic on page 257 for $k = 2$.

9.5 THE ANALYSIS OF $r \times c$ TABLES

As we suggested earlier, the method by which we analyzed the example on page 257 lends itself also to the analysis of $r \times c$ **tables**, or *r*-by-*c* tables, that is, tables in which data are tallied into a two-way classification having r rows and c columns. Such tables arise in essentially two kinds of problems. First, we might again have samples from several populations, with the distinction that now each trial permits more than two possible outcomes. This might happen, for example, if persons belonging to different income groups are asked whether they favor a certain political candidate, whether they are against him, or whether they are indifferent or undecided. The other situation giving rise to an $r \times c$ table is one in which we sample from one population but classify each item with respect to two (usually qualitative) categories. This might happen, for example, if a consumer testing service rates cars as excellent, superior, average, or poor with regard to performance and also with regard to appearance. Each car tested would then fall into one of the 16 cells of a 4×4 table, and it is mainly in connection with problems of this kind that $r \times c$ tables are referred to as **contingency tables**.

The essential difference between the two kinds of situations giving rise to $r \times c$ tables is that in the first case the column totals (the sample sizes) are fixed, while in the second case only the **grand total** (the total for the entire table) is fixed. As a result, there are also differences in the null hypotheses we shall want to test. In the first case we want to test whether the probability of obtaining an observation in the ith row is the same for each column; symbolically, we shall want to test the null hypothesis

$$p_{i1} = p_{i2} = \ldots = p_{ic} \quad \text{for } i = 1, 2, \ldots, r$$

where p_{ij} is the probability of obtaining an observation belonging to the ith row and the jth column, and $\sum_{i=1}^{r} p_{ij} = 1$ for each column. The alternative hypothesis is that the p's are not all equal for at least one row. In the second case we shall want to test the null hypothesis that the random variables represented by the two classifications are independent, so that p_{ij} is the product of the probability of getting a value belonging to the ith row and the probability of getting a value belonging to the jth column. The alternative hypothesis is that the two random variables are not independent.

In spite of the differences we have described, the analysis of an $r \times c$ table is the same for both cases. First we calculate the expected cell frequencies e_{ij} as on page 257, namely, by multiplying the totals of the respective rows and columns and then dividing by the grand total. In practice, we make use of the fact that the observed frequencies and the expected frequencies total the same for each row and column, so that only $(r - 1)(c - 1)$ of the e_{ij} have to be calculated directly, while the others can be obtained by subtraction from appropriate row or column totals. We then substitute into the formula

Statistic for analysis of r × c table

$$\chi^2 = \sum_{i=1}^{r} \sum_{j=1}^{c} \frac{(o_{ij} - e_{ij})^2}{e_{ij}}$$

and we reject the null hypothesis if the value of this statistic exceeds χ_α^2 for $(r - 1)(c - 1)$ degrees of freedom. This expression for the number of degrees of freedom is justified by the above observation that after we determine $(r - 1)(c - 1)$ of the expected cell frequencies, the others are automatically determined, that is, they may be obtained by subtraction from appropriate row or column totals.

EXAMPLE To determine whether there really is a relationship between an employee's performance in the company's training program and his or her ultimate success in the job, it takes a sample of 400 cases from its very extensive files and obtains the results shown in the following table:

		Performance in training program			
		Below average	*Average*	*Above average*	*Total*
Success in job (Employer's rating)	*Poor*	23	60	29	112
	Average	28	79	60	167
	Very good	9	49	63	121
	Total	60	188	152	400

Use the 0.01 level of significance to test the null hypothesis that performance in the training program and success in the job are independent.

Solution

1. *Null hypothesis*: Performance in training program and success in job are independent.

   Alternative hypothesis: Performance in training program and success in job are not independent.
2. *Level of significance*: $\alpha = 0.01$
3. *Criterion*: Reject the null hypothesis if $\chi^2 > 13.277$, the value of $\chi^2_{0.01}$ for $(3 - 1)(3 - 1) = 4$ degrees of freedom, where χ^2 is given by the formula on page 263.
4. *Calculations*: Calculating first the expected cell frequencies for the first two cells of the first two rows, we get

$$e_{11} = \frac{112 \cdot 60}{400} = 16.8, \qquad e_{12} = \frac{112 \cdot 188}{400} = 52.6,$$

$$e_{21} = \frac{167 \cdot 60}{400} = 25.0, \qquad e_{22} = \frac{167 \cdot 188}{400} = 78.5.$$

Then, by subtraction, we find that the expected frequencies for the third cell of the first two rows are 42.6 and 63.5, and those for the third row are 18.2, 56.9, and 45.9. Thus,

$$\chi^2 = \frac{(23 - 16.8)^2}{16.8} + \frac{(60 - 52.6)^2}{52.6} + \frac{(29 - 42.6)^2}{42.6}$$

$$+ \frac{(28 - 25.0)^2}{25.0} + \frac{(79 - 78.5)^2}{78.5} + \frac{(60 - 63.5)^2}{63.5}$$

$$+ \frac{(9 - 18.2)^2}{18.2} + \frac{(49 - 56.9)^2}{56.9} + \frac{(63 - 45.9)^2}{45.9}$$

$$= 20.34$$

5. *Decision*: Since $\chi^2 = 20.34$ exceeds 13.277, the null hypothesis must be rejected; we conclude that there is a dependence between an employee's performance in the training program and his or her success in the job.

9.6 GOODNESS OF FIT

We speak of **goodness of fit** when we try to compare an observed frequency distribution with the corresponding values of an expected, or theoretical, distribution. To illustrate, suppose that during 400 five-minute intervals the air-traffic control of an airport received 0, 1, 2, . . . , or 13 radio messages with respective frequencies of 3, 15, 47, 76, 68, 74, 46, 39, 15, 9, 5, 2, 0, and 1. Suppose, furthermore, that we want to check whether these data substantiate the claim that the number of radio messages which they receive during a 5-minute interval may be looked upon as a random variable having the Poisson distribution with $\lambda = 4.6$. Looking up the corresponding Poisson probabilities in Table 2 and multiplying them by 400 to get the expected frequencies, we arrive at the result shown in the following table together with the original data:

Number of radio messages	Observed frequencies	Poisson probabilities	Expected frequencies
0	3 ⎫ 18	0.010	4.0 ⎫ 22.4
1	15 ⎭	0.046	18.4 ⎭
2	47	0.107	42.8
3	76	0.163	65.2
4	68	0.187	74.8
5	74	0.173	69.2
6	46	0.132	52.8
7	39	0.087	34.8
8	15	0.050	20.0
9	9	0.025	10.0
10	5 ⎫	0.012	4.8 ⎫
11	2 ⎪ 8	0.005	2.0 ⎪ 8.0
12	0 ⎪	0.002	0.8 ⎪
13	1 ⎭	0.001	0.4 ⎭
	400		400.0

Note that we combined some of the data so that none of the expected frequencies is less than 5.

To test whether the discrepancies between the observed and expected frequencies can be attributed to chance, we use the statistic

Statistic for test of goodness of fit

$$\chi^2 = \sum_{i=1}^{k} \frac{(o_i - e_i)^2}{e_i}$$

where the o_i and e_i are the observed and expected frequencies. The sampling distribution of this statistic is approximately the chi-square distribution with $k - m$ degrees of freedom, where k is the number of terms in the formula for χ^2 and m is the number of quantities, obtained from the observed data, that are needed to calculate the expected frequencies.

EXAMPLE With reference to the illustration on page 265, test at the 0.01 level of significance whether the data can be looked upon as values of a random variable having the Poisson distribution with $\lambda = 4.6$.

Solution

1. *Null hypothesis*: Random variable has Poisson distribution with $\lambda = 4.6$.
   Alternative hypothesis: Random variable does not have Poisson distribution with $\lambda = 4.6$.
2. *Level of significance*: $\alpha = 0.01$
3. *Criterion*: Reject the null hypothesis if $\chi^2 > 16.919$, the value of $\chi^2_{0.01}$ for $k - m = 10 - 1 = 9$ degrees of freedom, where χ^2 is given by the formula above. (The number of degrees of freedom is $10 - 1 = 9$, since only one quantity, the total frequency of 400, is needed from the observed data to calculate the expected frequencies.)
4. *Calculations*: Substitution into the formula for χ^2 yields

$$\chi^2 = \frac{(18 - 22.4)^2}{22.4} + \frac{(47 - 42.8)^2}{42.8} + \cdots$$

$$+ \frac{(9 - 10.0)^2}{10.0} + \frac{(8 - 8.0)^2}{8.0}$$

$$= 6.749$$

5. *Decision*: Since $\chi^2 = 6.749$ does not exceed 16.919, the null hypothesis cannot be rejected; we conclude that the Poisson distribution with $\lambda = 4.6$ provides a good fit.

EXERCISES

9.43 The results of polls conducted two weeks and four weeks before a gubernatorial election are shown in the following table:

	Two weeks before election	Four weeks before election
For Republican candidate	79	91
For Democratic candidate	84	66
Undecided	37	43

Use the 0.05 level of significance to test whether there has been a change in opinion during the 2 weeks between the two polls.

9.44 Suppose that in Exercise 9.31 on page 260 we had been interested also in how many of the tires lasted more than 30,000 miles, and obtained the results shown in the following table:

	Brand A	Brand B	Brand C	Brand D
Failed to last 20,000 miles	26	23	15	32
Lasted from 20,000 to 30,000	118	93	116	121
Lasted more than 30,000 miles	56	84	69	47

Use the 0.01 level of significance to test the null hypothesis that there is no difference in the quality of the four kinds of tires with regard to their durability.

9.45 A large electronics firm which hires many handicapped workers wants to determine whether their handicaps affect such workers' performance. Use the level of significance $\alpha = 0.05$ to decide on the basis of the sample data shown in the following table whether it is reasonable to maintain that the handicaps have no effect on the workers' performance:

	Performance		
	Above average	*Average*	*Below average*
Blind	21	64	17
Deaf	16	49	14
No handicap	29	93	28

9.46 Tests of the fidelity and the selectivity of 190 radio receivers produced the results shown in the following table:

		Fidelity		
		Low	*Average*	*High*
Selectivity	*Low*	6	12	32
	Average	33	61	18
	High	13	15	0

Use the 0.01 level of significance to test whether there is a relationship (dependence) between fidelity and selectivity.

9.47 A quality control engineer takes daily samples of $n = 4$ tractors coming off an assembly line and on 200 consecutive working days he obtains the data summarized in the following table:

Number requiring adjustments	*Number of days*
0	101
1	79
2	19
3	1

To test the claim that 10% of all the tractors coming off this assembly line require adjustments, look up the corresponding probabilities in Table 1, calculate the expected frequencies, and perform the chi-square test at the 0.01 level of significance.

9.48 With reference to the preceding exercise, verify that the mean of the observed distribution is 0.60, corresponding to 15% of the tractors requiring adjustments. Then look up the probabilities for $n = 4$ and $p = 0.15$ in Table 1, calculate the expected frequencies, and test at the 0.01 level of significance whether the binomial distribution with $n = 4$ and $p = 0.15$ provides a suitable model for this situation.

9.49 Suppose that in the example on page 265 we had shown first that the mean of the distribution, rounded to one decimal, is 4.5, and then tested whether the Poisson distribution with $\lambda = 4.5$ provides a good fit. What would have been the number of degrees of freedom for the appropriate chi-square criterion?

9.50 The following is the distribution of the daily number of power failures reported in a western city on 300 days:

Number of power failures	Number of days
0	9
1	43
2	64
3	62
4	42
5	36
6	22
7	14
8	6
9	2

Test at the 0.05 level of significance whether the daily number of power failures in this city is a random variable having the Poisson distribution with $\lambda = 3.2$.

9.51 The following is the distribution of the hourly number of trucks arriving at a company's warehouse:

Trucks arriving per hour	Frequency
0	52
1	151
2	130
3	102
4	45
5	12
6	5
7	1
8	2

Find the mean of this distribution, and using it (rounded to one decimal) as the parameter λ, fit a Poisson distribution. Test for goodness of fit at the 0.05 level of significance.

9.52 Using any four columns of Table 7 (that is, a total of 200 random digits), construct a table showing how many times each of the digits 0, 1, . . . , and 9 occurred. Comparing the observed frequencies with the corresponding expected frequencies (based on the assumption that the digits are randomly generated), test at the 0.05 level of significance whether the assumption of randomness is tenable.

9.53 The following is the distribution of the sulfur oxides emission data on page 138, for which we showed that $\bar{x} = 18.85$ and $s = \sqrt{30.77} = 5.55$:

Class limits (tons)	Frequency
5.0– 8.9	3
9.0–12.9	10
13.0–16.9	14
17.0–20.9	25
21.0–24.9	17
25.0–28.9	9
29.0–32.9	2
	80

(a) Find the probabilities that a random variable having a normal distribution with $\mu = 18.85$ and $\sigma = 5.55$ takes on a value less than 8.95, between 8.95 and 12.95, between 12.95 and 16.95, between 16.95 and 20.95, between 20.95 and 24.95, between 24.95 and 28.95, and greater than 28.95.

(b) Multiply the probabilities obtained in part (a) by the total frequency, $n = 80$, thus getting the expected normal curve frequencies corresponding to the seven classes of the given distribution (with the first one changed to "8.9 or less" and the last one changed to "29.0 or more").

(c) Use the 0.05 level of significance to test the null hypothesis that the given data may be looked upon as a random sample from a normal population. Explain why the number of degrees of freedom for this χ^2 test is $k - 3$, where k is the number of terms in the χ^2 statistic.

9.54 Among 100 vacuum tubes used in an experiment, 46 had a service life of less than 20 hours, 19 had a service life of 20 or more but less than 40 hours, 17 had a service life of 40 or more but less than 60 hours, 12 had a service life of 60 or more but less than 80 hours, and 6 had a service life of 80 hours or more. Using steps similar to those outlined in the preceding exercise, test at the 0.01 level of significance whether the lifetimes may be regarded as a sample from an exponential population with $\mu = 40$ hours.

9.7 CHECK LIST OF KEY TERMS *(with page references)*

Contingency table, 262
Expected cell frequency, 257
Goodness of fit, 265
Observed cell frequency, 256

One-sided confidence interval, 248
r × c table, 262
Sample proportion, 243

10

Nonparametric Tests

Most of the methods of inference which we have studied are based on the assumption that the observations come from normal populations. If this is the case, these methods extract all the information that is available in a sample, and they usually attain the best possible precision. However, since there are many situations where it is doubtful whether the assumption of normality can be met, statisticians have developed alternative techniques based on less stringent assumptions, which have become known as non-parametric tests. Not only can these methods be used under more general conditions, but they are often easier to explain and understand; that is, they come under the heading of "quick and easy" or "shortcut" techniques.

Of the many nonparametric tests that have been developed in recent years, the sign test is presented in Section 10.2; tests based on rank sums are given in Section 10.3; a test of randomness is introduced in Section 10.4; and a goodness of fit test is given in Section 10.5.

10.1 INTRODUCTION

If there is a choice between several statistical methods which can all be used in a given situation, the criterion of **efficiency** is most commonly used as an appropriate guide. If we think of the methods based on the assumption of normality as being fully (100%) efficient, we can use this as a yardstick for measuring the merits of any other method. The most widely used measure of efficiency is based on the sample sizes required to give equally precise results by a given method and by the corresponding method which is fully efficient. For example, in estimating the mean of a normal population, the most efficient method involves use of the sample mean $\bar{x}$. If we wanted to use the median instead of the mean, we would have to take into account that the variance of the sampling distribution of the median is approximately $1.57\frac{\sigma^2}{n}$, so that the efficiency of the median is $1/1.57$ or approximately 64%. In other words, the median based on a sample of size 100 gives as reliable an estimate of the mean of a normal population as the mean based on a sample of size 64. Similarly, it can be shown that the efficiency of the range estimator for the standard deviation of a normal population decreases as the sample size increases; the efficiency is 100% for samples of size 2, 96% for samples of size 5, and 81% for samples of size 15.

The main advantage of **nonparametric tests** is that exact inferences can be made when the assumptions underlying so-called standard methods cannot all be met; their main disadvantage is that they may be wasteful of information and have a smaller efficiency when these assumptions can be met. However, if we say that the efficiency of a certain nonparametric method is 80%, we may actually be understating its relative worth, for the efficiency of the corresponding "standard" method will be somewhat less than 100% if all assumptions are not exactly met.

10.2 THE SIGN TEST

In this section we shall describe a nonparametric alternative to the one-sample t test, the paired-sample t test, and corresponding large-sample tests. As an alternative to the one-sample t test or the corresponding large-sample test, the **sign test** applies when we sample a continuous symmetrical population, so that the probability of getting a sample value less than the mean and the probability of getting a sample value greater than the mean are both $\frac{1}{2}$. To test the null hypothesis $\mu = \mu_0$ against an appropriate alternative on the basis of a random sample of size n, we replace each sample value greater than μ_0 with a plus sign and each sample value less than μ_0 with a minus sign, and then we test the null hypothesis that these plus and minus signs are the outcomes of binomial trials with $p = \frac{1}{2}$. If a sample value equals μ_0, which may well happen since the values of continuous random valuables are virtually always rounded, it is discarded.

To perform this kind of test when the sample is small, we refer directly to a table of binomial probabilities such as Table 1 at the end of the book; when the sample is large we use the test described in Section 9.3.

EXAMPLE The following data constitute a random sample of 15 measurements of the octane rating of a certain kind of gasoline:

$$99.0 \quad 102.3 \quad 99.8 \quad 100.5 \quad 99.7 \quad 96.2 \quad 99.1 \quad 102.5$$
$$103.3 \quad 97.4 \quad 100.4 \quad 98.9 \quad 98.3 \quad 98.0 \quad 101.6$$

Test the null hypothesis $\mu = 98.0$ against the alternative hypothesis $\mu > 98.0$ at the 0.01 level of significance.

> *Solution* Since one of the sample values equals 98.0 and must be discarded, the sample size for the sign test is only $n = 14$.
>
> 1. *Null hypothesis*: $\mu = 98.0\ (p = \frac{1}{2})$
>   *Alternative hypothesis*: $\mu > 98.0\ (p > \frac{1}{2})$
> 2. *Level of significance*: $\alpha = 0.01$
> 3. *Criterion*: The criterion may be based on the number of plus signs or the number of minus signs. Using the number of plus signs, denoted by x, reject the null hypothesis if the probability of getting x or more plus signs is less than or equal to 0.05.
> 4. *Calculations*: Replacing each value greater than 98.0 with a plus sign and each value less than 98.0 with a minus sign, the 14 sample values yield
>
> $$+ + + + + - + + + - + + + +$$
>
> Thus, $x = 12$ and Table 1 shows that for $n = 14$ and $p = 0.50$ the probability of $x \geq 12$ is $1 - 0.9935 = 0.0065$.
> 5. *Decision*: Since 0.0065 is less than 0.01, the null hypothesis must be rejected; we conclude that the mean octane rating of the given kind of gasoline exceeds 98.0.

The sign test can also be used as a nonparametric alternative to the paired-sample *t* test or the corresponding large-sample test. In such problems, each pair of sample values is replaced with a plus sign if the first value is greater than the second, with a minus sign if the first value is smaller than the second, or it is discarded if the two values are equal. The procedure is the same as before.

EXAMPLE   With reference to the example on page 223, which dealt with the effectiveness of an industrial safety program, use the sign test at the 0.05 level of significance to test whether the safety program is effective.

> ***Solution***
>
> 1. *Null hypothesis*: $\mu_1 - \mu_2 = 0$ $(p = \frac{1}{2})$
>   *Alternative hypothesis*: $\mu_1 - \mu_2 > 0$ $(p > \frac{1}{2})$
> 2. *Level of significance*: $\alpha = 0.05$
> 3. *Criterion*: If x is the number of plus signs, reject the null hypothesis if the probability of getting x or more plus signs is less than or equal to 0.05.
> 4. *Calculations*: Replacing each pair of values with a plus sign if the first value is greater than the second or with a minus sign if the first value is smaller than the second, the 10 sample pairs yield
>
> $$+ + + + - + + + + +$$
>
> Thus, $x = 9$ and Table 1 shows that for $n = 10$ and $p = 0.50$ the probability of $x \geq 9$ is $1 - 0.9990 = 0.0010$.
> 5. *Decision*: Since 0.0010 is less than 0.05, the null hypothesis must be rejected; we conclude that the safety program is effective.

10.3   RANK-SUM TESTS

In this section we shall introduce two tests based on **rank sums**—the U **test** will be presented as a nonparametric alternative to the two-sample t test, and the H **test** will be presented as a nonparametric alternative to the one-way analysis of variance, which we shall study in Chapter 12. In other words, the H test serves to test the null hypothesis that k samples come from identical populations against the alternative that the populations have unequal means.

To illustrate how the U test (also called the **Wilcoxon test** or the **Mann–Whitney test**, named after the statisticians who contributed to its development) is performed, suppose that in a study of sedimentary rocks, the following diameters (in millimeters) were obtained for two kinds of sand:

> *Sand I:*   0.63,   0.17,   0.35,   0.49,   0.18,   0.43,   0.12,   0.20,
>   0.47,   1.36,   0.51,   0.45,   0.84,   0.32,   0.40
> *Sand II:* 1.13,   0.54,   0.96,   0.26,   0.39,   0.88,   0.92,   0.53,
>   1.01,   0.48,   0.89,   1.07,   1.11,   0.58

The means of these two samples are 0.46 and 0.77, and the problem is to decide whether their difference is significant.

We begin the U test by ranking the data jointly, as if they comprise one sample, in an increasing order of magnitude, and for our data we get

0.12	0.17	0.18	0.20	0.26	0.32	0.35	0.39	0.40	0.43
I	I	I	I	II	I	I	II	I	I

0.45	0.47	0.48	0.49	0.51	0.53	0.54	0.58	0.63	0.84
I	I	II	I	I	II	II	II	I	I

0.88	0.89	0.92	0.96	1.01	1.07	1.11	1.13	1.36
II	II	II	II	II	II	II	II	I

Note that we indicated for each value whether it is a measurement of Sand I or Sand II. Assigning the data in this order the ranks $1, 2, 3, \ldots$, and 29, we find that the values of the first sample (Sand I) occupy ranks 1, 2, 3, 4, 6, 7, 9, 10, 11, 12, 14, 15, 19, 20, and 29, while those of the second sample (Sand II) occupy ranks 5, 8, 13, 16, 17, 18, 21, 22, 23, 24, 25, 26, 27, and 28. There are no ties here among values belonging to different samples, but if there were, we would assign to each of the tied observations the mean of the ranks which they jointly occupy. (For instance, if the third and fourth values are identical we would assign each the rank $\frac{3 + 4}{2} = 3.5$, and if the ninth, tenth, and eleventh values are identical we would assign each the rank $\frac{9 + 10 + 11}{3} = 10$.)

The null hypothesis we shall want to test is that the two samples come from identical populations, and it stands to reason that in that case the means of the ranks assigned to the values of the two samples should be more or less the same. Instead of the means, we can also compare the sums of the ranks assigned to the values of the two samples, suitably accounting for a possible difference in their size. For our two samples, the sums of the ranks are $R_1 = 162$ and $R_2 = 273$, and it remains to be seen whether their difference is large enough to reject the null hypothesis.

When the use of rank sums was first proposed as a nonparametric alternative to the two-sample t test, the decision was based on R_1 or R_2, but now the decision is based on either of the related statistics

U_1 and U_2 statistics

$$U_1 = n_1 n_2 + \frac{n_1(n_1 + 1)}{2} - R_1$$

or

$$U_2 = n_1 n_2 + \frac{n_2(n_2 + 1)}{2} - R_2$$

or on the statistic U, which always equals the smaller of the two. The sizes of the two samples are n_1 and n_2, and as it does not matter how we number the samples, we shall use here the statistic U_1.†

Under the null hypothesis that the two samples come from identical populations, it can be shown that the mean and the variance of the sampling distribution of U_1 are

Mean and variance of U_1 statistic

$$\mu_{U_1} = \frac{n_1 n_2}{2}$$

and

$$\sigma^2_{U_1} = \frac{n_1 n_2 (n_1 + n_2 + 1)}{12}$$

If there are ties in rank, these formulas provide only approximations, but if the number of ties is small, these approximations will generally be good.

Since numerical studies have shown that the sampling distribution of U_1 can be approximated closely by a normal distribution when n_1 and n_2 are both greater than 8, the test of the null hypothesis that the two samples come from identical populations can be based on

Statistic for large-sample U test

$$z = \frac{U_1 - \mu_{U_1}}{\sigma_{U_1}}$$

which is a value of a random variable having approximately the standard normal distribution. For small samples, we can base the test on special tables; for instance, on those in the handbook by D. B. Owen listed in the Bibliography.

Note that when we test the null hypothesis that the two samples come from identical populations against the alternative hypothesis $\mu_1 > \mu_2$, we reject the null hypothesis if $z < -z_\alpha$ since small values of U_1 correspond to large values of R_1; correspondingly, if the alternative hypothesis is $\mu_1 < \mu_2$, we reject the null hypothesis if $z > z_\alpha$ since large values of U_1 correspond to small values of R_1.

---

†The tests based on U_1 or U_2 are equivalent to those based on R_1 or R_2, but they have the advantage that they lend themselves more readily to the construction of tables of critical values. Not only do U_1 and U_2 take on values on the interval from 0 to $n_1 n_2$— indeed, their sum is always equal to $n_1 n_2$—but their sampling distributions are symmetrical about $\frac{n_1 n_2}{2}$. The use of U, which always equals the smaller of the values of U_1 and U_2, has the added advantage that the resulting test is one-tailed regardless of the alternative hypothesis, and hence easier to tabulate.

EXAMPLE With reference to the grain-size data on page 274, use the U test at the 0.01 level of significance to test the null hypothesis that the two samples come from identical populations against the alternative hypothesis that they have unequal means.

   Solution

   1. *Null hypothesis*: Populations are identical.
      Alternative hypothesis: $\mu_1 \neq \mu_2$
   2. *Level of significance*: $\alpha = 0.01$
   3. *Criterion*: Reject the null hypothesis if $z < -2.575$ or $z > 2.575$, where z is given by the above formula.
   4. *Calculations*: Since $n_1 = 15$, $n_2 = 14$, and we have already shown that $R_1 = 162$, we find that

$$U_1 = 15 \cdot 14 + \frac{15 \cdot 16}{2} - 162 = 168$$

$$\mu_{U_1} = \frac{15 \cdot 14}{2} = 105$$

   and

$$\sigma^2_{U_1} = \frac{15 \cdot 14 \cdot 30}{12} = 525$$

   and it follows that

$$z = \frac{168 - 105}{\sqrt{525}} = 2.74$$

   5. *Decision*: Since $z = 2.74$ exceeds 2.575, the null hypothesis must be rejected; we conclude that there is a difference in the true mean grain size of the two kinds of sand.

The **H test**, or **Kruskal–Wallis test**, is a generalization of the U test in that it enables us to test the null hypothesis that k independent random samples come from identical populations. As in the U test, all the observations are ranked jointly, and if R_i is the sum of the ranks occupied by the n_i observations of the ith sample and $n_1 + n_2 + \ldots + n_k = n$, the test is based on the statistic

Statistic for
H test

$$H = \frac{12}{n(n+1)} \sum_{i=1}^{k} \frac{R_i^2}{n_i} - 3(n+1)$$

When $n_i > 5$ for all i and the null hypothesis is true, the sampling distribution of the H statistic is well approximated by the chi-square distribution with $k - 1$ degrees of freedom. There exist special tables of critical values for the H test for selected small values of the n_i and k.

EXAMPLE   An experiment designed to compare three preventive methods against corrosion yielded the following maximum depths of pits (in thousandths of an inch) in pieces of wire subjected to the respective treatments:

> *Method A:*   77,   54,   67,   74,   71,   66
> *Method B:*   60,   41,   59,   65,   62,   64,   52
> *Method C:*   49,   52,   69,   47,   56

Use the 0.05 level of significance to test the null hypothesis that the three samples come from identical populations.

Solution

1. *Null hypothesis*: Populations are identical.
   Alternative hypothesis: Population means are not all equal.
2. *Level of significance*: $\alpha = 0.05$
3. *Criterion*: Reject the null hypothesis if $H > 5.991$, the value of $\chi^2_{0.05}$ for 2 degrees of freedom, where H is given by the formula on page 277.
4. *Calculations*: Ranking these measurements jointly from smallest to largest, we find that those of the first sample occupy ranks 6, 13, 14, 16, 17, and 18; those of the second sample occupy ranks 1, 4.5, 8, 9, 10, 11, and 12; and those of the third sample occupy ranks 2, 3, 4.5, 7, and 15. Thus, $R_1 = 84$, $R_2 = 55.5$, $R_3 = 31.5$, and substitution into the formula for H yields

$$H = \frac{12}{18 \cdot 19}\left(\frac{84^2}{6} + \frac{55.5^2}{7} + \frac{31.5^2}{5}\right) - 3 \cdot 19$$

$$= 6.7$$

5. Since $H = 6.7$ exceeds 5.991, the null hypothesis must be rejected; we conclude that the three preventive methods against corrosion are not equally effective.

EXERCISES

10.1 In a laboratory experiment, 18 determinations of the coefficient of friction between leather and metal yielded the following results: 0.59, 0.56, 0.49, 0.55, 0.65, 0.55, 0.51, 0.60, 0.56, 0.47, 0.58, 0.61, 0.54, 0.68, 0.56, 0.50, 0.57, and 0.53. Use the sign test at the 0.05 level of significance to test the null hypothesis $\mu = 0.55$ against the alternative hypothesis $\mu \neq 0.55$.

10.2 A random sample of nine women buying new eyeglasses tried on 12, 11, 14, 15, 10, 14, 11, 8, and 12 different frames. Use the sign test at the 0.05 level of significance to test the null hypothesis $\mu = 10$ (that on the average a woman buying new eyeglasses tries on 10 different frames) against the alternative hypothesis $\mu > 10$.

10.3 With reference to Exercise 5.15, which pertained to the boiling point of a certain silicon compound, use the sign test at the 0.05 level of significance to test the null hypothesis $\mu = 158$ (degrees Celsius) against the alternative hypothesis $\mu \neq 158$.

10.4 The quality control department of a large manufacturer obtained the following sample data (in pounds) on the breaking strength of a certain kind of 2-inch cotton ribbon: 153, 159, 144, 160, 158, 153, 171, 162, 159, 137, 159, 159, 148, 162, 154, 159, 160, 157, 140, 168, 163, 148, 151, 153, 157, 155, 148, 168, 162, and 149. Use the sign test at the 0.01 level of significance to test the null hypothesis $\mu = 150$ against the alternative hypothesis $\mu > 150$.

10.5 With reference to Exercise 5.9 on page 149, which pertained to the ignition times of certain upholstery materials, use the sign test at the 0.01 level of significance to test the null hypothesis $\mu = 6.50$ seconds against the alternative hypothesis $\mu < 6.50$ seconds.

10.6 The following are the number of speeding tickets issued by two policemen on 17 days: 7 and 10, 11 and 13, 14 and 14, 11 and 15, 12 and 9, 6 and 10, 9 and 13, 8 and 11, 10 and 11, 11 and 15, 13 and 11, 7 and 10, 8 and 8, 11 and 12, 9 and 14, 10 and 9, 13 and 16. Use the sign test at the 0.05 level of significance to test the null hypothesis that on the average the two policemen issue equally many speeding tickets per day against the alternative hypothesis that on the average the second policeman issues more than the first.

10.7 With reference to Exercise 7.72 on page 227, use the sign test at the 0.10 level of significance to check whether there is a systematic difference between weights obtained with the two scales.

10.8 With reference to Exercise 7.73 on page 227, use the sign test at the 0.01 level of significance to check whether the prescribed program of exercises is effective.

10.9 The following are the number of minutes it took a sample of 15 men and 12 women to complete the application form for a position:

Men: 16.5, 20.0, 17.0, 19.8, 18.5, 19.2, 19.0, 18.2, 20.8, 18.7, 16.7, 18.1, 17.9, 16.4, 18.9

Women: 18.6, 17.8, 18.3, 16.6, 20.5, 16.3, 19.3, 18.4, 19.7, 18.8, 19.9, 17.6

Use the U test at the 0.05 level of significance to test the null hypothesis that the two samples come from identical populations against the alternative that the two populations have unequal means.

10.10 With reference to part (a) of Exercise 7.71 on page 226, use the U test at the 0.05 level of significance to test whether the true average hardness of the two alloys is the same.

10.11 Comparing two kinds of emergency flares, a consumer testing service obtained the following burning times (rounded to the nearest tenth of a minute):

Brand C: 19.4, 21.5, 15.3, 17.4, 16.8, 16.6, 20.3, 22.5, 21.3, 23.4, 19.7, 21.0

> *Brand D:*   16.5, 15.8, 24.7, 10.2, 13.5, 15.9, 15.7, 14.0, 12.1, 17.4, 15.6, 15.8

Use the *U* test at the 0.01 level of significance to check whether it is reasonable to say that there is no difference between the true average burning times of the two kinds of flares.

10.12 To find the best arrangement of instruments on a control panel of an airplane, two different arrangements were compared by simulating an emergency condition and measuring the reaction time required to correct the condition. The reaction times (in tenths of a second) of 20 pilots (randomly assigned to the two different arrangements) were as follows:

> *Arrangement 1:*   8, 15, 10, 13, 17, 10, 9, 11, 12, 15
> *Arrangement 2:*   12, 7, 13, 8, 14, 6, 16, 7, 10, 9

Use the *U* test at the 0.05 level of significance to check the claim that the second arrangement is better than the first.

10.13 The following are the scores which random samples of students from two minority groups obtained on a current events test:

> *Minority Group 1:*   73, 82, 39, 68, 91, 75, 89, 67, 50, 86, 57, 65, 70
> *Minority Group 2:*   51, 42, 36, 53, 88, 59, 49, 66, 25, 64, 18, 76, 74

Use the *U* test at the 0.05 level of significance to test whether or not students from the two minority groups can be expected to score equally well on the test.

10.14 The following are data on the breaking strength (in pounds) of two kinds of material:

> *Material 1:*   144, 181, 200, 187, 169, 171, 186, 194, 176, 182, 133, 183, 197, 165, 180, 198
> *Material 2:*   175, 164, 172, 194, 176, 198, 154, 134, 169, 164, 185, 159, 161, 189, 170, 164

Use the *U* test at the 0.05 level of significance to test the claim that Material 1 is on the average stronger than Material 2.

10.15 The following are the numbers of misprints counted on pages selected at random from three Sunday editions of a newspaper:

> *April 11:*   4, 10, 2, 6, 4, 12
> *April 18:*   8, 5, 13, 8, 8, 10
> *April 25:*   7, 9, 11, 2, 14, 7

Use the *H* test at the 0.05 level of significance to test the null hypothesis that the three samples come from identical populations.

10.16 The following are the miles per gallon which a test driver got for 10 tankfuls each of three brands of gasoline:

Brand 1: 22, 25, 32, 18, 23, 15, 30, 27, 19, 23
Brand 2: 19, 22, 18, 29, 28, 32, 17, 33, 28, 20
Brand 3: 30, 29, 25, 24, 15, 27, 30, 27, 18, 32

Use the *H* test at the 0.05 level of significance to test whether there is a difference in the true average performance of the three brands of gasoline.

10.17 So-called Franklin tests were performed to determine the insulation properties of grain-oriented silicon steel specimens that were annealed in five different atmospheres with the following results:

Atmosphere	*Test results (amperes)*						
1	0.58	0.61	0.69	0.79	0.61	0.59	
2	0.37	0.37	0.58	0.40	0.28	0.44	0.35
3	0.29	0.19	0.34	0.17	0.29	0.16	
4	0.81	0.69	0.75	0.72	0.68	0.85	0.57	0.77
5	0.26	0.34	0.29	0.47	0.30	0.42	

Use the *H* test at the 0.05 level of significance to decide whether or not these five samples can be assumed to come from identical populations.

10.18 A panel of seven experts was asked to rate each of five industries on the likelihood that technological changes would produce improvement in environmental pollution over the next 10 years. Their ratings (in the form of judgmental probabilities) are as follows:

			Industry		
Expert	*A*	*B*	*C*	*D*	*E*
1	0.15	0.75	0.10	0.00	0.30
2	0.30	0.60	0.20	0.05	0.25
3	0.20	0.80	0.30	0.00	0.50
4	0.00	0.50	0.25	0.10	0.60
5	0.10	0.55	0.15	0.15	0.40
6	0.25	0.70	0.35	0.25	0.45
7	0.40	0.95	0.45	0.20	0.35

(a) Test at the 0.05 level of significance whether the average ratings given by the experts are significantly different.

(b) Test at the 0.01 level of significance whether the average ratings given to the industries are significantly different. What argument(s) can you make against the validity of this test?

10.4 TESTS OF RANDOMNESS

When we discussed random sampling in Chapter 6, we gave several methods which provide some assurance in advance that a sample taken will be random. Since there are also situations in which we have no control over the way in which the data are selected, it is useful to have a technique for testing whether a sample may be looked upon as random after it has actually been obtained. One such technique is based on the order in which the sample values were obtained; more specifically, it is based on the number of **runs** exhibited in the sample results.

Given a sequence of two symbols, such as *H* and *T* (which might represent the occurrence of heads and tails in repeated tosses of a coin), a run is defined as a succession of identical symbols contained between different symbols or none at all. For example, the sequence

$$\underline{T\ T}\ \ \underline{H\ H}\ \ \underline{T\ T}\ \ \underline{H\ H\ H}\ \ \underline{T}\ \ \underline{H\ H\ H}\ \ \underline{T\ T\ T\ T}\ \ \underline{H\ H\ H}$$

contains 8 runs, as indicated by the underlines. The total number of runs in a sequence of *n* trials often serves as an indication that the arrangement is not random. For instance, if there had been only two runs consisting of 10 heads followed by 10 tails, we might have suspected that the probability of a success did not remain constant from trial to trial. On the other hand, had the sequence of 20 tosses consisted of alternating heads and tails, we might have suspected that the trials were not independent. In either case, there are grounds to suspect a lack of randomness. Note that our suspicion is not aroused by the numbers of *H*'s and *T*'s, but by the order in which they appeared.

If a sequence contains n_1 symbols of one kind and n_2 of another kind (and neither n_1 nor n_2 is less than 10), the sampling distribution of the **total number of runs**, *u*, can be approximated closely by a normal distribution with

Mean and standard deviation of u

$$\mu_u = \frac{2n_1 n_2}{n_1 + n_2} + 1 \quad and \quad \sigma_u = \sqrt{\frac{2n_1 n_2 (2n_1 n_2 - n_1 - n_2)}{(n_1 + n_2)^2 (n_1 + n_2 - 1)}}$$

Thus, the test of the null hypothesis that the arrangement of the symbols (and, hence, the sample) is random can be based on the statistic

Statistic for test of randomness

$$z = \frac{u - \mu_u}{\sigma_u}$$

which has approximately the standard normal distribution. Special tables for performing the test when n_1, n_2, or both are small may be found in the handbook by D. B. Owen listed in the Bibliography.

EXAMPLE The following is the arrangement of defective, *d*, and nondefective, *n*, pieces produced in the given order by a certain machine:

$$\underline{n\,n\,n\,n\,n}\;\underline{d\,d\,d\,d}\;\underline{n\,n\,n\,n\,n\,n\,n\,n\,n}\;\underline{d\,d}\;\underline{n\,n}\;\underline{d\,d\,d\,d}$$

Test for randomness at the 0.01 level of significance.

Solution

1. *Null hypothesis*: Arrangement is random.
   Alternative hypothesis: Arrangement is not random.
2. *Level of significance*: $\alpha = 0.01$
3. *Criterion*: Reject the null hypothesis if $z < -2.575$ or $z > 2.575$, where z is given by the above formula.
4. *Calculations*: Since $n_1 = 10$, $n_2 = 17$, and $u = 6$, we get

$$\mu_u = \frac{2\cdot 10\cdot 17}{10 + 17} + 1 = 13.59$$

$$\sigma_u = \sqrt{\frac{2\cdot 10\cdot 17(2\cdot 10\cdot 17 - 10 - 17)}{(10 + 17)^2(10 + 17 - 1)}} = 2.37$$

and

$$z = \frac{6 - 13.59}{2.37} = -3.20$$

5. *Decision*: Since $z = -3.20$ is less than -2.575, the null hypothesis must be rejected; we conclude that the arrangement is not random. Indeed, the total number of runs is much smaller than expected and there is a strong indication that the defective pieces appear in clusters or groups; the reason for this will have to be uncovered by an engineer who is familiar with the process.

The run test can be used also to test the randomness of samples consisting of numerical data by counting **runs above and below the median**. Denoting an observation exceeding the median of the sample by the letter *a* and an observation less than the median by the letter *b*, we can use the resulting sequence of *a*'s and *b*'s to test for randomness by the method just indicated. A frequent application of this test is in quality control, where the means of successive small samples are exhibited on a graph in chronological order. The run test can then be used to check whether there might be a trend in the data, so that it is possible to adjust a machine setting or some other process variable before any serious damage occurs.

EXAMPLE   An engineer is concerned about the possibility that too many changes are being made in the settings of an automatic lathe. Given the following mean diameters (in inches) of 40 successive shafts turned on the lathe

0.261 0.258 0.249 0.251 0.247 0.256 0.250 0.247 0.255 0.243
0.252 0.250 0.253 0.247 0.251 0.243 0.258 0.251 0.245 0.250
0.248 0.252 0.254 0.250 0.247 0.253 0.251 0.246 0.249 0.252
0.247 0.250 0.253 0.247 0.249 0.253 0.246 0.251 0.249 0.253

use the 0.01 level of significance to test the null hypothesis of randomness against the alternative that there is a frequently alternating pattern.

Solution

1. *Null hypothesis*: Arrangement of sample values is random.
   Alternative hypothesis: There is a frequently alternating pattern.
2. *Level of significance*: $\alpha = 0.01$
3. *Criterion*: Reject the null hypothesis if $z > 2.33$, where z is obtained by means of the formula on page 282 for the total number of runs above and below the median.
4. *Calculations*: The median of the 40 measurements is 0.250, so that we get the following arrangement of values above and below 0.250:

   $a\ a\ b\ a\ b\ a\ b\ a\ b\ a\ b\ a\ a\ b\ a\ b\ a\ b\ a\ a\ b\ b\ a\ a\ b\ a\ a\ b\ b\ a\ b\ a\ b\ b\ a\ b\ a\ b\ a$

   Thus, $n_1 = 19$, $n_2 = 16$, and $u = 27$, so that

   $$\mu_u = \frac{2 \cdot 19 \cdot 16}{35} + 1 = 18.37$$

   $$\sigma_u = \sqrt{\frac{2 \cdot 19 \cdot 16(2 \cdot 19 \cdot 16 - 19 - 16)}{(19 + 16)^2(19 + 16 - 1)}} = 2.89$$

   and

   $$z = \frac{27 - 18.37}{2.89} = 2.98$$

5. *Decision*: Since $z = 2.98$ exceeds 2.33, the null hypothesis of randomness must be rejected; since the number of runs is much larger than one might expect due to chance, it is reasonable to conclude that the lathe is being adjusted too often.

10.5   THE KOLMOGOROV–SMIRNOV TESTS

The **Kolmogorov–Smirnov tests** are nonparametric tests for differences between cumulative distributions. The one-sample test concerns the agreement between an observed cumulative distribution of sample values and a specified continuous distribution function; thus, it is a test of goodness of fit. The two-sample test concerns the agreement between two observed cumulative distributions; it tests the hypothesis whether two independent samples come from identical continuous distributions, and it is sensitive to population differences with respect to location, dispersion, or skewness.

The Kolmogorov–Smirnov one-sample test is generally more efficient than the chi-square test for goodness of fit for small samples, and it can be used for very small samples where the chi-square test does not apply. It must be remembered, however, that the chi-square test of Section 9.6 can be used in connection with discrete distributions whereas the Kolmogorov–Smirnov test cannot.

The one-sample test is based on the maximum absolute difference D between the values of the cumulative distribution of a random sample of size n and a specified theoretical distribution. To determine whether this difference is larger than can reasonably be expected for a given level of significance, we look up the critical value of D in Table 10.

EXAMPLE It is desired to check whether pinholes in electrolytic tin plate are uniformly distributed across a plated coil on the basis of the following distances in inches of 10 pinholes from one edge of a long strip of tin plate 30 inches wide:

$$4.8 \quad 14.8 \quad 28.2 \quad 23.1 \quad 4.4 \quad 28.7 \quad 19.5 \quad 2.4 \quad 25.0 \quad 6.2$$

Test this null hypothesis at the 0.05 level of significance.

Solution

1. *Null hypothesis*:

$$F(x) = \begin{cases} 0 & \text{for } x \leq 0 \\ \dfrac{x}{30} & \text{for } 0 < x < 30 \\ 1 & \text{for } x \geq 30 \end{cases}$$

where x is the distance of a pinhole from the edge.
Alternative hypothesis: The pinholes are not uniformly distributed across the tin plate.

2. *Level of significance*: $\alpha = 0.05$

3. *Criterion*: Reject the null hypothesis if $D > 0.410$, where D is the maximum difference between the observed cumulative distribution and the cumulative distribution assumed under the null hypothesis.

4. *Calculations*: Plotting the two cumulative distributions as in Figure 10.1, we find that the difference is greatest at $x = 6.2$. and that its value is $D = 0.40 - \dfrac{6.2}{30} = 0.193.$

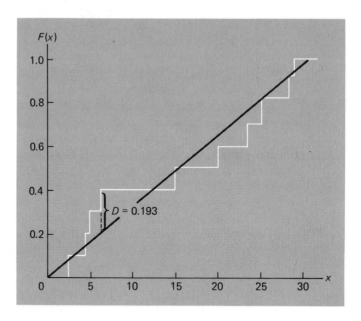

Figure 10.1 Diagram for Kolmogorov-Smirnov test.

5. *Decision*: Since $D = 0.193$ does not exceed 0.410, the null hypothesis (that the pinholes are uniformly distributed across the tin plate) cannot be rejected.

EXERCISES

10.19 The following arrangement indicates whether 60 consecutive cars which went by the toll booth of a bridge had local plates, L, or out-of-state plates, O:

$$L\,L\,O\,L\,L\,L\,L\,O\,O\,L\,L\,L\,L\,O\,L\,O\,O\,L\,L\,L\,L\,O\,L\,O\,O\,L\,L\,L\,L\,L$$
$$O\,L\,L\,L\,O\,L\,O\,L\,O\,L\,L\,L\,L\,O\,O\,L\,O\,O\,O\,O\,L\,L\,L\,L\,O\,L\,O\,O\,L\,L\,L\,O$$

Test at the 0.05 level of significance whether this arrangement of L's and O's may be regarded as random.

10.20 The following arrangement indicates whether 50 consecutive persons interviewed by a pollster are for, F, or against, A, an increase in the state gasoline tax to build more roads:

$$A\,A\,A\,F\,A\,F\,A\,A\,A\,A\,F\,F\,A\,A\,F\,A\,A\,A\,A\,A\,F\,A\,A\,F\,F$$
$$A\,A\,F\,A\,A\,A\,A\,F\,A\,F\,F\,A\,A\,A\,A\,A\,F\,A\,A\,F\,A\,A\,A\,A\,F$$

Test at the 0.05 level of significance whether this arrangement of A's and F's may be regarded as random.

10.21 To check the randomness of the digits in Table 7, take those in any five rows, replace the odd digits by the letter O and the even digits by the letter E, and base your decision on the total number of runs in the resulting sequence of O's and E's. Use the 0.01 level of significance.

10.22 To test whether radio signals from deep space contain a message, an interval of time could be subdivided into a number of very short intervals and it could then be determined whether the signal strength exceeded a certain level (background noise) in each short interval. Suppose that the following is part of such a record, where H denotes a high signal strength and L denotes that the signal strength does not exceed a given noise level.

$$L L H L H L H L H L H H H H L H H H H L H H H H L H L H L H L H L L H L$$
$$L L H L H L H L H H H H L H H H H L H H H H L H L H L H L H L L L L L$$

Test this sequence for randomness (using the 0.05 level of significance) and ascertain whether it is reasonable to assume that the signal contains a message.

10.23 The following are the number of defective pieces turned out by a machine during 24 consecutive shifts: 15, 11, 17, 14, 16, 12, 19, 17, 21, 15, 17, 19, 21, 14, 22, 16, 19, 12, 16, 14, 18, 17, 24, and 13. Test for randomness at the 0.01 level of significance.

10.24 The following are 50 consecutive downtimes of a machine (in minutes) which were observed during a certain period of time: 22, 29, 32, 25, 33, 34, 38, 34, 29, 25, 27, 33, 34, 28, 39, 41, 24, 31, 34, 29, 34, 25, 30, 37, 40, 39, 35, 24, 32, 43, 44, 34, 40, 38, 39, 43, 46, 34, 39, 45, 42, 39, 54, 50, 38, 41, 43, 46, 52, and 55. Use the method of runs above and below the median and the 0.05 level of significance to test the null hypothesis of randomness against the alternative that there is a trend.

10.25 Using the result of the illustration given in Exercise 5.50 on page 162, namely, that the median of the sulfur oxides emission data on page 138 is 19.0 (rounded to one decimal), test the original data for randomness at the 0.05 level of significance.

10.26 The total number of retail stores opening for business and also quitting business within the calendar years 1948–1980 in a large city were 108, 103, 109, 107, 125, 142, 147, 122, 116, 153, 144, 162, 143, 126, 145, 129, 134, 137, 143, 150, 148, 152, 125, 106, 112, 139, 132, 122, 138, 148, 155, 146, and 158. Making use of the fact that the median is 138, test at the 0.05 level of significance whether there is a significant trend.

10.27 In a vibration study, certain airplane components were subjected to severe vibrations until they showed structural failures. Given the following failure times (in minutes), test whether they can be looked upon as a sample from an exponential population with the mean $\mu = 10$:

$$1.5 \quad 10.3 \quad 3.6 \quad 13.4 \quad 18.4 \quad 7.7 \quad 24.3 \quad 10.7 \quad 8.4$$
$$15.4 \quad 4.9 \quad 2.8 \quad 7.9 \quad 11.9 \quad 12.0 \quad 16.2 \quad 6.8 \quad 14.7$$

Use the 0.05 level of significance.

10.28 Use the Kolmogorov–Smirnov test at the 0.01 level of significance to test the null hypothesis that the boiling points of Exercise 5.15 on page 150 come from a normal population with $\mu = 160$ degrees Celsius and $\sigma = 10$ degrees Celsius. [*Hint*: Use arithmetic probability paper.]

10.29 Survival times (days) of eight cancer-bearing mice that have been treated with a certain anticancer drug are as follows:

$$16, \quad 11, \quad 24, \quad 18, \quad 31, \quad 15, \quad 12, \quad 21$$

Test at the 0.01 level of significance whether these data are consistent with the assumption of a log-normal distribution of survival times.

10.6 *CHECK LIST OF KEY TERMS* (*with page references*)

Efficiency, 272
H test, 277
Kolmogorov–Smirnov test, 284
Kruskal–Wallis test, 277
Mann–Whitney test, 274
Nonparametric tests, 272
Rank sums, 274

Run, 282
Runs above and below the median, 283
Sign test, 272
Total number of runs, 282
U test, 274
Wilcoxon test, 274

11

Curve Fitting

The main objective of many statistical investigations is to make predictions, preferably on the basis of mathematical equations. For instance, an engineer may wish to predict the amount of oxide that will form on the surface of a metal baked in an oven for a specified amount of time at 200 degrees Celsius, or the amount of deformation of a ring subjected to a compressive force of 1,000 pounds, or the time between recappings of a tire having a given tread thickness and composition. Usually, such predictions require that a formula be found which relates the dependent variable (whose value one wants to predict) to one or more independent variables.

Problems relating to predictions which are based on the known value of one variable are treated in Sections 11.1 through 11.3 and 11.5; the case where predictions are based on the known values of several variables is treated in Section 11.4.

11.1 THE METHOD OF LEAST SQUARES

In this section we shall begin our study of the case where a dependent variable is to be predicted in terms of a single independent variable. In many problems of this kind the independent variable is observed without error, or with an error which is negligible when compared with the error (chance variation) in the dependent variable. For example, in measuring the amount of oxide on the surface of a metal specimen, the baking temperature can usually be controlled with good precision, but the oxide-thickness measurement may be subject to considerable chance variation. Thus, even though the independent variable may be fixed at x, repeated measurements of the dependent variable may lead to y-values which differ considerably. Such differences among y-values can be attributed to several causes, chiefly to errors of measurement and to the existence of other, uncontrolled variables which may influence the value of y when x is fixed. Thus, measurements of the thickness of oxide layers may vary over several specimens baked for the same length of time at the same temperature because of the difficulty in measuring thickness as well as possible differences in the composition of the oven atmosphere, surface conditions of the specimens, and the like.

It should be apparent from this discussion that in this context y is the value of a random variable whose distribution depends on x. In most situations of this sort we are interested mainly in the relationship between x and the mean of the corresponding distribution of the y's, and we refer to this relationship as the **regression curve of y on x**. (For the time being we shall consider the case where x is fixed, that is, not random; in Section 11.5 we shall consider the case where x and y are both values of random variables.)

Let us first treat the case where the regression curve of y on x is **linear**, that is, where for any given x, the mean of the distribution of the y's is given by $\alpha + \beta x$. In general, an observed y will differ from this mean, and we shall denote this difference by ϵ, writing

$$y = \alpha + \beta x + \epsilon$$

Thus, ϵ is a value of a random variable and we can always choose α so that the mean of the distribution of this random variable is equal to zero. The value of ϵ for any given observation will depend on a possible error of measurement and on the values of variables other than x which might have an influence on y.

To give an example where the regression curve of y on x can reasonably be assumed to be linear, suppose that a tensile ring is to be calibrated by measuring the deflection at various loads. In the following table, giving the results of 12 measurements, the x's are the load forces in thousands of pounds and the y's are the corresponding deflections in thousandths of an inch:

x	1	2	3	4	5	6	7	8	9	10	11	12
y	16	35	45	74	86	96	106	124	134	156	164	182

It is apparent from Figure 11.1, where these data have been plotted, that it is reasonable to assume that the relationship (regression curve) is linear; that is, a straight line gives a very good approximation over the range of the available data. This kind of diagram, which shows how the data points are scattered, is called a **scattergram**.

Figure 11.1 Scattergram indicating linear regression.

Now we face the problem of using the observed data to estimate the parameters α and β of the regression line, and this is equivalent to finding the equation of the straight line which somehow provides the best fit. In connection with Figure 11.1, it may well be satisfactory to do this by eye. If different experimenters were to fit a line in this way, they would probably all predict that for a load of 7,500 pounds the deflection should be close to 0.115 inch. However, if we have to deal with data such as those plotted in Figure 11.2, the problem of finding a best-fitting line is not so obvious. To handle problems of this kind, we must seek a nonsubjective method for fitting straight lines which reflects some desirable statistical properties.

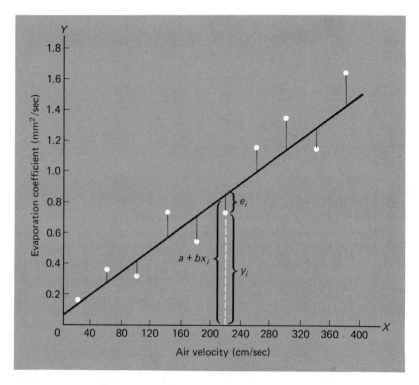

Figure 11.2 Diagram for least-squares criterion.

To state the problem formally, we have n paired observations (x_i, y_i) for which it is reasonable to assume that the regression of y on x is linear, and we want to determine the line (that is, the equation of the line) which in some sense provides the best fit. There are several ways in which we can interpret the word "best," and the meaning we shall give it here may be explained as follows. If we predict y by means of the equation

$$\hat{y} = a + bx$$

where a and b are constants, then e_i, the error in predicting the value of y corresponding to the given x_i, is

$$e_i = y_i - \hat{y}_i$$

and we shall want to determine a and b so that these errors are in some sense as small as possible.

Since we cannot minimize each of the e_i individually, it suggests itself that we might try to make their sum $\sum_{i=1}^{n} e_i$ as close as possible to zero. However, since this sum can be made equal to zero by many choices of totally unsuitable lines for which the positive and negative errors cancel, we shall minimize the sum of the squares of the e_i (for the same reason we worked with the squares of the deviations from the mean in the definition of the standard deviation). In other words, we shall choose a and b so that

$$\sum_{i=1}^{n} [y_i - (a + bx_i)]^2$$

is a minimum. Note from Figure 11.2 that this is equivalent to minimizing the sum of the squares of the vertical distances from the points to the line.

This method of finding the equation of the line which best fits a given set of paired data, called the **method of least squares**, yields values for a and b (estimates of α and β) that have many desirable properties; some of these are mentioned on page 295.

A necessary condition for a relative minimum is the vanishing of the partial derivatives with respect to a and b. We thus have

$$2 \sum_{i=1}^{n} [y_i - (a + bx_i)](-1) = 0$$

$$2 \sum_{i=1}^{n} [y_i - (a + bx_i)](-x_i) = 0$$

and we can rewrite these two equations as

Normal equations

$$\sum_{i=1}^{n} y_i = an + b \sum_{i=1}^{n} x_i$$

$$\sum_{i=1}^{n} x_i y_i = a \sum_{i=1}^{n} x_i + b \sum_{i=1}^{n} x_i^2$$

This set of two linear equations in the unknowns a and b, called the **normal equations**, gives the values of a and b for the line which provides the best fit to a given set of paired data in accordance with the criterion of least squares.

EXAMPLE The following are measurements of the air velocity and evaporation coefficient of burning fuel droplets in an impulse engine:

Air velocity (cm/sec) x	Evaporation coefficient (mm²/sec) y
20	0.18
60	0.37
100	0.35
140	0.78
180	0.56
220	0.75
260	1.18
300	1.36
340	1.17
380	1.65

Fit a straight line to these data by the method of least squares, and use it to estimate the evaporation coefficient of a droplet when the air velocity is 190 cm/sec.

Solution The quantities needed for substitution into the two normal equations are $n = 10$ and

$$\sum_{i=1}^{n} x_i = 2,000, \qquad \sum_{i=1}^{n} x_i^2 = 532,000$$

$$\sum_{i=1}^{n} y_i = 8.35, \qquad \sum_{i=1}^{n} x_i y_i = 2,175.40$$

and we get

$$8.35 = 10a + 2,000b$$
$$2,175.40 = 2,000a + 532,000b$$

Solving this system of equations by use of determinants or the method of elimination (or a statistical calculator which directly yields the values of a and b after we have punched in the data), we obtain $a = 0.069$ and $b = 0.0038$. Thus, the equation of the straight line which best fits the given data in the sense of least squares is

$$\hat{y} = 0.069 + 0.0038x$$

and for $x = 190$ we predict that the evaporation coefficient will be $\hat{y} = 0.069 + 0.0038(190) = 0.79$ mm²/sec.

It is impossible to make any exact statements about the "goodness" of an estimate like this unless we make some assumptions about the underlying distributions of the random variables with which we are concerned and about the true nature of the regression. Looking upon the values of a and b obtained from the normal equations as estimates of the actual regression coefficients

α and β, the reader will be asked to show in Exercises 11.22 and 11.23 on pages 305 and 306 that these estimates are linear in the observations y_i and that they are values of unbiased estimators of α and β. With these properties, we can refer to the remarkable **Gauss–Markov theorem**, which states that among all unbiased estimators for α and β which are linear in the y_i, the least-squares estimators have the smallest variance. In other words, the least-squares estimators are the most reliable in the sense that they are subject to the smallest chance variations. A proof of the Gauss–Markov theorem may be found in the book by H. Scheffe referred to in the Bibliography.

11.2 INFERENCES BASED ON THE LEAST-SQUARES ESTIMATORS

The method of the preceding section is used when the relationship between x and the mean of y is linear, or close enough to a straight line so that the least-squares line yields reasonably good predictions. In what follows we shall assume that the regression *is* linear and furthermore that the n random variables having the values y_i $(i = 1, \ldots, n)$ are independently normally distributed with the means $\alpha + \beta x_i$ and the common variance σ^2. If we write

$$y_i = \alpha + \beta x_i + \epsilon_i$$

it follows from these assumptions that the ϵ_i are values of independent normally distributed random variables having zero means and the common variance σ^2. The various assumptions we have made here are illustrated in Figure 11.3, showing the distributions of values of y_i for several values of the x_i. Note that these additional assumptions are required to discuss the goodness of predictions based on least-squares equations, the properties of a and b as estimates of α and β, and so on; they were not required to obtain the original estimates based on the method of least squares.

Before we state a theorem concerning the distribution of the least-squares estimators of α and β, it will be convenient to introduce some special notation. The following expressions pertaining to the sample values (x_i, y_i) occur so often that it is convenient to write them as

$$S_{xx} = n \sum_{i=1}^{n} x_i^2 - \left(\sum_{i=1}^{n} x_i \right)^2$$

$$S_{yy} = n \sum_{i=1}^{n} y_i^2 - \left(\sum_{i=1}^{n} y_i \right)^2$$

$$S_{xy} = n \sum_{i=1}^{n} x_i y_i - \left(\sum_{i=1}^{n} x_i \right)\left(\sum_{i=1}^{n} y_i \right)$$

In Exercise 11.24 on page 306 the reader will be asked to show that, in this

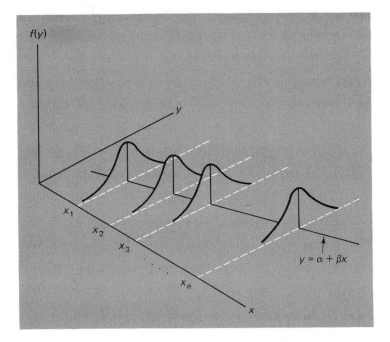

Figure 11.3 Diagram showing assumptions underlying Theorem 11.1.

notation, the solutions of the two normal equations on page 293 can be written as

Solutions of
normal equations

$$a = \bar{y} - b \cdot \bar{x} \quad \text{and} \quad b = \frac{S_{xy}}{S_{xx}}$$

where $\bar{x}$ and $\bar{y}$ are, respectively, the means of the x's and the y's. Note also the close relationship between S_{xx} and S_{yy} and the respective sample variances of the x's and the y's; in fact, $s_x^2 = S_{xx}/n(n-1)$ and $s_y^2 = S_{yy}/n(n-1)$, and we shall sometimes use this alternative notation.

The variance σ^2 defined on page 295 is usually estimated in terms of the vertical deviations of the sample points from the least-squares line. The ith such deviation is $y_i - \hat{y}_i = y_i - (a + bx_i)$ and the estimate of σ^2 is

$$s_e^2 = \frac{1}{n-2} \sum_{i=1}^{n} [y_i - (a + bx_i)]^2$$

where s_e is, traditionally, referred to as the **standard error of estimate**; also, the sum of squares given by $(n-2)s_e^2$ is referred to as the **residual sum of squares** or the **error sum of squares**. An equivalent formula for this estimate of σ^2, which is more convenient to use in actual applications, is given by

Estimate of σ^2

$$s_e^2 = \frac{S_{xx}S_{yy} - (S_{xy})^2}{n(n-2)S_{xx}}$$

In these formulas the divisor $n-2$ is used to make the resulting estimator for σ^2 unbiased. The "loss" of two degrees of freedom is explained by the fact that the two regression coefficients α and β had to be replaced by their least-squares estimates. It can also be shown that under the given assumptions $(n-2)s_e^2/\sigma^2$ is a value of a random variable having the chi-square distribution with $n-2$ degrees of freedom.

Based on the assumptions made concerning the distribution of the y's, one can prove the following theorem concerning the distributions of the least-squares estimators of the regression coefficients α and β:

***Statistics
for inferences
about α and β***

Theorem 11.1 *Under the assumptions given on page 295, the statistics*

$$t = \frac{(a - \alpha)}{s_e}\sqrt{\frac{nS_{xx}}{S_{xx} + (n\bar{x})^2}}$$

and

$$t = \frac{(b - \beta)}{s_e}\sqrt{\frac{S_{xx}}{n}}$$

are values of random variables having the t distribution with $n-2$ degrees of freedom.

To construct confidence intervals for the regression coefficients α and β, we substitute for the middle term of $-t_{\alpha/2} < t < t_{\alpha/2}$ the appropriate t statistic of Theorem 11.1. Then, simple algebra leads to

***Confidence limits
for regression
coefficients***

$$a \pm t_{\alpha/2} \cdot s_e\sqrt{\frac{S_{xx} + (n\bar{x})^2}{nS_{xx}}}$$

and

$$b \pm t_{\alpha/2} \cdot s_e\sqrt{\frac{n}{S_{xx}}}$$

EXAMPLE   With reference to the example on page 294, construct a 95% confidence interval for the regression coefficient α.

Solution   Using the numerical results on page 294 together with

$$\sum_{i=1}^{n} y_i^2 = 9.1097$$

we first obtain

$$S_{xx} = 10(532,000) - (2,000)^2 = 1,320,000$$
$$S_{yy} = 10(9.1097) - (8.35)^2 = 21.3745$$
$$S_{xy} = 10(2,175.40) - (2,000)(8.35) = 5,054.00$$

and, hence,

$$s_e^2 = \frac{(1,320,000)(21.3745) - (5,054.00)^2}{(10)(8)(1,320,000)} = 0.0253$$

Since $t_{0.025} = 2.306$ for $10 - 2 = 8$ degrees of freedom, we get the 95 % confidence limits

$$0.069 \pm (2.306)(0.159)\sqrt{\frac{1,320,000 + (2,000)^2}{10(1,320,000)}}$$

and, hence, the 95 % confidence interval

$$-0.164 < \alpha < 0.302$$

In connection with tests of hypotheses concerning the regression coefficients α and β, those concerning β are of special importance because β is the **slope of the regression line**; that is, β is the change in the mean of the y's corresponding to a unit increase in x. If $\beta = 0$, the regression line is horizontal and the mean of the y's does not depend linearly on x. For tests of the null hypothesis $\beta = \beta_0$ we use the second statistic of Theorem 11.1 and the criteria are like those in the table on page 208 with t and β substituted for z and μ.

EXAMPLE   With reference to the example on page 294, test the null hypothesis $\beta = 0$ against the alternative hypothesis $\beta \neq 0$ at the 0.05 level of significance.

Solution
1. *Null hypothesis*: $\beta = 0$
   Alternative hypothesis: $\beta \neq 0$
2. *Level of significance*: $\alpha = 0.05$
3. *Criterion*: Reject the null hypothesis if $t < -2.306$ or $t > 2.306$, where 2.306 is the value of $t_{0.025}$ for $10 - 2 = 8$ degrees of freedom, and t is given by the second formula of Theorem 11.1.

4. *Calculations*: Using the quantities obtained on pages 294 and 298, we get

$$t = \frac{0.0038 - 0}{0.159} \sqrt{\frac{1,320,000}{10}} = 8.36$$

5. *Decision*: Since $t = 8.36$ exceeds 2.306, the null hypothesis must be rejected; we conclude that there is a relationship between air velocity and the average evaporation coefficient. (The relationship is linear by the assumptions which underly the test.)

Another problem, closely related to the problem of estimating the regression coefficients α and β, is that of estimating $\alpha + \beta x$, namely, the mean of the distribution of the y's for a given value of x. If x is held fixed at x_0, the quantity we want to estimate is $\alpha + \beta x_0$ and it would seem reasonable to use $a + b x_0$, where a and b are again the values obtained by the method of least squares. In fact, it can be shown that this estimator is unbiased, has the variance

$$\sigma^2 \left[\frac{1}{n} + \frac{n(x_0 - \bar{x})^2}{S_{xx}} \right]$$

and that $(1 - \alpha)100\%$ confidence limits for $\alpha + \beta x_0$ are given by

Confidence limits for $\alpha + \beta x_0$

$$(a + b x_0) \pm t_{\alpha/2} \cdot s_e \sqrt{\frac{1}{n} + \frac{n(x_0 - \bar{x})^2}{S_{xx}}}$$

where the number of degrees of freedom for $t_{\alpha/2}$ is $n - 2$.

EXAMPLE   With reference to the example on page 294, construct a 95% confidence interval for the mean evaporation coefficient when the air velocity is 190 cm/sec.

Solution   Substituting the various quantities already calculated on pages 294 and 298 into the above formula, we get

$$0.79 \pm (2.306)(0.159) \sqrt{\frac{1}{10} + \frac{(10)(190 - 200)^2}{1,320,000}}$$

and, hence,

$$0.67 < \alpha + 190\beta < 0.91$$

Of even greater importance than the estimation of $\alpha + \beta x_0$ is the prediction of a future value of y when $x = x_0$, where x_0 is within the range of experimentation. (We added "within the range of experimentation" because, here as elsewhere extrapolation is risky and a linear relationship we have observed may not continue beyond that range.) Thus, for our example we already showed on page 294 that for an air velocity of 190 cm/sec, a value well within the range of experimentation, the predicted evaporation coefficient is 0.79 mm²/sec. Now let us indicate a method of constructing an interval in which a future observation, y, can be expected to lie with a given probability (or confidence) when $x = x_0$. If α and β were known, we could use the fact that y is a value of a random variable having a normal distribution with the mean $\alpha + \beta x_0$ and the variance σ^2 (or that $y - \alpha - \beta x_0$ is a value of a random variable having a normal distribution with zero mean and the variance σ^2). However, if α and β are not known, we must consider the quantity $y - a - b x_0$, where y, a, and b are all values of random variables, and the resulting theory leads to the following **limits of prediction** for y when $x = x_0$:

Limits of prediction

$$(a + b x_0) \pm t_{\alpha/2} \cdot s_e \sqrt{1 + \frac{1}{n} + \frac{n(x_0 - \bar{x})^2}{S_{xx}}}$$

The number of degrees of freedom for $t_{\alpha/2}$ is again $n - 2$.

EXAMPLE   With reference to the example on page 294, find 95 % limits of prediction for an observation of the evaporation coefficient when the air velocity is 190 cm/sec.

    Solution   Adding a 1 under the radical sign in the preceding example, we get

$$0.79 \pm (2.306)(0.159)\sqrt{1 + \frac{1}{10} + \frac{(10)(190 - 200)^2}{1,320,000}}$$

and, hence, the limits of prediction are $0.79 - 0.39 = 0.40$ and $0.79 + 0.39 = 1.18$.

Note that although the mean of the distribution of y's when $x = 190$ can be estimated fairly closely, the value of a single future observation cannot be predicted with much precision. Indeed, even as $n \to \infty$ the difference between the limits of prediction does not approach zero; the limiting width of the interval of prediction depends on s_e, which measures the inherent variability of the data.

Note, further, that if we do wish to extrapolate, the interval of prediction (and also the confidence interval for $\alpha + \beta x_0$) becomes increasingly wide.

EXAMPLE   With reference to the example on page 294, assume that the linear relationship continues beyond the range of experimentation and find 95 % limits of prediction for an observation of the evaporation coefficient when the air velocity is 450 cm/sec.

Solution Substituting the various quantities already calculated on pages 294 and 298 into the formula on page 300, we get

$$1.78 \pm (2.306)(0.159)\sqrt{1 + \frac{1}{10} + \frac{10(450 - 200)^2}{1,320,000}}$$

and, hence, the limits of prediction are $1.78 - 0.46 = 1.32$ and $1.78 + 0.46 = 2.24$ mm²/sec. The width of this interval of prediction is $2(0.46) = 0.92$ compared to the width of $2(0.39) = 0.78$ obtained in the preceding example.

EXERCISES

11.1 A chemical company, wishing to study the effect of extraction time on the efficiency of an extraction operation, obtained the data shown in the following table:

Extraction time (minutes) x	Extraction efficiency (%) y
27	57
45	64
41	80
19	46
35	62
39	72
19	52
49	77
15	57
31	68

(a) Draw a scattergram to verify that a straight line will provide a good fit to the data, draw a straight line by eye, and use it to predict the extraction efficiency one can expect when the extraction time is 35 minutes.
(b) Fit a straight line to the given data by the method of least squares and use it to predict the extraction efficiency one can expect when the extraction time is 35 minutes.

11.2 Solve the normal equations on page 293 symbolically to show that

$$a = \frac{(\sum x^2)(\sum y) - (\sum x)(\sum xy)}{n(\sum x^2) - (\sum x)^2}$$

$$b = \frac{n(\sum xy) - (\sum x)(\sum y)}{n(\sum x^2) - (\sum x)^2}$$

where the subscripts and limits of summation have been omitted for simplicity. Also use these formulas to verify the values obtained for a and b in part (b) of Exercise 11.1.

11.3 In the accompanying table, x is the tensile force applied to a steel specimen in thousands of pounds, and y is the resulting elongation in thousandths of an inch:

x	1	2	3	4	5	6
y	14	33	40	63	76	85

(a) Graph the data to verify that it is reasonable to assume that the regression of y on x is linear.

(b) Use the formulas of the preceding exercise to find the equation of the least-squares line, and use it to predict the elongation when the tensile force is 3.5 thousand pounds.

11.4 With reference to the preceding exercise, construct a 95% confidence interval for β, the elongation per thousand pounds of tensile stress.

11.5 With reference to Exercise 11.3, find 95% limits of prediction for the elongation of a specimen when $x = 3.5$ thousand pounds.

11.6 The following table shows how many weeks a sample of six persons have worked at an automobile inspection station and the number of cars each one inspected between noon and 2 P.M. on a given day:

Number of weeks employed x	Number of cars inspected y
2	13
7	21
9	23
1	14
5	15
12	21

(a) Use the formulas of Exercise 11.2 or directly solve the normal equations to find the equation of the least-squares line which will enable us to predict y in terms of x.

(b) Use the result of part (a) to estimate how many cars someone who has been working at the inspection station for 8 weeks can be expected to inspect during the given 2-hour period.

11.7 With reference to the preceding exercise, test the null hypothesis $\beta = 1.2$ against the alternative hypothesis $\beta < 1.2$ at the 0.05 level of significance.

11.8 With reference to Exercise 11.6 find

(a) a 95% confidence interval for the average number of cars inspected in the given period of time by a person who has been working at the inspection station for 8 weeks;

(b) 95% limits of prediction for the number of cars that will be inspected in the given period of time by a person who has worked at the inspection station for 8 weeks.

11.9 The following data pertain to the chlorine residual in a swimming pool at various times after it has been treated with chemicals:

Number of hours	Chlorine residual (parts per million)
2	1.8
4	1.5
6	1.4
8	1.1
10	1.1
12	0.9

(a) Fit a least-squares line from which we can predict the chlorine residual in terms of the number of hours since the pool has been treated with chemicals.

(b) Use the equation of the least-squares line to estimate the chlorine residual in the pool 5 hours after it has been treated with chemicals.

11.10 With reference to the preceding exercise, construct a 95% confidence interval for α.

11.11 With reference to Exercise 11.9 test the null hypothesis $\beta = -0.12$ against the alternative hypothesis $\beta > -0.12$ at the 0.01 level of significance.

11.12 Raw material used in the production of a synthetic fiber is stored in a place which has no humidity control. Measurements of the relative humidity in the storage place and the moisture content of a sample of the raw material (both in percentages) on 12 days yielded the following results:

Humidity x	Moisture content y
42	12
35	8
50	14
43	9
48	11
62	16
31	7
36	9
44	12
39	10
55	13
48	11

(a) Plot a scattergram to verify that it is reasonable to assume that the regression of y on x is linear.

(b) Fit a straight line by the method of least squares.

(c) Find a 99% confidence interval for the mean moisture content of the raw material when the humidity of the storage place is 40%.

11.13 With reference to the preceding exercise, find 95% limits of prediction for the moisture content of the raw material when the humidity of the storage place is 40%. Also indicate to what extent the width of the interval is affected by the size of the sample and to what extent it is affected by the inherent variability of the data.

11.14 The following show the improvement (gain in reading speed) of eight students in a speed-reading program, and the number of weeks they have been in the program:

Number of weeks	Speed gain (words per minute)
3	86
5	118
2	49
8	193
6	164
9	232
3	73
4	109

(a) Plot a scattergram to verify that it is reasonable to assume that the regression of speed gain on the number of weeks is linear.

(b) Fit a straight line by the method of least squares.

11.15 With reference to the preceding exercise, find a 90% confidence interval for β.

11.16 With reference to Exercise 11.1, express 95% limits of prediction for the extraction efficiency in terms of the extraction time x_0. Choosing suitable values of x_0, sketch graphs of the loci of the upper and lower limits of prediction on the diagram of part (a) of Exercise 11.1. Note that since any two sets of limits of prediction obtained from these bands are dependent, they should be used only once for one extraction time x_0.

11.17 In Exercises 11.3 and 11.14 it would have been entirely reasonable to impose the condition $\alpha = 0$ before fitting a straight line by the method of least squares.

(a) Use the method of least squares to derive a formula for estimating β when the regression line has the form $y = \beta x$.

(b) With reference to Exercise 11.3, use the formula obtained in part (a) to estimate β and compare the result with the estimate previously obtained without the condition that the line must pass through the origin.

(c) With reference to Exercise 11.14, use the formula obtained in part (a) to estimate β and compare the result with the estimate previously obtained without the condition that the line must pass through the origin.

11.18 The cost of manufacturing a lot of a certain product depends on the lot size as shown by the following sample data:

Cost (dollars)	30	70	140	270	530	1,010	2,500	5,020
Lot size	1	5	10	25	50	100	250	500

(a) Draw a scattergram to verify the assumption that the relationship is linear, letting lot size be x and cost y.
(b) Fit a straight line to these data by the method of least squares, using lot size as the independent variable, and draw its graph on the diagram obtained in part (a).

11.19 With reference to the preceding exercise, find a 90% confidence interval for α, which can be interpreted here as the fixed overhead cost of manufacturing.

11.20 With reference to Exercise 11.18, fit a straight line to the data by the method of least squares, using cost as the independent variable, and draw its graph on the diagram obtained in part (a) of Exercise 11.18. Note that the two estimated regression lines do not coincide.

11.21 When the sum of the x values is equal to zero, the calculation of the coefficients of the regression line of y on x is greatly simplified; in fact, their estimates are given by

$$a = \frac{\sum y}{n} \quad \text{and} \quad b = \frac{\sum xy}{\sum x^2}$$

This simplification can also be attained when the x's are equally spaced, that is, when they are in arithmetic progression. We then code the data by substituting for the x's the values . . . , $-2, -1, 0, 1, 2, . . .$, when n is odd, or the values . . . , $-3, -1, 1, 3, . . .$, when n is even. The above formulas are then used in connection with the coded data.
(a) During its first 7 years of operation, a company's gross income from sales was 1.4, 2.1, 2.6, 3.5, 3.7, 4.9, and 5.5 million dollars. Fit a least-squares line and, assuming that the trend continues, predict the company's gross income from sales during the eighth year of operation.
(b) At the ends of the years 1977–1984 a manufacturing company had the following net investments in plants and equipments: 1.0, 1.7, 2.3, 3.1, 3.5, 3.4, 3.9, and 4.7 million dollars. Fit a least-squares line and, assuming that the trend continues, predict the company's net investment in plant and equipment at the end of 1986.

11.22 Using the formulas obtained in Exercise 11.2, show that
(a) the expression for a is linear in the y_i;
(b) a is an unbiased estimate of α.

11.23 Using the formulas obtained in Exercise 11.2, show that
(a) the expression for b is linear in the y_i;
(b) b is an unbiased estimate of β.

11.24 Show that the least-squares estimates of the coefficients of the regression line of y on x can be written in the form

$$a = \bar{y} - b \cdot \bar{x} \quad \text{and} \quad b = \frac{S_{xy}}{S_{xx}}$$

11.3 CURVILINEAR REGRESSION

So far we have studied only the case where the regression curve of y on x is linear; that is, where for any given x, the mean of the distribution of the y's is given by $\alpha + \beta x$. In this section we shall first investigate cases where the regression curve is nonlinear, but where the methods of Section 11.1 can nevertheless be applied; then we shall take up the problem of polynomial regression, that is, problems where for any given x the mean of the distribution of the y's is given by

$$\beta_0 + \beta_1 x + \beta_2 x^2 + \ldots + \beta_p x^p$$

Polynomial curve fitting is also used to obtain approximations when the exact functional form of the regression curve is unknown.

It is common practice for engineers to plot paired data on various kinds of graph paper, in order to determine whether for suitably transformed scales the points will fall close to a straight line. If that is the case, the nature of the transformation used leads to a functional form of the regression equation, and the necessary constants (parameters) can be determined by applying the method of Section 11.1 to the transformed data. For instance, if a set of paired data consisting of n points (x_i, y_i) "straightens out" when plotted on semilog paper, this indicates that the regression curve of y on x is **exponential**, namely, that for any given x, the mean of the distribution of the y's is given by $\alpha \cdot \beta^x$. If we take logarithms to the base 10 (or any other convenient base), the predicting equation $y = \alpha \cdot \beta^x$ becomes

$$\log y = \log \alpha + x \cdot \log \beta$$

and we can now get estimates of $\log \alpha$ and $\log \beta$, and hence of α and β, by applying the method of Section 11.1 to the n pairs of values $(x_i, \log y_i)$.

EXAMPLE The following are data on the percentage of the radial tires made by a certain manufacturer that are still usable after having been driven for the given numbers of miles:

Miles driven (thousands) x	Percentage usable y
1	98.2
2	91.7
5	81.3
10	64.0
20	36.4
30	32.6
40	17.1
50	11.3

(a) Plot the given data on semilog paper to verify that it is reasonable to assume that the relationship is exponential.

(b) Fit an exponential curve by applying the method of least squares to the data points $(x_i, \log y_i)$.

(c) Use the result of part (b) to estimate what percentage of the manufacturer's radial tires will last at least 25,000 miles.

Solution (a) As can be seen from Figure 11.4, the overall pattern is linear and this justifies fitting an exponential curve. (b) Determining first the logarithms of the eight y's, we get 1.9921, 1.9624, 1.9101, 1.8062, 1.5611, 1.5132, 1.2330, and 1.0531, and the summations required for substitution into the normal equations are $\sum x = 158$, $\sum x^2 = 5{,}530$, $\sum \log y = 13.0312$, and $\sum x \cdot \log y = 212.1224$, where the subscripts and limits of summation have been omitted for simplic-

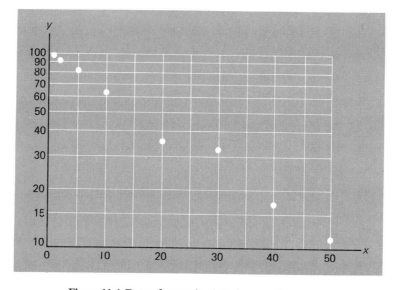

Figure 11.4 Data of example plotted on semilog paper.

ity. Again using a and b for the least-squares estimates of α and β, we obtain the normal equations

$$13.0312 = 8(\log a) + 158(\log b)$$
$$212.1224 = 158(\log a) + 5,530(\log b)$$

whose solution is $\log a = 2.002$ and $\log b = -0.0188$, and hence $a = 100.0$ and $b = 0.96$. Thus, the equation of the estimated regression curve can be written as

$$\log \hat{y} = 2.0002 - 0.0188x$$

in **logarithmic form**, or as

$$\hat{y} = 100.0(0.96)^x$$

in **exponential form**. (c) Using the logarithmic form, which is more convenient, we get

$$\log \hat{y} = 2.002 - 0.0188(25)$$
$$= 1.5302$$

and, hence, $\hat{y} = 33.9\%$.

Two other relationships that frequently arise in engineering applications, and can be fitted by the method of Section 11.1 after suitable transformations, are the **reciprocal function** $y = \dfrac{1}{\alpha + \beta x}$, and the **power function** $y = \alpha \cdot x^\beta$. The first of these represents a linear relationship between x and $\dfrac{1}{y}$, namely

$$\frac{1}{y} = \alpha + \beta x$$

and we obtain estimates of α and β by applying the method of Section 11.1 to the points $\left(x_i, \dfrac{1}{y_i} \right)$. The second represents a linear relationship between $\log x$ and $\log y$, namely,

$$\log y = \log \alpha + \beta \cdot \log x$$

and we obtain estimates of $\log \alpha$ and β, and hence of α and β, by applying the method of Section 11.1 to the points $(\log x_i, \log y_i)$. Examples of other curves that can be fitted by the method of least squares after a suitable transformation are given in Exercises 11.32 and 11.34 on page 318.

If there is no clear indication about the functional form of the regression of y on x, we often assume that the underlying relationship is at least "well behaved" to the extent that it has a Taylor series expansion and that the first few terms of this expansion will yield a fairly good approximation.

We thus fit to our data a **polynomial**, that is, a predicting equation of the form

$$y = \beta_0 + \beta_1 x + \beta_2 x^2 + \ldots + \beta_p x^p$$

where the degree is determined by inspection of the data or by a more rigorous method to be discussed below.

Given a set of data consisting of n points (x_i, y_i), we estimate the coefficients $\beta_0, \beta_1, \beta_2, \ldots, \beta_p$ of the pth-degree polynomial by minimizing

$$\sum_{i=1}^{n} [y_i - (\beta_0 + \beta_1 x_i + \beta_2 x_i^2 + \ldots + \beta_p x_i^p)]^2$$

In other words, we are now applying the least-squares criterion by minimizing the sum of the squares of the vertical distances from the points to the curve (see Figure 11.5). Differentiating partially with respect to $\beta_0, \beta_1, \beta_2, \ldots, \beta_p$, equating these partial derivatives to zero, rearranging some of the terms, and letting b_i be the estimate of β_i, we obtain the $p + 1$ normal equations

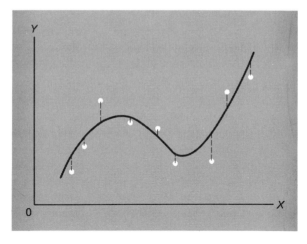

Figure 11.5 Least-squares criterion for polynomial curve fitting.

Normal equations for polynomial regression

$$\sum y = nb_0 + b_1 \sum x + \ldots + b_p \sum x^p$$
$$\sum xy = b_0 \sum x + b_1 \sum x^2 + \ldots + b_p \sum x^{p+1}$$
$$\cdots\cdots\cdots\cdots\cdots\cdots\cdots\cdots\cdots\cdots\cdots\cdots\cdots\cdots$$
$$\sum x^p y = b_0 \sum x^p + b_1 \sum x^{p+1} + \ldots + b_p \sum x^{2p}$$

where the subscripts and limits of summation are omitted for simplicity. Note that this is a system of $p + 1$ linear equations in the $p + 1$ unknowns

$b_0, b_1, b_2, \ldots,$ and b_p. Unless the various summations have very unusual values, this system of equations will have a unique solution.

EXAMPLE   The following are data on the drying time of a certain varnish and the amount of an additive which is intended to reduce the drying time:

Amount of varnish additive (grams) x	Drying time (hours) y
0	12.0
1	10.5
2	10.0
3	8.0
4	7.0
5	8.0
6	7.5
7	8.5
8	9.0

(a) Draw a scattergram on ordinary graph paper to verify that it is reasonable to assume that the relationship is parabolic.

(b) Fit a second-degree polynomial by the method of least squares.

(c) Use the result of part (b) to predict the drying time of the varnish when 6.5 grams of the additive is being used.

Solution   (a) As can be seen from Figure 11.6, the overall pattern suggests fitting a second-degree polynomial having one relative minimum. (b) The summations required for substitution into the normal

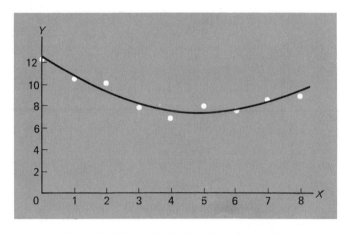

Figure 11.6 Parabola fitted to data of example.

equations are $\sum x = 36$, $\sum x^2 = 204$, $\sum x^3 = 1,296$, $\sum x^4 = 8,772$, $\sum y = 80.5$, $\sum xy = 299.0$, and $\sum x^2 y = 1,697.0$, and we thus have to solve the following system of three linear equations in the unknowns b_0, b_1, and b_2:

$$80.5 = 9b_0 + 36b_1 + 204b_2$$

$$299.0 = 36b_0 + 204b_1 + 1296b_2$$

$$1697.0 = 204b_0 + 1296b_1 + 8772b_2$$

Getting $b_0 = 12.2$, $b_1 = -1.85$, and $b_2 = 0.183$, we find that the equation of the least-squares polynomial is

$$\hat{y} = 12.2 - 1.85x + 0.183x^2$$

(c) Substituting $x = 6.5$ into this equation, we get

$$\hat{y} = 12.2 - 1.85(6.5) + 0.183(6.5)^2$$
$$= 7.9$$

that is, a predicted drying time of 7.9 hours.

Note that it would have been rather dangerous in the preceding example to predict the drying time which corresponds to, say, 24.5 grams of the additive. The risks inherent in extrapolation, discussed on page 300 in connection with fitting straight lines, increase greatly when polynomials are used to approximate unknown regression functions.

In actual practice, it may be difficult to determine the degree of the polynomial to fit to a given set of paired data. As it is always possible to find a polynomial of degree at most $n - 1$ that will pass through each of n points corresponding to n distinct values of x, it should be clear that what we actually seek is a polynomial of lowest possible degree that "adequately" describes the data. As we did in our example, it is often possible to determine the degree of the polynomial by inspection of the data.

There also exists a more rigorous method for determining the degree of the polynomial to be fitted to a given set of data. Essentially, it consists of first fitting a straight line as well as a second-degree polynomial and testing the null hypothesis $\beta_2 = 0$, namely, that nothing is gained by including the quadratic term. If this null hypothesis can be rejected, we then fit a third-degree polynomial and test the hypothesis $\beta_3 = 0$, namely, that nothing is gained by including the cubic term. This procedure is continued until the null hypothesis $\beta_i = 0$ cannot be rejected in two successive steps and there is, thus, no apparent advantage to carrying the extra terms. Note that in order to perform these tests it is necessary to impose the assumptions of normality, independence, and equal variances introduced in Section 11.2. Also, these tests should never be used "blindly," that is, without inspection of the overall pattern of the data.

The use of this technique is fairly tedious and we shall not illustrate it in the text. In Exercise 11.38 on page 319 the reader will be given detailed instructions to apply it to the varnish-additive, drying-time data in order to check whether it was really worthwhile to carry the quadratic term.

11.4 MULTIPLE REGRESSION

Before we extend the methods of the preceding sections to problems involving more than one independent variable, let us point out that the curves obtained (and the surfaces we will obtain) are not used only to make predictions. They are often used also for purposes of optimization namely, to determine for what values of the independent variable (or variables) the dependent variable is a maximum or minimum. For instance, in the example on page 310 we might use the polynomial fitted to the data to conclude that the drying time is a minimum when the amount of varnish additive used is 5.1 grams (see Exercise 11.39 on page 320).

Statistical methods of prediction and optimization are often referred to under the general heading of **response surface analysis**. Within the scope of this text, we shall be able to introduce two further methods of response surface analysis: **multiple regression** here and related problems of **factorial experimentation** in Chapter 13.

In multiple regression, we deal with data consisting of n $(r + 1)$-tuples $(x_{1i}, x_{2i}, \ldots, x_{ri}, y_i)$, where the x's are again assumed to be known without error while the y's are values of random variables. Data of this kind arise, for example, in studies designed to determine the effect of various climatic conditions on a metal's resistance to corrosion; the effect of kiln temperature, humidity, and iron content on the strength of a ceramic coating; or the effect of factory production, consumption level, and stocks in storage on the price of a product.

As in the case of one independent variable, we shall first treat the problem where the regression equation is linear, namely, where for any given set of values $x_1, x_2, \ldots,$ and x_r the mean of the distribution of the y's is given by

$$\beta_0 + \beta_1 x_1 + \beta_2 x_2 + \ldots + \beta_r x_r$$

For two independent variables, this is the problem of fitting a plane to a set of n points with coordinates (x_{1i}, x_{2i}, y_i) as is illustrated in Figure 11.7.

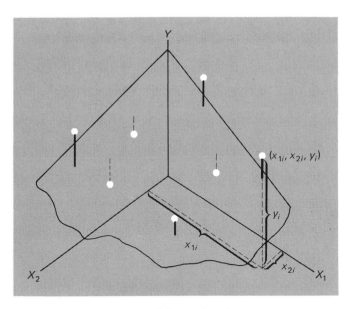

Figure 11.7 Regression plane.

Applying the method of least squares to obtain estimates of the coefficients β_0, β_1, and β_2, we minimize the sum of the squares of the vertical distances from the points to the plane (see Figure 11.7); symbolically, we minimize

$$\sum_{i=1}^{n} [y_i - (\beta_0 + \beta_1 x_{1i} + \beta_2 x_{2i})]^2$$

and it will be left to the reader to verify in Exercise 11.40 on page 320 that the resulting normal equations are

Normal equation for multiple regression with $r = 2$

$$\sum y = nb_0 + b_1 \sum x_1 + b_2 \sum x_2$$
$$\sum x_1 y = b_0 \sum x_1 + b_1 \sum x_1^2 + b_2 \sum x_1 x_2$$
$$\sum x_2 y = b_0 \sum x_2 + b_1 \sum x_1 x_2 + b_2 \sum x_2^2$$

As before, we write the least-squares estimates of β_0, β_1, and β_2 as b_0, b_1, and b_2. Note that in the abbreviated notation $\sum x_1$ stands for $\sum_{i=1}^{n} x_{1i}$, $\sum x_1 x_2$ stands for $\sum_{i=1}^{n} x_{1i} x_{2i}$, $\sum x_1 y$ stands for $\sum_{i=1}^{n} x_{1i} y_i$, and so forth.

EXAMPLE The following are data on the numbers of twists required to break a certain kind of forged alloy bar and the percentages of two alloying elements present in the metal:

Number of twists y	Percent of element A x_1	Percent of element B x_2
38	1	5
40	2	5
85	3	5
59	4	5
40	1	10
60	2	10
68	3	10
53	4	10
31	1	15
35	2	15
42	3	15
59	4	15
18	1	20
34	2	20
29	3	20
42	4	20

Fit a least-squares regression plane and use its equation to estimate the number of twists required to break one of the bars when $x_1 = 2.5$ and $x_2 = 12$.

Solution Substituting $\sum x_1 = 40$, $\sum x_2 = 200$, $\sum x_1^2 = 120$, $\sum x_1 x_2 = 500$, $\sum x_2^2 = 3{,}000$, $\sum y = 733$, $\sum x_1 y = 1{,}989$, and $\sum x_2 y = 8{,}285$ into the normal equations, we get

$$733 = 16b_0 + 40b_1 + 200b_2$$
$$1{,}989 = 40b_0 + 120b_1 + 500b_2$$
$$8{,}285 = 200b_0 + 500b_1 + 3{,}000b_2$$

The unique solution of this system of equations is $b_0 = 48.2$, $b_1 = 7.83$, $b_2 = -1.76$, and the equation of the estimated regression plane is

$$\hat{y} = 48.2 + 7.83x_1 - 1.76x_2$$

Finally, substituting $x_1 = 2.5$ and $x_2 = 12$ into this equation, we get

$$\hat{y} = 48.2 + 7.83(2.5) - 1.76(12)$$
$$= 46.7$$

Note that b_1 and b_2 are estimates of the average change in y resulting from a unit increase in the corresponding independent variable when the other independent variable is held fixed.

When a regression problem involves many variables and (or) the fitting of polynomials, the necessary calculations can become quite tedious. However, if the values of the independent variables can be controlled, it is possible to attain considerable simplifications by using appropriate spacings and subsequent coding (similar to that used in Exercise 11.21 on page 305). Although we shall not illustrate it here, matrix theory can also be used to great advantage.

Since the x_1's as well as the x_2's are equally spaced in the above example and each value (level) of x_1 occurs once in combination with each value (level) of x_2, coding makes possible a substantial reduction in the labor required to obtain and solve the normal equations.

EXAMPLE Rework the preceding example, coding the x_1 values as well as the x_2 values -3, -1, 1, and 3, and calling them z_1 and z_2.

Solution With this coding, the data become

y	z_1	z_2
38	-3	-3
40	-1	-3
85	1	-3
59	3	-3
40	-3	-1
60	-1	-1
68	1	-1
53	3	-1
31	-3	1
35	-1	1
42	1	1
59	3	1
18	-3	3
34	-1	3
29	1	3
42	3	3

and we now get $\sum z_1 = 0$, $\sum z_2 = 0$, $\sum z_1^2 = 80$, $\sum z_1 z_2 = 0$, $\sum z_2^2 = 80$, $\sum y = 733$, $\sum z_1 y = 313$, and $\sum z_2 y = -351$. Thus, the normal equations become

$$733 = 16b_0$$
$$313 = 80b_1$$
$$-351 = 80b_2$$

and, in terms of the coded variables, the estimated regression equation is

$$\hat{y} = 45.8 + 3.91z_1 - 4.39z_2$$

To estimate y for $x_1 = 2.5$ and $x_2 = 12$, we substitute $z_1 = 0$ and $z_2 = -0.2$, and get

$$\hat{y} = 45.8 + 3.91(0) - 4.39(-0.2)$$
$$= 46.7$$

as before.

Had there been five values of x_1 in this example, we would have coded them -2, -1, 0, 1, and 2, as in Exercise 11.21 on page 305; in this kind of coding the values must always be equally spaced and their sum must equal zero.

The same coding would have produced even greater simplifications if we had wanted to fit a second-degree polynomial to our data, that is, if we had used an equation of the form

$$y = \beta_0 + \beta_1 x_1 + \beta_2 x_2 + \beta_3 x_1 x_2 + \beta_4 x_1^2 + \beta_5 x_2^2$$

Without coding, the method of least squares would have led to a system of normal equations consisting of six simultaneous linear equations in six unknowns. As the reader will be asked to verify in Exercise 11.47 on page 321, the same coding as used above reduces this problem, essentially, to that of solving three simultaneous linear equations in three unknowns.

The example of this section served to illustrate how a careful choice of the values of the independent variables can simplify the calculations required in a problem of multiple regression. The principles and procedures for making this choice, whenever possible, are part of the important subject of **experimental design**. In Chapter 12 we shall consider some of the simpler, but nevertheless widely used, experimental designs. We shall focus our attention on what is called the **analysis of variance**, where tests are made concerning the significance of the effects of certain combinations of levels of the independent variables. In Chapter 13 we shall introduce the related problem of **factorial experimentation**.

EXERCISES

11.25 Fit an exponential curve to the data of Exercise 11.9 on page 303.

11.26 The following data pertain to the growth of a colony of bacteria in a culture medium:

| *Days since* | *Bacteria* |
| *inoculation* | *count* |
x	*y*
3	115,000
6	147,000
9	239,000
12	356,000
15	579,000
18	864,000

(a) Plot the data on semilog graph paper to verify that it is reasonable to fit an exponential curve.

(b) Fit an exponential curve to the given data.

(c) Use the result obtained in part (b) to estimate the bacteria count at the end of 20 days.

11.27 The following data pertain to the cosmic ray doses measured at various altitudes:

| *Altitude* | *Dose rate* |
| *(feet)* | *(mrem/year)* |
x	*y*
50	28
450	30
780	32
1,200	36
4,400	51
4,800	58
5,300	69

(a) Fit an exponential curve.

(b) Use the result obtained in part (a) to estimate the mean dose at an altitude of 3,000 feet.

11.28 With reference to the preceding exercise, change the equation obtained in part (a) to the form $\hat{y} = a \cdot e^{-cx}$, and use the result to rework part (b).

11.29 The following data pertain to the demand for a product (in thousands of units) and its price (in cents) charged in five different market areas:

| *Price* | *Demand* |
x	*y*
20	22
16	41
10	120
11	89
14	56

Fit a power function and use it to estimate the demand when the price of the product is 12 cents.

11.30 In an experiment designed to determine the specific heat ratio γ for a certain gas, the gas was compressed adiabatically to several predetermined volumes V, and the corresponding pressure p was measured with the following results:

p (lb/in^2)	16.6	39.7	78.5	115.5	195.3	546.1
V (in^3)	50	30	20	15	10	5

Assuming the ideal gas law $p \cdot V^\gamma = C$, use these data to estimate γ for this gas.

11.31 With reference to the preceding exercise, use the method of Section 11.2 to construct a 95% confidence interval for γ. State what assumptions will have to be made.

11.32 Fit a **Gompertz curve** of the form

$$y = e^{e^{\alpha x + \beta}}$$

to the data of Exercise 11.27 on page 317.

11.33 Plot the curve obtained in the preceding exercise and the one obtained in Exercise 11.27 on one diagram and compare the fit of these two curves.

11.34 The rise of current in an inductive circuit having the time constant τ is given by

$$I = 1 - e^{-t/\tau}$$

where t is the time measured from the instant the switch is closed, and I is the ratio of the current at time t to the full value of the current given by Ohm's law. Given the measurements

I	0.073	0.220	0.301	0.370	0.418	0.467	0.517	0.578
t (sec)	0.1	0.2	0.3	0.4	0.5	0.6	0.7	0.8

estimate the time constant of this circuit from the experimental results given. [*Hint*: Note that the relationship between $\ln(1 - I)$ and t is linear.]

11.35 The number of inches which a newly built structure is settling into the ground is given by

$$y = 3 - 3e^{-\alpha x}$$

where x is its age in months.

x	2	4	6	12	18	24
y	1.07	1.88	2.26	2.78	2.97	2.99

Use the method of least squares to estimate α. [*Hint*: Note that the relationship between $\ln(3 - y)$ and x is linear.]

11.36 The following data pertain to the amount of hydrogen present (y, in parts per million) in core drillings made at 1-foot intervals along the length of a vacuum-cast ingot (x, core location in feet from base):

x	1	2	3	4	5	6	7	8	9	10
y	1.28	1.53	1.03	0.81	0.74	0.65	0.87	0.81	1.10	1.03

(a) Draw a scattergram to check whether it is reasonable to fit a parabola to the given data.

(b) Fit a parabola by the method of least squares.

(c) Use the equation obtained in part (b) to estimate the amount of hydrogen present at $x = 7.5$.

11.37 In the years 1973–1978 the production of uranium in the United States was 13,235, 11,528, 11,600, 12,747, 14,940, and 18,400 short tons (according to the *1979 Statistical Abstract of the United States*). Fit a parabola by the method of least squares and plot it together with the original data on ordinary graph paper.

11.38 When fitting a polynomial to a set of paired data, we usually begin by fitting a straight line and using the method on page 298 to test the null hypothesis $\beta_1 = 0$. Then we fit a second-degree polynomial and test whether it is worthwhile to carry the quadratic term by comparing $\hat{\sigma}_1^2$, the **residual variance** after fitting the straight line, with $\hat{\sigma}_2^2$, the residual variance after fitting the second-degree polynomial. Each of these residual variances is given by the formula

$$\frac{\Sigma(y - \hat{y})^2}{\text{degrees of freedom}}$$

with $\hat{y}$ determined, respectively, from the equation of the line and the equation of the second-degree polynomial. The decision whether to carry the quadratic term is based on the statistic

$$F = \frac{\hat{\sigma}_1^2 - \hat{\sigma}_2^2}{\hat{\sigma}_2^2}$$

which (under the assumptions of Section 11.2) is a value of a random variable having the F distribution with 1 and $n - 3$ degrees of freedom.

(a) Fit a straight line to the varnish-additive and drying-time data on page 310, test the null hypothesis $\beta_1 = 0$ at the 0.05 level of significance, and calculate $\hat{\sigma}_1^2$.

(b) Using the result on page 311 calculate $\hat{\sigma}_2^2$ for the given data and test at the 0.05 level whether we should carry the quadratic term. (Note that we could continue this procedure and test whether to carry a cubic term by means of a corresponding comparison of residual variances. Then we could test whether to carry a fourth-degree term, and so on. It is customary to terminate this procedure after two successive steps have not produced significant results.)

11.39 With reference to the example on page 310, verify that the drying time is a minimum when the amount of additive used is 5.1 grams.

11.40 Verify that the system of normal equations on page 313 leads to the system of normal equations on page 315.

11.41 Twelve specimens of cold-reduced sheet steel, having different copper contents and annealing temperatures, are measured for hardness with the following results:

Hardness (Rockwell 30-T)	Copper content (%)	Annealing temperature (degrees F)
78.9	0.02	1000
65.1	0.02	1100
55.2	0.02	1200
56.4	0.02	1300
80.9	0.10	1000
69.7	0.10	1100
57.4	0.10	1200
55.4	0.10	1300
85.3	0.18	1000
71.8	0.18	1100
60.7	0.18	1200
58.9	0.18	1300

Fit an equation of the form $y = \beta_0 + \beta_1 x_1 + \beta_2 x_2$, where x_1 represents the copper content, x_2 represents the annealing temperature, and y represents the hardness.

11.42 Rework the preceding exercise, coding the x_1 values as $z_1 = -1, 0, +1$, and the x_2 values $z_2 = -3, -1, +1, +3$.

11.43 With reference to the two preceding exercises, estimate the hardness of a sheet of steel with a copper content of 0.05% and an annealing temperature of 1150 degrees Fahrenheit, using the estimated regression equation obtained in
(a) Exercise 11.41;   (b) Exercise 11.42.

11.44 The following are data on the ages and incomes of five executives working for the same company, and the number of years they went to college

Age x_1	Years college x_2	Income (dollars) y
37	4	51,200
45	0	46,800
38	5	55,000
42	2	50,300
31	4	45,400

Fit an equation of the form $y = \beta_0 + \beta_1 x_1 + \beta_2 x_2$ to the given data, and use it to estimate how much on the average an executive working for this company will make if he is 40 years old and has had 4 years of college.

11.45 The following sample data were collected to determine the relationship between two processing variables and the current gain of a certain kind of transistor:

Diffusion time (hours) x_1	*Sheet resistance (ohm-cm)* x_2	*Current gain* y
1.5	66	5.3
2.5	87	7.8
0.5	69	7.4
1.2	141	9.8
2.6	93	10.8
0.3	105	9.1
2.4	111	8.1
2.0	78	7.2
0.7	66	6.5
1.6	123	12.6

Fit a regression plane and use its equation to estimate the expected current gain when the diffusion time is 2.2 hours and the sheet resistance is 90 ohm-cm.

11.46 The following are sample data provided by a moving company on the weights of six shipments, the distances they were moved, and the damage that was incurred:

Weight (1,000 pounds) x_1	*Distance (1,000 miles)* x_2	*Damage (dollars)* y
4.0	1.5	160
3.0	2.2	112
1.6	1.0	69
1.2	2.0	90
3.4	0.8	123
4.8	1.6	186

(a) Fit an equation of the form $y = \beta_0 + \beta_1 x_1 + \beta_2 x_2$.

(b) Use the equation obtained in part (a) to estimate the damage when a shipment weighing 2,400 pounds is moved 1,200 miles.

11.47 Verify that if the x's are coded as on page 315, the normal equations for fitting a second-degree polynomial of the form

$$y = \beta_0 + \beta_1 x_1 + \beta_2 x_2 + \beta_3 x_1 x_2 + \beta_4 x_1^2 + \beta_5 x_2^2$$

to the data on page 314 dealing with the number of twists required to break certain forged alloy bars, become

$$733 = 16b_0 \qquad\qquad\qquad + 80b_4 + 80b_5$$
$$315 = \qquad 80b_1$$
$$351 = \qquad\qquad 80b_2$$
$$-79 = \qquad\qquad\qquad 400b_3$$
$$3453 = 80b_0 \qquad\qquad\qquad + 656b_4 + 400b_5$$
$$3493 = 80b_0 \qquad\qquad\qquad + 400b_4 + 656b_5$$

11.5 CORRELATION

So far in this chapter, we have studied problems where the independent variable (or variables) was assumed to be known without error. Although this applies to many experimental situations, there are also problems where the x's as well as the y's are values assumed by random variables. This would be the case, for instance, if we studied the relationship between rainfall and the yield of a certain crop, the relationship between the tensile strength and the hardness of aluminum, or the relationship between impurities in the air and the incidence of a certain disease. Problems like these are referred to as problems of **correlation analysis** where it is assumed that the data points (x_i, y_i) for $i = 1, 2, \ldots, n$ are values of a pair of random variables whose joint density is given by $f(x, y)$.

The bivariate density which is most commonly used in problems of correlation analysis is the **bivariate normal distribution**. We shall introduce it here in terms of the conditional density $g_2(y|x)$ and the marginal density $f_1(x)$, as defined in Section 4.10. So far as $g_2(y|x)$ is concerned, the conditions we shall impose are practically identical with the ones we used in connection with the sampling theory of Section 11.2. For any given x, it will be assumed that $g_2(y|x)$ is a normal distribution with the mean $\alpha + \beta x$ and the variance σ^2. Thus, the regression of y on x is linear and the variance of the conditional density does not depend on x. Furthermore, we shall assume that the marginal density $f_1(x)$ is normal with the mean μ_1 and the variance σ_1^2. Making use of the relationship $f(x, y) = f_1(x) \cdot g_2(y|x)$ given on page 130, we thus obtain

$$f(x, y) = \frac{1}{\sqrt{2\pi}\sigma_1} e^{-(x-\mu_1)^2/2\sigma_1^2} \cdot \frac{1}{\sqrt{2\pi}\sigma} e^{-[y-(\alpha+\beta x)]^2/2\sigma^2}$$

$$= \frac{1}{2\pi \cdot \sigma \cdot \sigma_1} e^{-\{[y-(\alpha+\beta x)]^2/2\sigma^2 + (x-\mu_1)^2/2\sigma_1^2\}}$$

for $-\infty < x < \infty$ and $-\infty < y < \infty$. Note that this joint distribution involves the *five* parameters μ_1, σ_1, α, β, and σ.

For reasons of symmetry and other considerations to be explained below, it is customary to express the bivariate normal density in terms of the parameters $\mu_1, \sigma_1, \mu_2, \sigma_2$, and ρ. Here μ_2 and σ_2^2 are the mean and the variance of the marginal distribution $f_2(y)$, while ρ (*rho*), called the **population correlation coefficient**, is defined by

$$\rho^2 = 1 - \frac{\sigma^2}{\sigma_2^2}$$

with ρ taken to be positive when β is positive, and negative when β is negative. Leaving it to the reader to show in Exercise 11.59 on page 329 that

$$\mu_2 = \alpha + \beta\mu_1 \quad \text{and} \quad \sigma_2^2 = \sigma^2 + \beta^2\sigma_1^2$$

we then substitute into the above expression for $f(x, y)$ and obtain the following form of the bivariate normal distribution:

$$f(x, y) = \frac{1}{2\pi \cdot \sigma_1\sigma_2\sqrt{1 - \rho^2}} e^{-\left[\left(\frac{x-\mu_1}{\sigma_1}\right)^2 - 2\rho\left(\frac{x-\mu_1}{\sigma_1}\right)\left(\frac{y-\mu_2}{\sigma_2}\right) + \left(\frac{y-\mu_2}{\sigma_2}\right)^2\right]/2(1-\rho^2)}$$

for $-\infty < x < \infty$ and $-\infty < y < \infty$ (see Exercise 11.60 on page 329).

Concerning the correlation coefficient ρ, note that $-1 \le \rho \le +1$ since $\sigma_2^2 = \sigma^2 + \beta^2\sigma_1^2$ and, hence, $\sigma_2^2 \ge \sigma^2$. Furthermore, ρ can equal -1 or $+1$ only when $\sigma^2 = 0$, which represents the degenerate case where all the probability is concentrated along the line $y = \alpha + \beta x$ and there is, thus, a perfect linear relationship between the two random variables. (That is, for a given value of x, y must equal $\alpha + \beta x$.) The correlation coefficient is equal to zero if and only if $\sigma^2 = \sigma_2^2$, and it follows from the identity $\sigma_2^2 = \sigma^2 + \beta^2\sigma_1^2$ that this can happen only when $\beta = 0$. Thus, $\rho = 0$ implies that the regression line of y on x is a horizontal line and, hence, that knowledge of x does not help in the prediction of y. Thus, when $\rho = \pm 1$ we say that there is a perfect linear correlation (relationship, or association) between the two random variables; when $\rho = 0$ we say that there is no correlation (relationship, or association) between the two random variables. In fact, $\rho = 0$ implies for the bivariate normal density that the two random variables are independent (see Exercise 11.61 on page 329).

For values between 0 and $+1$ or 0 and -1, we interpret ρ by referring back to the identity

$$\rho^2 = 1 - \frac{\sigma^2}{\sigma_2^2} = \frac{\sigma_2^2 - \sigma^2}{\sigma_2^2}$$

given above. Since σ^2 is a measure of the variation of the y's when x is known while σ_2^2 is a measure of the variation of the y's when x is unknown, $\sigma_2^2 - \sigma^2$ measures the variation of the y's that is accounted for by the linear relationship with x, and ρ^2 **tells us what proportion of the variation of the y's can be attributed to the linear relationship with x.**

The same argument applies also to r^2, the square of the **sample correlation coefficient**, given by

Sample correlation coefficient

$$r = \frac{S_{xy}}{\sqrt{S_{xx} \cdot S_{yy}}}$$

for any n pairs of values (x_i, y_i); S_{xy}, S_{xx}, and S_{yy} are as defined on page 295. This estimator is not unbiased and, except for the factor $\sqrt{\frac{n-1}{n-2}}$, it may be obtained by substituting for σ_2^2 the sample variance of the y's and for σ^2 the square of the standard error of estimate, s_e^2 (see also Exercise 11.62 on page 330).

EXAMPLE The following are the numbers of minutes it took 10 mechanics to assemble a piece of machinery in the morning, x, and in the late afternoon, y:

x	y
11.1	10.9
10.3	14.2
12.0	13.8
15.1	21.5
13.7	13.2
18.5	21.1
17.3	16.4
14.2	19.3
14.8	17.4
15.3	19.0

Calculate r.

Solution First we determine the necessary summations, getting $\sum x = 142.3$, $\sum y = 166.8$, $\sum x^2 = 2,085.31$, $\sum xy = 2,434.69$, $\sum y^2 = 2,897.80$. Then, substituting into the formulas on page 295, we get

$$S_{xx} = 10(2,085.31) - (142.3)^2 = 603.81$$
$$S_{xy} = 10(2,434.69) - (142.3)(166.8) = 611.26$$
$$S_{yy} = 10(2,897.80) - (166.8)^2 = 1,155.76$$

and, hence,

$$r = \frac{611.26}{\sqrt{(603.81)(1,155.76)}} = 0.73$$

The result we have obtained here implies that $100r^2 = 53\%$ of the variation among the afternoon times is explained by (is accounted for or may be attributed to) corresponding differences among the morning times.

Whenever a value of r is based on a random sample from a bivariate normal population, we can perform a test of significance (a test of the null hypothesis $\rho = \rho_0$) or construct a confidence interval for ρ on the basis of the following transformation:

Fisher Z transformation

$$Z = \frac{1}{2} \ln \frac{1+r}{1-r}$$

This statistic is a value of a random variable having approximately a normal distribution with the mean $\mu_z = \frac{1}{2} \ln \frac{1+\rho}{1-\rho}$ and the variance $\frac{1}{n-3}$. Thus, we can base inferences about ρ on

Statistic for inferences about ρ

$$z = \frac{Z - \mu_z}{1/\sqrt{n-3}} = \frac{\sqrt{n-3}}{2} \cdot \ln \frac{(1+r)(1-\rho)}{(1-r)(1+\rho)}$$

which is a value of a random variable having approximately the standard normal distribution.

In particular, we can test the null hypothesis of no correlation, namely, the null hypothesis $\rho = 0$, with the statistic

Statistic for test of null hypothesis ρ = 0

$$z = \sqrt{n-3} \cdot Z = \frac{\sqrt{n-3}}{2} \cdot \ln \frac{1+r}{1-r}$$

These tests are facilitated by the use of Table 11, which gives the values of Z corresponding to $r = 0.00, 0.01, 0.02, \ldots$, and 0.99. When r is negative, we look up the value of Z corresponding to $-r$, and then take $-Z$.

EXAMPLE   With reference to the preceding example, where we had $n = 10$ and $r = 0.73$, test the null hypothesis $\rho = 0$ against the alternative hypothesis $\rho \neq 0$ at the 0.05 level of significance.

   Solution
   1. *Null hypothesis*: $\rho = 0$
      Alternative hypothesis: $\rho \neq 0$
   2. *Level of significance*: $\alpha = 0.05$

3. *Criterion*: Reject the null hypothesis if $z < -1.96$ or $z > 1.96$, where $z = \sqrt{n-3} \cdot Z$.

4. *Calculations*: The value of Z corresponding to $r = 0.73$ is 0.929 according to Table 11, so that

$$z = \sqrt{10-3} \cdot (0.929) = 2.46$$

5. *Decision*: Since $z = 2.46$ exceeds 1.96, the null hypothesis must be rejected; we conclude that there is a relationship between the morning and late afternoon time it takes a mechanic to assemble the given kind of machinery.

To construct a confidence interval for ρ, we first construct a confidence interval for μ_Z, the mean of the sampling distribution of Z, and then convert to r and ρ by means of Table 11. Making use of the theory on page 325, we can write the first of these confidence intervals as

Confidence interval for μ_Z

$$Z - \frac{z_{\alpha/2}}{\sqrt{n-3}} < \mu_Z < Z + \frac{z_{\alpha/2}}{\sqrt{n-3}}$$

EXAMPLE   If $r = 0.70$ for the mathematics and physics grades of 30 students, construct a 95% confidence interval for the population correlation coefficient.

Solution   Reading the value of Z which corresponds to $r = 0.70$ from Table 11, and substituting it together with $n = 30$ and $z_{0.025} = 1.96$ into the above confidence interval formula for μ_Z, we get

$$0.867 - \frac{1.96}{\sqrt{27}} < \mu_Z < 0.867 + \frac{1.96}{\sqrt{27}}$$

or

$$0.490 < \mu_Z < 1.244$$

Then, looking up the values of r for which Z is closest to 0.490 and 1.244 in Table 11, we get the 95% confidence interval

$$0.45 < \rho < 0.85$$

for the true strength of the linear relationship between grades of students in the two given subjects.

EXAMPLE If $r = 0.20$ for a random sample of $n = 40$ paired data, construct a 95% confidence interval for ρ.

> **Solution** Reading the value of Z which corresponds to $r = 0.20$ from Table 11, and substituting it together with $n = 40$ and $z_{0.025} = 1.96$ into the above confidence-interval formula for μ_Z, we get

$$0.203 - \frac{1.96}{\sqrt{37}} < \mu_z < 0.203 + \frac{1.96}{\sqrt{37}}$$

or

$$-0.119 < \mu_z < 0.525$$

> Then, looking up the values of r for which Z is closest to 0.119 and 0.525 in Table 11, we get the 95% confidence interval

$$-0.12 < \rho < 0.48$$

for the population correlation coefficient.

Note that in both of these examples the confidence intervals for ρ are fairly wide; this illustrates the fact that correlation coefficients based on relatively small samples are generally not very reliable.

There are several serious pitfalls in the interpretation of the coefficient of correlation. First, it must be emphasized that r is an estimate of the strength of the linear relationship between the values of two random variables; thus, as is shown in Figure 11.8, r may be close to 0 when there is actually a strong (but nonlinear) relationship. Second, and perhaps of greatest

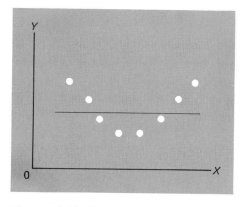

Figure 11.8 Nonlinear relationship where $r = 0$.

importance, a significant correlation does not necessarily imply a causal relationship between the two random variables. Although it would not be surprising, for example, to obtain a high positive correlation between the annual sales of chewing gum and the incidence of crime in the United States, one cannot conclude that crime might be reduced by prohibiting the sale of chewing gum. Both variables depend upon the size of the population, and it is this mutual relationship with a third variable (population size) which produces the positive correlation.

EXERCISES

11.48 Calculate r for the air velocities and evaporation coefficients of the example on page 294. Also, assuming that the necessary assumptions can be met, test the null hypothesis $\rho = 0$ against the alternative hypothesis $\rho \neq 0$ at the 0.05 level of significance.

11.49 The following data pertain to the resistance (ohms) and the failure time (minutes) of certain overloaded resistors:

Resistance	Failure time	Resistance	Failure time
43	32	36	36
29	20	39	33
44	45	36	21
33	35	47	44
33	22	28	26
47	46	40	45
34	28	42	39
31	26	33	25
48	37	46	36
34	33	28	25
46	47	48	45
37	30	45	36

Calculate r.

11.50 With reference to the preceding exercise, test for significance at $\alpha = 0.01$.

11.51 Calculate r for the extraction times and extraction efficiencies of Exercise 11.1 on page 301. Assuming that the necessary assumptions can be met, test the null hypothesis $\rho = 0.75$ against the alternative hypothesis $\rho > 0.75$ at the 0.05 level of significance.

11.52 Calculate r for the humidities and moisture contents of Exercise 11.12 on page 303. Assuming that the necessary assumptions can be met, construct a 95% confidence interval for the population correlation coefficient ρ.

11.53 The following are measurements of the carbon content and the permeability index of 22 sinter mixtures:

Carbon content (%)	Permeability index	Carbon content (%)	Permeability index
4.4	12	4.1	13
5.5	14	4.9	19
4.2	18	4.7	22
3.0	35	5.0	20
4.5	23	4.6	16
4.9	29	3.6	27
4.6	16	4.9	21
5.0	12	5.1	13
4.7	18	4.8	18
5.1	21	5.2	17
4.4	27	5.2	11

(a) Calculate r.
(b) Find 99% confidence limits for ρ.

11.54 If $r = 0.41$ for one set of paired data and $r = 0.29$ for another, compare the strengths of the two relationships.

11.55 If data on the ages and prices of 25 pieces of equipment yielded $r = -0.58$, test the null hypothesis $\rho = -0.40$ against the alternative hypothesis $\rho < -0.40$ at the 0.05 level of significance.

11.56 If for certain paired data $n = 18$ and $r = 0.44$, test the null hypothesis $\rho = 0.30$ against the alternative hypothesis $\rho > 0.30$ at the 0.01 level of significance.

11.57 Assuming that the necessary assumptions are met, construct 95% confidence intervals for ρ when
(a) $r = 0.78$ and $n = 15$; (c) $r = 0.17$ and $n = 35$.
(b) $r = -0.62$ and $n = 32$;

11.58 Assuming that the necessary assumptions are met, construct 99% confidence intervals for ρ when
(a) $r = -0.87$ and $n = 19$; (c) $r = 0.16$ and $n = 40$.
(b) $r = 0.39$ and $n = 24$;

11.59 Evaluating the necessary integrals, verify the identities

$$\mu_2 = \alpha + \beta\mu_1 \quad \text{and} \quad \sigma_2^2 = \sigma^2 + \beta^2\sigma_1^2$$

on page 323.

11.60 Substitute $\mu_2 = \alpha + \beta\mu_1$ and $\sigma_2^2 = \sigma^2 + \beta^2\sigma_1^2$ into the formula for the bivariate density given on page 322, and show that this gives the final form shown on page 323.

11.61 Show that for the bivariate normal distribution
(a) independence implies zero correlation;
(b) zero correlation implies independence.

11.62 By substituting the sample variance of the y's for $\sigma_{\hat{2}}^2$ and s_e^2 for σ^2 in the formula

$$\rho^2 = 1 - \frac{\sigma^2}{\sigma_{\hat{2}}^2}$$

verify the formula for r on page 324 (except for a multiplicative constant).

11.63 Instead of using the computing formula on page 324, we can obtain the correlation coefficient r with the formula

$$r = \pm\sqrt{1 - \frac{\sum (y - \hat{y})^2}{\sum (y - \bar{y})^2}}$$

which is analogous to the formula used to define ρ. Although the computations required by the use of this formula are tedious, the formula has the advantage that it can be used also to measure the strength of nonlinear relationships or relationships in several variables. For instance, in the multiple linear regression example on page 314, one could calculate the predicted values by means of the equation

$$\hat{y} = 48.2 + 7.83x_1 - 1.76x_2$$

and then determine r as a measure of how strongly y, the twist required to break one of the forged alloy bars, depends on both percentages of alloying elements present.

(a) Using the data on page 314, find $\sum (y - \bar{y})^2$ by means of the formula $\sum (y - \bar{y})^2 = \sum y^2 - n\bar{y}^2$.

(b) Using the regression equation obtained on page 314, calculate $\hat{y}$ for the sixteen points and then determine $\sum (y - \hat{y})^2$.

(c) Substitute the results obtained in (a) and (b) into the above formula for r. The result is called the **multiple correlation coefficient**.

11.64 With reference to Exercise 11.45 on page 321, use the theory of the preceding exercise to calculate the multiple correlation coefficient (which measures how strongly the current gain is related to the two independent variables).

11.65 With reference to Exercise 11.46 on page 321, use the theory of Exercise 11.63 to calculate the multiple correlation coefficient (which measures how strongly the damage is related to both weight and distance).

11.66 When we substitute for the actual values of paired data the ranks which the values occupy in the respective samples, the correlation coefficient can be written in the form

$$r_S = 1 - \frac{6 \cdot \sum d_i^2}{n(n^2 - 1)}$$

in which it is called the **rank-correlation coefficient** or **Spearman's rank-correlation coefficient**. In this formula, d_i is the difference between the ranks of the paired observations (x_i, y_i) and n is the number of pairs in the sample. The formula for r_S follows from that for r when there are no ties in rank; if there are ties, we substitute for each of the tied observations the mean of the ranks which they jointly occupy, and there may be a small difference between the values of r and r_S obtained for the two sets of ranks.

(a) Calculate r_S for the example on page 324, which deals with the time it takes mechanics to assemble a certain piece of machinery in the morning and in the late afternoon. Compare the result with the value of r obtained for the original data.

(b) Calculate r_S for the data of Exercise 11.49 on page 328 and compare the result with the value of r obtained for the original data.

(c) Calculate r_S for the air velocities and evaporation coefficients of the example on page 294 and compare the result with the value of r obtained for the original data in Exercise 11.48 on page 328.

11.67 Since the assumptions underlying the significance test of Section 11.5 are rather stringent, it is often desirable to use a nonparametric alternative based on the rank-correlation coefficient defined in the preceding exercise. It is based on the theory that if there is no relationship between the x's and the y's (in fact, if they are randomly matched), the sampling distribution of r_S can be approximated closely with a normal distribution having zero mean and the standard deviation

$$\sigma_{r_S} = \frac{1}{\sqrt{n-1}}$$

Use the 0.01 level of significance to test the significance of the values of r_S obtained in the three parts of Exercise 11.66.

11.68 If a sample of $n = 18$ pairs of data yielded $r_S = 0.39$, is this rank-correlation coefficient significant at the 0.05 level of significance? [*Hint*: Refer to the two preceding exercises.]

11.69 If a sample of $n = 40$ pairs of data yielded $r_S = 0.48$, is this rank-correlation coefficient significant at the 0.01 level of significance? [*Hint*: Refer to Exercises 11.66 and 11.67.]

11.6 CHECK LIST OF KEY TERMS (with page references)

12

Analysis of Variance

Some of the examples of Chapter 11 have already taught us that considerable economies in calculation can result by appropriately planning an experiment in advance. What is even more important, proper experimental planning can give a reasonable assurance that the results of experiment will provide clear-cut answers to questions under investigation. While it is impossible to give in this chapter a complete discussion of experimental design, including the many pitfalls to which the experimenter is exposed, we shall begin by presenting some of the general principles of experimental design. Several of the designs most frequently used in engineering and other applied research will be given in subsequent sections.

In Sections 12.2 and 12.3 we shall discuss the often used one- and two-way classification designs, in Section 12.5 we introduce Latin-square and Graeco-Latin-square designs, and in the remainder of the chapter we introduce tests for comparing several means as well as the analysis of a balanced experiment in the presence of a concomitant variable, or covariate.

Many of the most important aspects of **experimental design** can be illustrated by means of an example drawn from the important field of engineering measurement.

Suppose that a steel mill supplies tin plate to three can manufacturers, the major specification being that the tin-coating weight should be at least 0.25 pound per base box. The mill and each can manufacturer has a laboratory where measurements are made of the tin-coating weights of samples taken from each shipment. Suppose, also, that some disagreement has arisen about the actual tin-coating weights of the tin plate being shipped, and it is decided to plan an experiment to determine whether the four laboratories are making consistent measurements. A complicating factor is that part of the measuring process consists of the chemical removal of the tin from the surface of the base metal; thus, it is impossible to have the identical sample measured by each laboratory to determine how closely the measurements correspond.

One possibility is to send several samples (in the form of circular discs having equal areas) to each of the laboratories. Although these discs may not actually have identical tin-coating weights, it is hoped that such differences will be small and that they will more or less "average out." In other words, it will be assumed that whatever differences there may be among the means of the four samples can be attributed to no other causes but systematic differences in measuring techniques and chance variability. This would make it possible to determine whether the results produced by the laboratories are consistent by comparing the variability of the four sample means with an appropriate measure of chance variation.

Now there remains the problem of deciding how many discs are to be sent to each laboratory and how the discs are actually to be selected. The question of sample size can be answered in many different ways, one of which is to use the formula on page 219 for the standard deviation of the sampling distribution of the difference between two means. Substituting known values of σ_1 and σ_2 and specifying what differences between the true means of any two of the laboratories should be detected with a probability of at least 0.95 (or 0.98, or 0.99), it is possible to determine $n_1 = n_2 = n$ (see Exercise 12.12 on page 347). Suppose that this method and, perhaps, also considerations of cost and availability of the necessary specimens lead to the decision to send a sample of 12 discs to each laboratory.

The problem of selecting the required 48 discs and allocating 12 to each laboratory is not as straightforward as it may seem at first. To begin with, suppose that a sheet of tin plate, sufficiently long and sufficiently wide, is selected and that the 48 discs are cut as shown in Figure 12.1. The 12 discs cut from strip 1 are sent to the first laboratory, the 12 discs from strip 2 are sent to the second laboratory, and so forth. If the four mean coating weights subsequently obtained were then found to differ significantly, would this

Figure 12.1 Numbering of tin-plate samples.

allow us to conclude that these differences can be attributed to lack of consistency in the measuring techniques? Suppose, for instance, that additional investigation shows that the amount of tin deposited electrolytically on a long sheet of steel has a distinct and repeated pattern of variation perpendicular to the direction in which it is rolled. (Such a pattern might be caused by the placement of electrodes, "edge effects," and so forth.) Thus, even if all four laboratories measured the amount of tin consistently and without error, there could be cause for differences in the tin-coating weight determinations. The allocation of an entire strip of discs to each laboratory is such that inconsistencies among the laboratories' measuring techniques are inseparable from (or **confounded** with) whatever differences there may be in the actual amount of tin deposited perpendicular to the direction in which the sheet of steel is rolled.

One way to avoid this kind of confounding is to number the discs and allocate them to the four laboratories at random, such as in the following arrangement, which was obtained with the aid of a table of random numbers:

Laboratory A: 3, 38, 17, 32, 24, 30, 48, 19, 11, 31, 22, 41
Laboratory B: 44, 20, 15, 25, 45, 4, 14, 5, 39, 7, 40, 34
Laboratory C: 12, 21, 42, 8, 27, 16, 47, 46, 18, 43, 35, 26
Laboratory D: 9, 2, 28, 23, 37, 1, 10, 6, 29, 36, 33, 13

If there were any actual pattern of tin-coating thickness on the sheet of tin plate, it would be "broken up" by the **randomization**.

Although we have identified and counteracted one possible systematic pattern of variation, there is no assurance that there can be no others. For instance, there may be systematic differences in the areas of the discs caused by progressive wear of the cutting instrument, or there may be scratches or other imperfections on one part of the sheet which could affect the measurements. Thus, there is always the possibility that differences in means attributed to inconsistencies among the laboratories are actually caused by some other uncontrolled variable, and it is the purpose of randomization to avoid confounding the variable under investigation with such other variables.

By distributing the 48 discs among the four laboratories entirely at random, we have no choice but to include whatever variation may be attributable to extraneous causes under the heading of "chance variation." This may give us an excessively large estimate of chance variation which, in turn, may make it difficult to detect differences between the true laboratory means. In order to avoid this, we could, perhaps, use only discs cut from the same strip (or from an otherwise homogeneous region). Unfortunately, this kind of **controlled experimentation** presents us with new complications. Of what use would it be, for example, to perform an experiment which allows us to conclude that the laboratories are consistent (or inconsistent), *if such a conclusion is limited to measurements made at a fixed distance from one edge of a sheet*? To consider a more poignant example, suppose that a manufacturer of plumbing materials wishes to compare the performance of several kinds of material to be used in underground water pipes. If such conditions as soil acidity, depth of pipe, and mineral content of water were all held fixed, any conclusions as to which material is best would be valid only for the given set of conditions. What the manufacturer really wants to know is which material is best over a fairly wide variety of conditions, and in designing a suitable experiment it would be advisable (indeed, necessary) to specify that pipe of each material be buried at each of several depths in each of several kinds of soil, and in locations where the water varies in hardness.

This example serves to illustrate that it is seldom desirable to hold all or most extraneous factors fixed throughout an experiment in order to obtain an estimate of chance variation that is not "inflated" by variations due to other causes. (In fact, it is rarely, if ever, possible to exercise such strict control, that is, to hold *all* extraneous variables fixed.) In actual practice, experiments should be planned so that known sources of variability are deliberately varied over as wide a range as necessary; furthermore, they should be varied in such a way that their variability can be eliminated from the estimate of chance variation. One way to accomplish this is to repeat the experiment in several **blocks**, where known sources of variability (that is, extraneous variables) are held fixed in each block, but vary from block to block.

In the tin-plating problem we might thus account for variations across the sheet of steel by randomly allocating three discs from each strip to each of the laboratories as in the following arrangement:

	Strip 1	*Strip 2*	*Strip 3*	*Strip 4*
Laboratory A:	8, 4, 10	23, 24, 19	26, 29, 35	37, 44, 48
Laboratory B:	2, 6, 12	21, 15, 22	34, 33, 32	45, 43, 46
Laboratory C:	1, 5, 11	16, 20, 13	36, 29, 30	41, 38, 47
Laboratory D:	7, 3, 9	17, 18, 14	28, 31, 25	39, 40, 42

In this experimental layout, the strips form the blocks, and if we base our estimate of chance variation on the variability *within* each of the 16 sets of three discs, this estimate will not be inflated by the extraneous variable, that is, differences among the strips. (Note also that, with this arrangement, differences among the means obtained for the four laboratories cannot be attributed to differences among the strips. This cannot be said for the arrangement on page 334.)

The analysis of experiments in which **blocking** is used to eliminate one source of variability will be discussed in Section 12.3. The analysis of experiments in which two or three sources of variability are thus eliminated will be treated in Section 12.5.

12.2 COMPLETELY RANDOMIZED DESIGNS

In this section we shall consider, in general, the statistical analysis of the **completely randomized design**, or **one-way classification**. We shall suppose that the experimenter has available the results of k independent random samples, each of size n, from k different populations (that is, data concerning k treatments, k groups, k methods of production, etc.); and he is concerned with testing the hypothesis that the means of these k populations are all equal. An example of such an experiment, with $k = 4$, is given by the layout on page 334. If we denote the jth observation in the ith sample by y_{ij}, the general schema for a one-way classification is as follows:

$$\begin{array}{llll} & & & \textit{Means} \\ \textit{Sample 1:} & y_{11}, y_{12}, \ldots, y_{1j}, \ldots, y_{1n} & & \bar{y}_1 \\ \textit{Sample 2:} & y_{21}, y_{22}, \ldots, y_{2j}, \ldots, y_{2n} & & \bar{y}_2 \\ & \qquad\qquad \cdot \quad\cdot\quad\cdot\quad\cdot\quad\cdot \\ \textit{Sample i:} & y_{i1}, y_{i2}, \ldots, y_{ij}, \ldots, y_{in} & & \bar{y}_i \\ & \qquad\qquad \cdot \quad\cdot\quad\cdot\quad\cdot\quad\cdot \\ \textit{Sample k:} & y_{k1}, y_{k2}, \ldots, y_{kj}, \ldots, y_{kn} & & \bar{y}_k \\ & & & \overline{\bar{y}.} \end{array}$$

With reference to the experimental layout on page 334, y_{ij} ($i = 1, 2, 3, 4$; $j = 1, 2, \ldots, 12$) is the jth tin-coating weight measured by the ith laboratory, $\bar{y}_i$ is the mean of the measurements obtained by the ith laboratory, and $\bar{y}.$ is the overall mean (or **grand mean**) of all 48 observations.

To be able to test the hypothesis that the samples were obtained from k populations with equal means, we shall make several assumptions. Specifically, it will be assumed that we are dealing with *normal populations* having *equal variances*. There exist methods for testing the reasonableness of this last assumption (see the book by A. M. Mood and F. A. Graybill mentioned in the Bibliography), but the methods we shall develop in this chapter are fairly

robust; that is, they are relatively insensitive to violations of the assumption of normality as well as the assumption of equal variances.

If μ_i denotes the mean of the ith population and σ^2 denotes the common variance of the k populations, we can express each observation y_{ij} as μ_i plus the value of a random component; that is, we can write

$$y_{ij} = \mu_i + \epsilon_{ij} \quad \text{for } i = 1, 2, \ldots, k; j = 1, 2, \ldots, n$$

In accordance with the above assumptions, the ϵ_{ij} are values of independent, normally distributed random variables with zero means and the common variance σ^2.†

To attain uniformity with corresponding equations for more complicated kinds of designs, it is customary to replace μ_i by $\mu + \alpha_i$, where μ is the mean of the μ_i and α_i is the **effect** of the ith treatment; hence, $\sum_{i=1}^{k} \alpha_i = 0$ (see Exercise 12.13 on page 347). Using these new parameters, we can write the model equation for the one-way classification as

Model equation for one-way classification

$$y_{ij} = \mu + \alpha_i + \epsilon_{ij} \quad \text{for } i = 1, 2, \ldots, k; j = 1, 2, \ldots, n$$

and the null hypothesis that the k population means are all equal can be replaced by the null hypothesis that $\alpha_1 = \alpha_2 = \ldots = \alpha_k = 0$. The alternative hypothesis that at least two of the population means are unequal is equivalent to the alternative hypothesis that $\alpha_i \neq 0$ for some i.

To test the null hypothesis that the k population means are all equal, we shall compare two estimates of σ^2—one based on the variation among the sample means, and one based on the variation within the samples. Since, by assumption, each sample comes from a population having the variance σ^2, this variance can be estimated by any one of the sample variances

$$s_i^2 = \sum_{j=1}^{n} \frac{(y_{ij} - \bar{y}_i)^2}{n-1}$$

and hence, also by their mean

$$\hat{\sigma}_W^2 = \sum_{i=1}^{k} \frac{s_i^2}{k} = \sum_{i=1}^{k} \sum_{j=1}^{n} \frac{(y_{ij} - \bar{y}_i)^2}{k(n-1)}$$

---

†Note that this equation, or model, can be regarded as a multiple regression equation; introducing the variables x_{il} which equal 0 or 1 depending on whether the two subscripts are unequal or equal, we can write

$$y_{ij} = \mu_1 x_{i1} + \mu_2 x_{i2} + \ldots + \mu_k x_{ik} + \epsilon_{ij}$$

The parameters μ_i can now be interpreted as regression coefficients, and they can be estimated by the least-squares methods of Chapter 11.

Note that each of the sample variances s_i^2 is based on $n - 1$ degrees of freedom ($n - 1$ independent deviations from $\bar{y}_i$) and, hence, $\hat{\sigma}_W^2$ is based on $k(n - 1)$ degrees of freedom. Now, the variance of the k sample means is given by

$$s_{\bar{x}}^2 = \sum_{i=1}^{k} \frac{(\bar{y}_i - \bar{y}.)^2}{k - 1}$$

and *if the null hypothesis is true* it estimates σ^2/n. Thus, an estimate of σ^2 based on the differences among the sample means is given by

$$\hat{\sigma}_B^2 = n \cdot s_{\bar{x}}^2 = n \cdot \sum_{i=1}^{k} \frac{(\bar{y}_i - \bar{y}.)^2}{k - 1}$$

and it is based on $k - 1$ degrees of freedom.

If the null hypothesis is true, it can be shown that $\hat{\sigma}_W^2$ and $\hat{\sigma}_B^2$ are independent estimates of σ^2, and it follows that

$$F = \frac{\hat{\sigma}_B^2}{\hat{\sigma}_W^2}$$

is a value of a random variable having the F distribution with $k - 1$ and $k(n - 1)$ degrees of freedom. Since the **between-sample variance**, $\hat{\sigma}_B^2$, can be expected to exceed the **within-sample variance**, $\hat{\sigma}_W^2$, when the null hypothesis is *false*, the null hypothesis will be rejected if F exceeds F_α, where F_α is obtained from Table 6 with $k - 1$ and $k(n - 1)$ degrees of freedom.

The preceding argument has shown how the test of the equality of k means can be based on the comparison of two variance estimates. More remarkable, perhaps, is the fact that the two estimates in question [except for the divisors $k - 1$ and $k(n - 1)$] can be obtained by "breaking up" or analyzing the total variance of all nk observations into two parts. The sample variance of all nk observations is given by

$$s^2 = \sum_{i=1}^{k} \sum_{j=1}^{n} \frac{(y_{ij} - \bar{y}.)^2}{nk - 1}$$

and with reference to its numerator, called the **total sum of squares**, we shall now prove the following theorem.

Identity for one-way analysis of variance

Theorem 12.1

$$\sum_{i=1}^{k} \sum_{j=1}^{n} (y_{ij} - \bar{y}.)^2 = \sum_{i=1}^{k} \sum_{j=1}^{n} (y_{ij} - \bar{y}_i)^2 + n \cdot \sum_{i=1}^{k} (\bar{y}_i - \bar{y}.)^2$$

The proof of this theorem is based on the identity

$$y_{ij} - \bar{y}.. = (y_{ij} - \bar{y}_i) + (\bar{y}_i - \bar{y}..)$$

Squaring both sides and summing on i and j, we obtain

$$\sum_{i=1}^{k} \sum_{j=1}^{n} (y_{ij} - \bar{y}..)^2 = \sum_{i=1}^{k} \sum_{j=1}^{n} (y_{ij} - \bar{y}_i)^2 + \sum_{i=1}^{k} \sum_{j=1}^{n} (\bar{y}_i - \bar{y}..)^2$$

$$+ 2 \sum_{i=1}^{k} \sum_{j=1}^{n} (y_{ij} - \bar{y}_i)(\bar{y}_i - \bar{y}..)$$

Next, we observe that

$$\sum_{i=1}^{k} \sum_{j=1}^{n} (y_{ij} - \bar{y}_i)(\bar{y}_i - \bar{y}..) = \sum_{i=1}^{k} (\bar{y}_i - \bar{y}..) \sum_{j=1}^{n} (y_{ij} - \bar{y}_i) = 0$$

since $\bar{y}_i$ is the mean of the ith sample and, hence, $\sum_{j=1}^{n} (y_{ij} - \bar{y}_i) = 0$ for all i.

To complete the proof of Theorem 12.1, we have only to observe that the summand of the second sum on the right-hand side of the above identity does not involve the subscript j and that, consequently,

$$\sum_{i=1}^{k} \sum_{j=1}^{n} (\bar{y}_i - \bar{y}..)^2 = n \cdot \sum_{i=1}^{k} (\bar{y}_i - \bar{y}..)^2$$

It is customary to denote the total sum of squares, the left-hand member of the identity of Theorem 12.1, by *SST*. The first term on the right-hand side is $\hat{\sigma}_W^2$ times its degrees of freedom, and we refer to this sum as the **error sum of squares**, *SSE*. The term "error sum of squares" expresses the idea that the quantity estimates random (or chance) error. The second term on the right-hand side of the identity of Theorem 12.1 is $\hat{\sigma}_B^2$ times its degrees of freedom, and we refer to its as the **between-samples sum of squares** or the **treatment sum of squares**, *SS(Tr)*. (Most of the early applications of this kind of analysis were in the field of agriculture, where the k populations represented different **treatments**, such as fertilizers, applied to agricultural plots.) Note that in this notation the F ratio on page 338 can be written

F ratio
for treatments

$$F = \frac{SS(Tr)/(k-1)}{SSE/k(n-1)}$$

The sums of squares required for substitution into this last formula are usually obtained by means of the following shortcut formulas, which the reader will be asked to verify in Exercise 12.14 on page 347. We first calculate *SST* and *SS(Tr)* by means of the formulas

Sums of squares—equal sample sizes

$$SST = \sum_{i=1}^{k} \sum_{j=1}^{n} y_{ij}^2 - C$$

$$SS(Tr) = \frac{\sum_{i=1}^{k} T_i^2}{n} - C$$

where C, called the **correction term,** *is given by*

$$C = \frac{T_{\cdot}^2}{kn}$$

In these formulas, T_i is the total of the *n* observations in the *i*th sample while $T_{\cdot}$ is the grand total of all *kn* observations. The error sum of squares, *SSE*, is then obtained by substraction; according to Theorem 12.1 we can write

Error sum of squares

$$SSE = SST - SS(Tr)$$

The results obtained in analyzing the total sum of squares into its components are conveniently summarized by means of the following kind of **analysis of variance table:**

Source of variation	Degrees of freedom	Sum of squares	Mean square	F
Treatments	$k - 1$	$SS(Tr)$	$MS(Tr)$ $= SS(Tr)/(k - 1)$	$\dfrac{MS(Tr)}{MSE}$
Error	$k(n - 1)$	SSE	MSE $= SSE/k(n - 1)$	
Total	$nk - 1$	SST		

Note that each **mean square** is obtained by dividing the corresponding sum of squares by its degrees of freedom.

EXAMPLE To illustrate the **analysis of variance** (as this technique is appropriately called) for a one-way classification, suppose that in accordance with the layout on page 334 each laboratory measures the tin-coating weights of 12 discs and that the results are as follows:

Laboratory A	Laboratory B	Laboratory C	Laboratory D
0.25	0.18	0.19	0.23
0.27	0.28	0.25	0.30
0.22	0.21	0.27	0.28
0.30	0.23	0.24	0.28
0.27	0.25	0.18	0.24
0.28	0.20	0.26	0.34
0.32	0.27	0.28	0.20
0.24	0.19	0.24	0.18
0.31	0.24	0.25	0.24
0.26	0.22	0.20	0.28
0.21	0.29	0.21	0.22
0.28	0.16	0.19	0.21

Construct an analysis-of-variance table.

Solution   The totals for the four samples are, respectively, 3.21, 2.72, 2.76, and 3.00, the grand total is 11.69, and the calculations required to obtain the necessary sums of squares are as follows:

$$C = \frac{(11.69)^2}{48} = 2.8470$$

$$SST = (0.25)^2 + (0.27)^2 + \ldots + (0.21)^2 - 2.8470 = 0.0809$$

$$SS(Tr) = \frac{(3.21)^2 + (2.72)^2 + (2.76)^2 + (3.00)^2}{12} - 2.8470 = 0.0130$$

$$SSE = 0.0809 - 0.0130 = 0.0679$$

Thus, we get the following analysis-of-variance table:

Source of variation	Degrees of freedom	Sum of squares	Mean square	F
Laboratories	3	0.0130	0.0043	2.87
Error	44	0.0679	0.0015	
Total	47	0.0809		

Since the value obtained for F exceeds 2.82, the value of $F_{.05}$ with 3 and 44 degrees of freedom, the null hypothesis can be rejected at the 0.05 level of significance; we conclude that the laboratories are *not* obtaining consistent results.

To estimate the parameters $\mu, \alpha_1, \alpha_2, \alpha_3$, and α_4 (or μ_1, μ_2, μ_3, and μ_4), we can use the method of least squares, minimizing

$$\sum_{i=1}^{k} \sum_{j=1}^{n} (y_{ij} - \mu - \alpha_i)^2$$

with respect to μ and the α_i, subject to the restriction that $\sum_{i=1}^{k} \alpha_i = 0$. This may be done by eliminating one of the α's or, better, by using the method of Lagrange multipliers, which is treated in most texts on advanced calculus. In either case we obtain the "intuitively obvious" estimates $\hat{\mu} = \bar{y}.$ and $\hat{\alpha}_i = \bar{y}_i - \bar{y}.$ for $i = 1, 2, \ldots, k$, and the corresponding estimates for the μ_i are given by $\hat{\mu}_i = \bar{y}_i.$

EXAMPLE   Estimate the parameters of the one-way classification model for the tin-coating weights given in the preceding example.

Solution   For the data from the four laboratories we get

$$\hat{\mu} = \frac{11.69}{48} = 0.244, \qquad \hat{\alpha}_1 = \frac{3.21}{12} - 0.244 = 0.024$$

thus,

$$\hat{\alpha}_2 = \frac{2.72}{12} - 0.244 = -0.017, \qquad \hat{\alpha}_3 = \frac{2.76}{12} - 0.244 = -0.014,$$

and

$$\hat{\alpha}_4 = \frac{3.00}{12} - 0.244 = 0.006.$$

The analysis of variance described in this section applies to one-way classifications where each sample has the same number of observations. If this is not the case and the sample sizes are $n_1, n_2, \ldots,$ and n_k we have only to substitute $N = \sum_{i=1}^{k} n_i$ for nk throughout and write the computing formulas for SST and $SS(Tr)$ as

Sums of squares—unequal sample sizes

$$SST = \sum_{i=1}^{k} \sum_{j=1}^{n_i} y_{ij}^2 - C$$

$$SS(Tr) = \sum_{i=1}^{k} \frac{T_i^2}{n_i} - C$$

Otherwise, the procedure is the same as before. (See also Exercise 12.15 on page 347.)

EXAMPLE As part of the investigation of the collapse of the roof of a building, a testing laboratory is given all the available bolts that connected the steel structure at three different positions on the roof. The forces required to shear each of these bolts (coded values) are as follows:

Position 1: 90, 82, 79, 98, 83, 91
Position 2: 105, 89, 93, 104, 89, 95, 86
Position 3: 83, 89, 80, 94

Perform an analysis of variance to test at the 0.05 level of significance whether the differences among the sample means at the three positions are significant.

Solution Using the same steps as in previous chapters for tests of hypotheses, we get

1. *Null hypothesis:* $\mu_1 = \mu_2 = \mu_3$
   Alternative hypothesis: The μ's are not all equal.
2. *Level of significance*: $\alpha = 0.05$
3. *Criterion*: Reject the null hypothesis if $F > 3.74$, the value of $F_{0.05}$ for $k - 1 = 3 - 1 = 2$ and $N - k = 17 - 3 = 14$ degrees of freedom, where F is to be determined by an analysis of variance; otherwise, accept it.
4. *Calculations*: Substituting $n_1 = 6, n_2 = 7, n_3 = 4, N = 17, T_1 = 523, \ T_2 = 661, \ T_3 = 346, \ T = 1530,$ and $\sum\sum y_{ij}^2 = 138{,}638$ into the computing formulas for the sums of squares, we get

$$SST = 138{,}638 - \frac{1530^2}{17} = 938$$

$$SS(Tr) = \frac{523^2}{6} + \frac{661^2}{7} + \frac{346^2}{4} - \frac{1530^2}{17}$$

$$= 234$$

and

$$SSE = 938 - 234 = 704$$

The remainder of the work is shown in the following analysis-of-variance table:

Source of variation	Degrees of freedom	Sum of squares	Mean square	F
Positions	2	234	117	2.33
Error	14	704	50.3	
Total	16	938		

5. *Decision*: Since $F = 2.33$ does not exceed 3.74, the value of $F_{0.05}$ for 2 and 14 degrees of freedom, the null hypothesis cannot be rejected; in other words, we cannot conclude that there is a difference in the mean shear strengths of the bolts at the three different positions on the roof.

EXERCISES

12.1 An experiment is performed to compare the cleansing action of two detergents, Detergent A and Detergent B. Twenty swatches of cloth are soiled with dirt and grease, each is washed with one of the detergents in an agitator-type machine, and then measured for "whiteness." Criticize the following aspects of the experiment:

(a) The entire experiment is performed with soft water.

(b) Fifteen of the swatches are washed with Detergent A and five with Detergent B.

(c) To accelerate the testing procedure, very hot water and 30-second washing times are used in the experiment.

(d) The "whiteness" readings of all swatches washed with Detergent A are taken first.

12.2 A certain *bon vivant*, wishing to ascertain the cause of his frequent hangovers, conducted the following experiment. On the first night, he drank nothing but whiskey and water; on the second night he drank vodka and water; on the third night he drank gin and water; and on the fourth night he drank rum and water. On each of the following mornings he had a hangover, and he concluded that it was the common factor, the water, that made him ill.

(a) This conclusion is obviously unwarranted, but can you state what principles of sound experimental design are violated?

(b) Give a less obvious example of an experiment having the same shortcoming.

(c) Suppose that our friend had modified his experiment so each of the four alcoholic beverages was used both with and without water, so that the experiment lasted eight nights. Could the results of this enlarged experiment serve to support or refute the hypothesis that water was the cause of the hangovers? Explain.

12.3 To compare the effectiveness of three methods of teaching the programming of a certain computer—Method A which is straight teaching-machine instruction, Method B which involves the personal attention of an instructor and some direct experience working with the computer, and Method C which involves the personal attention of an instructor but no work with the computer—random samples of size four are taken from large groups of persons taught by the three methods and the following are the scores which they obtained in an appropriate achievement test:

	Method A	Method B	Method C
	73	91	72
	77	81	77
	67	87	76
	71	85	79

(a) Without using the shortcut formulas, calculate

$$\sum_{i=1}^{k}\sum_{j=1}^{n}(y_{ij}-\bar{y}_{\cdot\cdot})^2, \quad \sum_{i=1}^{k}\sum_{j=1}^{n}(y_{ij}-\bar{y}_i)^2, \quad \text{and} \quad n\cdot\sum_{i=1}^{k}(\bar{y}_i-\bar{y}_{\cdot\cdot})^2$$

and verify the identity of Theorem 12.1.

(b) Verify the results obtained for the three sums of squares by using the shortcut formulas on page 340.

12.4 Using the sums of squares obtained in Exercise 12.3, test at the level of significance $\alpha = 0.05$ whether the differences among the means obtained for the three samples are significant.

12.5 The following are the numbers of mistakes made in five successive days by four technicians working for a photographic laboratory:

Technician I	Technician II	Technician III	Technician IV
6	14	10	9
14	9	12	12
10	12	7	8
8	10	15	10
11	14	11	11

Test at the level of significance $\alpha = 0.01$ whether the differences among the four sample means can be attributed to chance.

12.6 The following are the weight losses of certain machine parts (in milligrams) due to friction, when three different lubricants were used under controlled conditions:

Lubricant A: 12.2, 11.8, 13.1, 11.0, 3.9, 4.1, 10.3, 8.4
Lubricant B: 10.9, 5.7, 13.5, 9.4, 11.4, 15.7, 10.8, 14.0
Lubricant C: 12.7, 19.9, 13.6, 11.7, 18.3, 14.3, 22.8, 20.4

(a) Test at the 0.01 level of significance whether the differences among the sample means can be attributed to chance.

(b) Estimate the parameters of the model used in the analysis of this experiment.

12.7 To determine the effect on exit dust loading in a precipitator, the following measurements were made:

Total flow (cu ft/hr)	Exit dust loading (grains per cubic yard in flue gas)				
200	1.5	1.7	1.6	1.9	1.9
300	1.5	1.8	2.2	1.9	2.2
400	1.4	1.6	1.7	1.5	1.8
500	1.1	1.5	1.4	1.4	2.0

Use the level of significance $\alpha = 0.05$ to test whether the flow through the precipitator has an effect on the exit dust loading.

12.8 To study the performance of a newly designed motorboat it was timed over a marked course under various wind and water conditions. Use the following data (in minutes) to test the null hypothesis that the boat's performance is not affected by the differences in wind and water conditions:

Calm conditions:	20, 17, 14, 24
Moderate conditions:	21, 23, 16, 25, 18, 23
Choppy conditions:	26, 24, 23, 29, 21

Use the level of significance $\alpha = 0.05$.

12.9 To find the best arrangement of instruments on a control panel of an airplane, three different arrangements were tested by simulating an emergency condition and observing the reaction time required to correct the condition. The reaction times (in *tenths* of a second) of 28 pilots (randomly assigned to the different arrangements) were as follows:

Arrangement 1:	14, 13, 9, 15, 11, 13, 14, 11
Arrangement 2:	10, 12, 9, 7, 11, 8, 12, 9, 10, 13, 9, 10
Arrangement 3:	11, 5, 9, 10, 6, 8, 8, 7

Test at the level of significance $\alpha = 0.01$ whether we can reject the null hypothesis that the differences among the arrangements have no effect.

12.10 Several different aluminum alloys are under consideration for use in heavy-duty circuit wiring applications. Among the desired properties is low electrical resistance, and a number of specimens of each wire are tested by applying a fixed voltage to a given length of wire and measuring the current passing through the wire. Given the following results, would you conclude that these alloys differ in resistance? (Use the 0.01 level of significance.)

Alloy	Current (amperes)				
1	1.085	1.016	1.009	1.034	
2	1.051	0.993	1.022		
3	0.985	1.001	0.990	0.988	1.011
4	1.101	1.015			

12.11 Two tests are made of the compressive strength of each of six samples of poured concrete. The force required to crumble each cylindrical specimen, measured in kilograms, is as follows:

	\multicolumn{6}{c}{Sample}				
	A	*B*	*C*	*D*	*E*	*F*
Test 1	110	125	98	95	104	115
Test 2	105	130	107	92	96	121

Test at the 0.05 level of significance whether these samples differ in compressive strength.

12.12 Referring to the discussion on page 333, assume that the standard deviations of the tin-coating weights determined by any one of the three laboratories have the common value $\sigma = 0.012$, and that it is desired to be 95% confident of detecting a difference in means between any two of the laboratories in excess of 0.01 pound per base box. Show that these assumptions lead to the decision to send a sample of 12 discs to each laboratory.

12.13 Show that if $\mu_i = \mu + \alpha_i$, and μ is the mean of the μ_i, it follows that
$$\sum_{i=1}^{k} \alpha_i = 0.$$

12.14 Verify the shortcut formulas for computing SST and $SS(Tr)$ given on page 340.

12.15 State and prove a theorem analogous to Theorem 12.1 for the case where the size of the ith sample is n_i, that is, where the sample sizes are not necessarily equal.

12.16 Samples of peanut butter produced by two different manufacturers are tested for aflatoxin content, with the following results:

Aflatoxin content (ppb)

Brand A	Brand B
0.5	4.7
0.0	6.2
3.2	0.0
1.4	10.5
0.0	2.1
1.0	0.8
8.6	
2.9	

(a) Use analysis of variance to test whether the two brands differ in aflatoxin content.

(b) Test the same hypothesis using a two-sample *t* test.

(c) It can be shown that the *t* statistic with v degrees of freedom and the F statistic with 1 and v degrees of freedom are related by the formula

$$F(1, v) = t^2(v)$$

where $v =$ degrees of freedom. Using this result, prove that the analysis of variance and two-sample *t*-test methods are equivalent in this case.

12.3   RANDOMIZED-BLOCK DESIGNS

As we observed in Section 12.1, the estimate of chance variation (the experimental error) can often be reduced, that is, freed of variability due to extraneous causes, by dividing the observations in each classification into blocks. This is accomplished when known sources of variability (that is, extraneous variables) are fixed in each block, but vary from block to block.

In this section we shall suppose that the experimenter has available measurements pertaining to a treatments distributed over b blocks. First, we shall consider the case where there is exactly one observation from each treatment in each block; with reference to the illustration on page 335, this case would arise if each laboratory tested one disc from each strip. Letting y_{ij} denote the observation pertaining to the ith treatment and the jth block, $\bar{y}_{i.}$ the mean of the b observations for the ith treatment, $\bar{y}_{.j}$ the mean of the a observations in the jth block, and $\bar{y}_{..}$ the grand mean of all the ab observations, we shall use the following layout for this kind of **two-way classification**:

Blocks

	B_1	B_2	...	B_j	...	B_b	*Means*
Treatment 1:	$y_{11}, y_{12}, \ldots, y_{1j}, \ldots, y_{1b}$						$\bar{y}_1.$
Treatment 2:	$y_{21}, y_{22}, \ldots, y_{2j}, \ldots, y_{2b}$						$\bar{y}_2.$
Treatment i:	$y_{i1}, y_{i2}, \ldots, y_{ij}, \ldots, y_{ib}$						$\bar{y}_i.$
Treatment a:	$y_{a1}\ y_{a2}, \ldots, y_{aj}, \ldots, y_{ab}$						$\bar{y}_a.$
Means	$\bar{y}_{.1}\ \bar{y}_{.2}\ \ldots\ \bar{y}_{.j}\ \ldots\ \bar{y}_{.b}$						$\bar{y}_{..}$

This kind of arrangement is also called a **randomized-block design**, provided the treatments are allocated at random *within* each block. Note that when a dot is used in place of a subscript, this means that the mean is obtained by summing over that subscript.

The underlying model which we shall assume for the analysis of this kind of experiment with one observation per "cell" (that is, there is one observation corresponding to each treatment within each block) is given by

Model equation for randomized-block design

$$y_{ij} = \mu + \alpha_i + \beta_j + \epsilon_{ij} \quad \text{for } i = 1, 2, \ldots, a; j = 1, 2, \ldots, b$$

Here μ is the grand mean, α_i is the effect of the ith treatment, β_j is the effect of the jth block, and the ϵ_{ij} are values of *independent, normally distributed* random variables having *zero means* and the *common variance* σ^2. Analogous to the model for the one-way classification, we restrict the parameters by imposing the conditions that $\sum_{i=1}^{a} \alpha_i = 0$ and $\sum_{j=1}^{b} \beta_j = 0$ (see Exercise 12.28 on page 360).

In the analysis of a two-way classification where each treatment is represented once in each block, the major objective is to test for the significance of the difference among the $\bar{y}_{i.}$, that is, to test the null hypothesis

$$\alpha_1 = \alpha_2 = \ldots = \alpha_a = 0$$

In addition, it may also be desirable to test whether the blocking has been effective, that is, whether the null hypothesis

$$\beta_1 = \beta_2 = \ldots = \beta_b = 0$$

can be rejected. In either case, the alternative hypothesis is that at least one of the effects is different from zero.

As in the one-way analysis of variance, we shall base these significance tests on comparisons of estimates of σ^2—one based on the variation among treatments, one based on the variation among blocks, and one measuring the experimental error. Note that only the latter is an estimate of σ^2 when either (or both) of the null hypotheses do not hold. The required sums of squares are given by the three components into which the total sum of squares is partitioned by means of the following theorem:

Identity for analysis of two-way classification

Theorem 12.2

$$\sum_{i=1}^{a} \sum_{j=1}^{b} (y_{ij} - \bar{y}_{..})^2 = \sum_{i=1}^{a} \sum_{j=1}^{b} (y_{ij} - \bar{y}_{i.} - \bar{y}_{.j} + \bar{y}_{..})^2$$
$$+ b \sum_{i=1}^{a} (\bar{y}_{i.} - \bar{y}_{..})^2 + a \sum_{j=1}^{b} (\bar{y}_{.j} - \bar{y}_{..})^2$$

The left-hand side of this identity represents the total sum of squares, *SST*, and the terms on the right-hand side are, respectively, the error sum of squares, *SSE*, the treatment sum of squares, *SS(Tr)*, and the **block sum of squares** *SS(Bl)*. To prove this theorem, we make use of the identity

$$y_{ij} - \bar{y}_{..} = (y_{ij} - \bar{y}_{i.} - \bar{y}_{.j} + \bar{y}_{..}) + (\bar{y}_{i.} - \bar{y}_{..}) + (\bar{y}_{.j} - \bar{y}_{..})$$

and follow essentially the same argument as in the proof of Theorem 12.1.

In practice, we calculate the necessary sums of squares by means of shortcut formulas analogous to those on page 340, rather than by using the expressions which define these sums of squares in Theorem 12.2. We first calculate SST, $SS(Tr)$, and $SS(Bl)$ by means of the formulas

Sums of squares for two-way analysis of variance

$$SST = \sum_{i=1}^{a} \sum_{j=1}^{b} y_{ij}^2 - C$$

$$SS(Tr) = \frac{\sum_{i=1}^{a} T_{i.}^2}{b} - C$$

$$SS(Bl) = \frac{\sum_{j=1}^{b} T_{.j}^2}{a} - C$$

where C, the correction term, is given by

$$C = \frac{T_{..}^2}{ab}$$

In these formulas $T_{i.}$ is the sum of the b observations for the ith treatment, $T_{.j}$ is the sum of the a observations in the jth block, and $T_{..}$ is the grand total of all the observations. Note that the divisors for $SS(Tr)$ and $SS(Bl)$ are the number of observations in the respective totals, $T_{i.}$ and $T_{.j}$. The error sum of squares is then obtained by subtraction; according to Theorem 12.2 we can write

Error sum of squares

$$SSE = SST - SS(Tr) - SS(Bl)$$

In Exercise 12.29 on page 360, the reader will be asked to verify that all these computing formulas are, indeed, equivalent to the corresponding terms of the identity of Theorem 12.2.

Using these sums of squares, we can reject the null hypothesis that the α_i are all equal to zero at the level of significance α if

F-ratio for treatments

$$F_{Tr} = \frac{MS(Tr)}{MSE} = \frac{SS(Tr)/(a-1)}{SSE/(a-1)(b-1)}$$

exceeds F_α with $a - 1$ and $(a - 1)(b - 1)$ degrees of freedom. The null hypothesis that the β_j are all equal to zero can be rejected at the level of significance α if

F-ratio for blocks

$$F_{Bl} = \frac{MS(Bl)}{MSE} = \frac{SS(Bl)/(b-1)}{SSE/(a-1)(b-1)}$$

exceeds F_{α} with $b-1$ and $(a-1)(b-1)$ degrees of freedom. Note that the mean squares, $MS(Tr)$, $MS(Bl)$, and MSE, are again defined as the corresponding sums of squares divided by their degrees of freedom.

The results obtained in this analysis are summarized in the following analysis-of-variance table:

Source of variation	Degrees of freedom	Sum of squares	Mean square	F
Treatments	$a-1$	$SS(Tr)$	$MS(Tr) = \frac{SS(Tr)}{(a-1)}$	$F_{Tr} = \frac{MS(Tr)}{MSE}$
Blocks	$b-1$	$SS(Bl)$	$MS(Bl) = \frac{SS(Bl)}{(b-1)}$	$F_{Bl} = \frac{MS(Bl)}{MSE}$
Error	$(a-1)(b-1)$	SSE	$MSE = \frac{SSE}{(a-1)(b-1)}$	
Total	$ab-1$	SST		

EXAMPLE An experiment was designed to study the performance of four different detergents. The following "whiteness" readings were obtained with specially designed equipment for 12 loads of washing distributed over three different models of washing machines:

	Machine 1	Machine 2	Machine 3	Totals
Detergent A	45	43	51	139
Detergent B	47	46	52	145
Detergent C	48	50	55	153
Detergent D	42	37	49	128
Totals	182	176	207	565

Looking on the detergents as treatments and the machines as blocks, obtain the appropriate analysis-of-variance table and test at the 0.01 level of significance whether there are differences in the detergents or in the washing machines.

Solution

1. *Null hypotheses*: $\alpha_1 = \alpha_2 = \alpha_3 = \alpha_4 = 0$; $\beta_1 = \beta_2 = \beta_3 = 0$
   Alternative hypotheses: The α's are not all equal to zero; the β's are not all equal to zero.
2. *Level of significance*: $\alpha = 0.01$
3. *Criteria*: For treatments, reject the null hypothesis if $F > 9.78$, the value of $F_{0.01}$ with $a - 1 = 4 - 1 = 3$ and $(a - 1)(b - 1) = (4 - 1)(3 - 1) = 6$ degrees of freedom; for blocks, reject the null hypothesis if $F > 10.9$, the value of $F_{0.01}$ for $b - 1 = 3 - 1 = 2$ and $(a - 1)(b - 1) = (4 - 1)(3 - 1) = 6$ degrees of freedom.
4. *Calculations*: Substituting $a = 4$, $b = 3$, $T_{1.} = 139$, $T_{2.} = 145$, $T_{3.} = 153$, $T_{4.} = 128$, $T_{.1} = 182$, $T_{.2} = 176$, $T_{.3} = 207$, $T_{..} = 565$, and $\sum \sum y_{ij}^2 = 26{,}867$ into the formulas for the sums of squares, we get

$$C = \frac{(565)^2}{12} = 26{,}602$$

$$SST = 45^2 + 43^2 + \ldots + 49^2 = 26{,}867 - 26{,}602 = 265$$

$$SS(Tr) = \frac{139^2 + 145^2 + 153^2 + 128^2}{3} - 26{,}602 = 111$$

$$SS(Bl) = \frac{182^2 + 176^2 + 207^2}{4} - 26{,}602 = 135$$

$$SSE = 265 - 111 - 135 = 19$$

Then dividing the sums of squares by their respective degrees of freedom to obtain the appropriate mean squares, we get the results shown in the following analysis-of-variance table:

Source of variation	Degrees of freedom	Sum of squares	Mean square	F
Detergents	3	111	37.0	11.6
Machines	2	135	67.5	21.1
Error	6	19	3.2	
Total	11	265		

5. *Decisions*: Since $F_{Tr} = 11.6$ exceeds 9.78, the value of $F_{0.01}$ with 3 and 6 degrees of freedom, we conclude that there are differences in the effectiveness of the four detergents. Also, since $F_{Bl} = 21.1$ exceeds 10.9, the value of $F_{0.01}$ with 2 and 6 degrees of freedom, we conclude that the differences among the results obtained for the three washing machines are significant, namely, that the blocking has been effective. To make the effect of this blocking even more evident, the reader will be asked to verify in Exercise 12.24 on page 359 that the test for differences among the detergents would *not* yield significant results if we looked upon the data as a one-way classification.

The effect of the *i*th detergent can be estimated by means of the formula $\hat{\alpha}_i = \bar{y}_{i.} - \bar{y}_{..}$, which may be obtained by the method of least squares. The resulting estimates are

$$\hat{\alpha}_1 = 46.3 - 47.1 = -0.8, \qquad \hat{\alpha}_2 = 48.3 - 47.1 = 1.2,$$
$$\hat{\alpha}_3 = 51.0 - 47.1 = 3.9, \qquad \hat{\alpha}_4 = 42.7 - 47.1 = -4.4$$

Similar calculations lead to $\hat{\beta}_1 = -1.6$, $\hat{\beta}_2 = -3.1$, and $\hat{\beta}_3 = 4.7$ for the estimated effects of the washing machines.

It should be observed that a two-way classification automatically allows for repetitions of the experimental conditions; for example, in the preceding experiment each detergent was tested three times. Further repetitions may be handled in several ways, and care must be taken that the model used appropriately describes the situation. One way to provide further repetition in a two-way classification is to include additional blocks—for example, to test each detergent using several additional washing machines, randomizing the order of testing for each machine. Note that the model remains essentially the same as before, the only change being an increase in b, and a corresponding increase in the degrees of freedom for blocks and for error. The latter is important, because an increase in the degrees of freedom for error makes the test of the null hypothesis $\alpha_i = 0$ for all i *more sensitive* to small differences among the treatment means. In fact, the real purpose of this kind of repetition is to increase the degrees of freedom for error, thereby increasing the sensitivity of the F tests (see Exercise 12.27 on page 359).

A second method is to repeat the entire experiment, using a new pattern of randomization to obtain $a \cdot b$ additional observations. This is possible only if the blocks are *identifiable*, that is, if the conditions defining each block can be repeated. For example, in the tin-coating weight experiment described in Section 12.1, the blocks are strips across the rolling direction of a sheet of tin plate, and, given a new sheet, it is possible to identify which is strip 1, which is strip 2, and so forth. In the example of this section, this kind of repetition (usually called **replication**) would require that the settings of the machines be exactly duplicated. This kind of repetition will be used in connection with Latin-square designs in Section 12.5; see also Exercises 12.25 and 12.26 on pages 359.

A third method of repetition is to include *n* observations for each treatment in each block. When an experiment is designed in this way, the *n* observations in each "cell" are regarded as duplicates, and it is to be expected that their variability will be somewhat less than experimental error. To illustrate this point, suppose that the tin-coating weights of three discs from adjacent positions in a strip are measured in sequence by one of the laboratories, using the same chemical solutions. The variability of these measurements will probably be considerably less than that of three discs from the same strip measured in that laboratory at different times, using different chemical solutions, and perhaps different technicians. The analysis-of-variance appropriate for this kind of repetition reduces essentially to a two-way analysis of variance applied to the *means* of the *n* duplicates in the *a·b* cells; thus, *there would be no gain in degrees of freedom for error*, and, consequently, *no gain in sensitivity of the F tests*. It can be expected, however, that there will be some reduction in the error mean square, since it now measures the residual variance of the *means* of several observations.

12.4 MULTIPLE COMPARISONS

The *F* tests used so far in this chapter showed whether differences among several means are significant, but they did not tell us whether a given mean (or group of means) differs significantly from another given mean (or group of means). In actual practice, the latter is the kind of information an investigator really wants; for instance, having determined on page 341 that the means of the tin-coating weights obtained by the four laboratories differ significantly, it may be important to find out which laboratory (or laboratories) differs from which others.

If an experimenter is confronted with *k* means, it may seem reasonable at first to test for significant differences between all possible pairs, that is, to perform

$$\binom{k}{2} = \frac{k(k-1)}{2}$$

two-sample *t* tests as described on page 222. Aside from the fact that this would require a large number of tests even if *k* is relatively small, these tests would not be independent, and it would be virtually impossible to assign an overall level of significance to this procedure.

Several **multiple-comparisons tests** have been proposed to overcome these difficulties, among them the **Duncan multiple-range test**, which we shall study in this section. (References to other multiple-comparisons tests are mentioned in the book by W. T. Federer listed in the Bibliography.) The assumptions underlying the Duncan multiple-range test are, essentially, those of the one-way analysis of variance for which the sample sizes are equal. The test compares the range of any set of p means with an appropriate **least significant range**, R_p, given by

Least significant range

$$R_p = s_{\bar{x}} \cdot r_p$$

Here $s_{\bar{x}}$ is an estimate of $\sigma_{\bar{x}} = \sigma/\sqrt{n}$, and it is computed by means of the formula

Standard error of the mean

$$s_{\bar{x}} = \sqrt{\frac{MSE}{n}}$$

where MSE is the error mean square in the analysis of variance. The value of r_p depends on the desired level of significance and the number of degrees of freedom corresponding to MSE, and it may be obtained from Tables 12(a) and (b) for $\alpha = 0.05$ and 0.01, for $p = 2, 3, \ldots, 10$, and for various degrees of freedom from 1 to 120.

EXAMPLE   Referring to the tin-coating weight data on page 341, perform a Duncan multiple-range test to determine which laboratory means differ from which others, using the 0.05 level of significance.

Solution   First we arrange the four sample means as follows in an increasing order of magnitude:

Laboratory	B	C	D	A
Mean	0.227	0.230	0.250	0.268

Next, we compute $s_{\bar{x}}$, using the error mean square of 0.0015 in the analysis of variance on page 341, and we obtain

$$s_{\bar{x}} = \sqrt{\frac{0.0015}{12}} = 0.011$$

Then, we get (by linear interpolation) from Table 12(a) the following values of r_p for $\alpha = 0.05$ and 44 degrees of freedom:

p	2	3	4
r_p	2.85	3.00	3.09

Multiplying each value of r_p by $s_{\bar{x}} = 0.011$, we finally obtain

p	2	3	4
R_p	0.031	0.033	0.034

The range of *all four means* is $0.268 - 0.227 = 0.041$, which exceeds $R_4 = 0.034$, the least significant range. This result should have been expected, since the F test on page 341 showed that the differences among all four means are significant at $\alpha = 0.05$. To test for significant differences among *three adjacent means*, we obtain ranges of 0.038 and 0.023, respectively, for 0.230, 0.250, 0.268 and 0.227, 0.230, 0.250. Since the first of these values exceeds $R_3 = 0.033$, the differences observed in the first set are significant; since the second value does not exceed 0.033, the corresponding differences are not significant. Finally, for *adjacent pairs* of means we find that no adjacent pair has a range greater than the least significant range $R_2 = 0.031$. All these results can be summarized by writing

$$\underline{0.227 \quad 0.230 \quad 0.250} \quad 0.268$$

where *a line is drawn under any set of adjacent means for which the range is less than the appropriate value of R_p, that is, under any set of adjacent means for which differences are not significant.* We thus conclude in our example that Laboratory A averages higher tin-coating weights than Laboratories B and C.

If we apply this same method to the example of Section 12.3, where we compared the four detergents, we obtain (see also Exercise 12.30 on page 360)

Detergent

D	A	B	C
42.7	46.3	48.3	51.0

In other words, among triplets of adjacent means both sets of differences are significant. So far as pairs of adjacent means are concerned, we find that only the difference between 42.7 and 46.3 is significant. Interpreting these results, we conclude that detergent D is significantly inferior to any of the others, and detergent A is significantly inferior to detergent C.

EXERCISES

12.17 A laboratory technician measures the breaking strength of each of five kinds of linen threads by means of four different instruments, and obtains the following results (in ounces):

Measuring instrument

	I_1	I_2	I_3	I_4
Thread 1	20.6	20.7	20.0	21.4
Thread 2	24.7	26.5	27.1	24.3
Thread 3	25.2	23.4	21.6	23.9
Thread 4	24.5	21.5	23.6	25.2
Thread 5	19.3	21.5	22.2	20.6

Looking upon the threads as treatments and the instruments as blocks, perform an analysis of variance at the level of significance $\alpha = 0.01$.

12.18 Looking at the days (rows) as blocks, rework Exercise 12.5 on page 345 by the method of Section 12.3.

12.19 Four different, though supposedly equivalent, forms of a standardized reading achievement test were given to each of five students, and the following are the scores which they obtained:

	Student 1	*Student 2*	*Student 3*	*Student 4*	*Student 5*
Form A	75	73	59	69	84
Form B	83	72	56	70	92
Form C	86	61	53	72	88
Form D	73	67	62	79	95

Perform a two-way analysis of variance to test at the level of significance $\alpha = 0.01$ whether it is reasonable to treat the four forms as equivalent.

12.20 An experiment was performed to judge the effect of four different fuels and two different types of launchers on the range of a certain rocket. Test, on the basis of the following data (in nautical miles) whether there are significant differences (a) among the means obtained for the fuels, and (b) between the means obtained for the launchers:

	Fuel I	*Fuel II*	*Fuel III*	*Fuel IV*
Launcher X	62.5	49.3	33.8	43.6
Launcher Y	40.4	39.7	47.4	59.8

Use the level of significance $\alpha = 0.05$.

12.21 Samples of groundwater were taken from five different toxic-waste dump sites by each of three different agencies, the EPA, the company that owned each site, and an independent consulting engineer. Each sample was analyzed for the presence of a certain contaminant by whatever laboratory method was customarily used by the agency collecting the sample, with the following results:

Concentration (parts per million)

	Site A	Site B	Site C	Site D	Site E
Agent 1	23.8	7.6	15.4	30.6	4.2
Agent 2	19.2	6.8	13.2	22.5	3.9
Agent 3	20.9	5.9	14.0	27.1	3.0

Is there reason to believe that the agents are not consistent with one another in their measurements? Do the dump sites differ from one another in their level of contamination? Use the 0.05 level of significance.

12.22 An industrial engineer tests four different shop-floor layouts by having each of six work crews construct a subassembly and measuring the construction times (minutes) as follows:

	Layout 1	Layout 2	Layout 3	Layout 4
Crew A	48.2	53.1	51.2	58.6
Crew B	49.5	52.9	50.0	60.1
Crew C	50.7	56.8	49.9	62.4
Crew D	48.6	50.6	47.5	57.5
Crew E	47.1	51.8	49.1	55.3
Crew F	52.4	57.2	53.5	61.7

Test at the 0.01 level of significance whether the four floor layouts produce different assembly times and whether some of the work crews are consistently faster in constructing this subassembly than the others.

12.23 Referring to Exercise 12.7 on page 346, suppose that the data in each of the five columns were obtained from a different precipitator. Repeat the analysis of variance, treating the experiment as a two-way classification, and observe what change results in the error mean square.

12.24 To emphasize the importance of "blocking," reanalyze the "whiteness" data on page 351 as a one-way classification with the four detergents being the different treatments.

12.25 If, in a two-way classification, the entire experiment is repeated r times, the model becomes

$$y_{ijk} = \mu + \alpha_i + \beta_j + \rho_k + \epsilon_{ijk}$$

for $i = 1, 2, \ldots, a$, $j = 1, 2, \ldots, b$, and $k = 1, 2, \ldots, r$, where the sum of the α's, the sum of the β's, and the sum of the ρ's are equal to zero, and where the ρ's represent the effects of the replications. The ϵ_{ijk} are again values of independent normally distributed random variables with zero means and the common variance σ^2.

(a) Write down (but do not prove) an identity analogous to the one of Theorem 12.2, subdividing the total sum of squares into components attributable to treatments, blocks, replications, and error.

(b) Generalize the computing formulas on page 350 so that they apply to a replicated randomized-block design. Note that the divisor in each case equals the number of observations in the respective totals.

(c) If the number of degrees of freedom for the replicate sum of squares is $r - 1$, how many degrees of freedom are there for the error sum of squares?

12.26 The following are the number of defectives produced by four workmen operating, in turn, three different machines; in each case, the first figure represents the number of defectives produced on a Friday and the second figure represents the number of defectives produced on the following Monday:

	Workman			
	B_1	B_2	B_3	B_4
Machine A_1	37, 43	38, 44	38, 40	32, 36
Machine A_2	31, 36	40, 44	43, 41	31, 38
Machine A_3	36, 40	33, 37	41, 39	38, 45

Use the theory developed in Exercise 12.25 to analyze the combined figures for the two days as a two-way classification with replication. Use the level of significance $\alpha = 0.05$.

12.27 As was pointed out on page 353, two ways of increasing the size of a two-way classification experiment are (a) to double the number of blocks, and (b) to replicate the entire experiment. Discuss and compare the gain in degrees of freedom for the error sum of squares by the two methods.

12.28 Show that if $\mu_{ij} = \mu + \alpha_i + \beta_j$, the mean of the μ_{ij} (summed on j) is equal to $\mu + \alpha_i$, and the mean of the μ_{ij} (summed on i and j) is equal to μ, it follows that $\sum_{i=1}^{a} \alpha_i = \sum_{j=1}^{b} \beta_j = 0$.

12.29 Verify that the computing formulas for *SST*, *SS(Tr)*, *SS(Bl)*, and *SSE*, given on page 350, are equivalent to the corresponding terms of the identity of Theorem 12.2.

12.30 Verify the results of Duncan's test for the comparison of the four detergents, given on page 356.

12.31 Use the Duncan test with $\alpha = 0.05$ to compare the effectiveness of the three methods of teaching the programming of the computer in Exercise 12.3 on page 344.

12.32 Use the Duncan test with $\alpha = 0.05$ to compare the effectiveness of the three lubricants in Exercise 12.6 on page 345.

12.33 Use the Duncan test with $\alpha = 0.01$ to compare the strength of the five linen threads in Exercise 12.17 on page 356.

12.34 Does it make sense to use the Duncan test with $\alpha = 0.05$ to compare the results obtained with the four fuels in Exercise 12.20 on page 357? Why?

12.35 Use the Duncan test with $\alpha = 0.05$ to compare the pollution levels of the five sites in Exercise 12.21 on page 358.

12.36 Is one of the agencies in Exercise 12.21 on page 358 producing results that are consistently higher (or lower) than those of the other two? (Use $\alpha = 0.05$.)

12.37 Determine which of the shop-floor layouts of Exercise 12.22 on page 358 (if any) differ from the others in the time it takes the average crew to construct the given subassembly. (Use the 0.01 level of significance.)

12.5 SOME FURTHER EXPERIMENTAL DESIGNS

The randomized-block design of Section 12.4 is appropriate when one extraneous source of variability is to be eliminated in comparing a set of sample means. An important feature of this kind of design is its **balance**, achieved by assigning the same number of observations for each treatment to each block. (In this connection, see also the comment on page 336, where we pointed out that differences due to blocks will not affect the means obtained for the different treatments.) The same kind of balance can be attained in more complicated kinds of designs, where it is desired to eliminate the effect of several extraneous sources of variability. In this section we shall introduce two further balanced designs, the Latin-square design and the Graeco-Latin-square design, which are used to eliminate the effects of two and three extraneous sources of variability, respectively.

To introduce the **Latin-square design**, suppose that it is desired to compare three treatments, *A*, *B*, and *C*, in the presence of two other sources of variability. For example, the three treatments may be three methods for soldering copper electrical leads, and the two extraneous sources of variability may be (1) different operators doing the soldering, and (2) the use of different solder fluxes. If three operators and three fluxes are to be considered, the experiment might be arranged in the following pattern:

	Flux 1	*Flux 2*	*Flux 3*
Operator 1	*A*	*B*	*C*
Operator 2	*C*	*A*	*B*
Operator 3	*B*	*C*	*A*

Here each soldering method is applied once by each operator in conjunction with each flux, and if there are systematic effects due to differences among operators or differences among fluxes, these effects are present equally for each treatment, that is, for each method of soldering.

An experimental arrangement such as the one just described is called a **Latin square**. An $n \times n$ Latin square is a square array of n distinct letters, with each letter appearing once and only once in each row and in each column. Examples of Latin squares with $n = 4$ and $n = 5$ are shown in Figure 12.2, and larger ones may be obtained from the book by W. G. Cochran and G. M. Cox mentioned in the Bibliography. Note that in a Latin-square experiment involving n treatments, it is necessary to include n^2 observations, n for each treatment.

As we shall see on page 363, a Latin-square experiment without replication provides only $(n - 1)(n - 2)$ degrees of freedom for estimating the

Figure 12.2 Latin squares.

experimental error. Thus, such experiments are rarely run without replication if n is small, that is, without repeating the entire Latin-square pattern several times. If there is a total of r replicates, the analysis of the data presumes the following model, where $y_{ij(k)l}$ is the observation in the ith row and the jth column of the lth replicate, and the subscript k, in parentheses, indicates that it pertains to the kth treatment:

Model equation for Latin square

$$y_{ij(k)l} = \mu + \alpha_i + \beta_j + \gamma_k + \rho_l + \epsilon_{ij(k)l}$$

for $i, j, k = 1, 2, \ldots, n$, and $l = 1, 2, \ldots, r$, subject to the restrictions that

$$\sum_{i=1}^{n} \alpha_i = 0, \ \sum_{j=1}^{n} \beta_j = 0, \ \sum_{k=1}^{n} \gamma_k = 0 \ \text{ and } \ \sum_{l=1}^{r} \rho_l = 0$$

Here μ is the grand mean, α_i is the effect of the ith row, β_j is the effect of the jth column, γ_k is the effect of the kth treatment, ρ_l is the effect of the lth replicate, and the $\epsilon_{ij(k)l}$ are values of independent, normally distributed random variables with zero means and the common variance σ^2. Note that by "effects of the rows" and "effects of the columns" we mean the effects of the two extraneous variables, and that we are including replicate effects, since, as we shall see, replication can introduce a third extraneous variable. Note also that the subscript k is in parentheses in $y_{ij(k)l}$, because, for a given Latin-square design, k is automatically determined when i and j are known.

The main hypothesis we shall want to test is the null hypothesis $\gamma_k = 0$ for all k, namely, the null hypothesis that there is no difference in the effectiveness of the n treatments. However, we can also test whether the "cross blocking" of the Latin-square design has been effective; that is, we can test the two null hypotheses $\alpha_i = 0$ for all i and $\beta_j = 0$ for all j (against suitable alternatives) to see whether the two extraneous variables actually have an effect on the phenomenon under consideration. Furthermore, we can test the null hypothesis $\rho_l = 0$ for all l against the alternative that the ρ_l are not all equal to zero, and this test for effects of the replicates may be important if the parts of the experiment representing the individual Latin squares were performed on different days, by different technicians, at different temperatures, and so on.

The sums of squares required to perform these tests are usually obtained by means of the following shortcut formulas, where $T_{i..}$ is the total of the $r \cdot n$ observations in all of the ith rows, $T_{.j.}$ is the total of the $r \cdot n$ observations in all of the jth columns, $T_{..l}$ is the total of the n^2 observations in the lth replicate, $T_{(k)}$ is the total of all the $r \cdot n$ observations pertaining to the kth treatment, and $T_{...}$ is the grand total of all the $r \cdot n^2$ observations:

Sums of squares— Latin square

$$C = \frac{(T_{...})^2}{r \cdot n^2}$$

$$SS(Tr) = \frac{1}{r \cdot n} \sum_{k=1}^{n} T_{(k)}^2 - C$$

$$SSR = \frac{1}{r \cdot n} \sum_{i=1}^{n} T_{i..}^2 - C \quad \text{(for rows)}$$

$$SSC = \frac{1}{r \cdot n} \sum_{j=1}^{n} T_{.j.}^2 - C \quad \text{(for columns)}$$

$$SS(Rep) = \frac{1}{n^2} \sum_{l=1}^{r} T_{..l}^2 - C \quad \text{(for replicates)}$$

$$SST = \sum_{i=1}^{n} \sum_{j=1}^{n} \sum_{l=1}^{r} y_{ij(k)l}^2 - C$$

$$SSE = SST - SS(Tr) - SSR - SSC - SS(Rep)$$

Note that again each divisor equals the number of observations in the corresponding squared totals. Finally, the results of the analysis are as shown in the following analysis-of-variance table:

Source of variation	Degrees of freedom	Sum of squares	Mean square	F
Treatments	$n - 1$	$SS(Tr)$	$MS(Tr)$ $= \frac{SS(Tr)}{n - 1}$	$\frac{MS(Tr)}{MSE}$
Rows	$n - 1$	SSR	MSR $= \frac{SSR}{n - 1}$	$\frac{MSR}{MSE}$
Columns	$n - 1$	SSC	MSC $= \frac{SSC}{n - 1}$	$\frac{MSC}{MSE}$
Replicates	$r - 1$	$SS(Rep)$	$MS(Rep)$ $= \frac{SS(Rep)}{r - 1}$	$\frac{MS(Rep)}{MSE}$
Error	$(n - 1)(rn + r - 3)$	SSE	MSE $= \frac{SSE}{(n - 1)(rn + r - 3)}$	
Total	$rn^2 - 1$	SST		

As before, the degrees of freedom for the total sum of squares equals the *sum* of the degrees of freedom for the individual components; thus, the degrees of freedom for error are usually found last, by subtraction.

EXAMPLE Suppose that two replicates of the aforementioned soldering experiment were run, using the following arrangement:

	Replicate I flux			Replicate II flux		
	1	*2*	*3*	*1*	*2*	*3*
Operator 1	A	B	C	C	B	A
Operator 2	C	A	B	A	C	B
Operator 3	B	C	A	B	A	C

The results, showing the number of pounds tensile force required to separate the soldered leads, were as follows:

Replicate I			Replicate II		
14.0	16.5	11.0	10.0	16.5	13.0
9.5	17.0	15.0	12.0	12.0	14.0
11.0	12.0	13.5	13.5	18.0	11.5

Analyze this experiment as a Latin square and test at the 0.01 level of significance whether there are differences in the methods, the operators, the fluxes, or the replicates.

Solution

1. *Null hypotheses*: $\alpha_1 = \alpha_2 = \alpha_3 = 0$; $\beta_1 = \beta_2 = \beta_3 = 0$; $\gamma_1 = \gamma_2 = \gamma_3 = 0$; $\rho_1 = \rho_2 = 0$.

   Alternate hypotheses: The α's are not all equal to zero; the β's are not all equal to zero; the γ's are not all equal to zero; the ρ's are not all equal to zero.

2. *Levels of significance*: $\alpha = 0.01$ for each test.

3. *Criteria*: For treatments, rows, or columns, reject the null hypothesis if $F > 7.56$, the value of $F_{0.01}$ for $n - 1 = 3 - 1$

$= 2$ and $(n-1)(rn+r-3) = (3-1)(2 \times 3+2-3) = 10$ degrees of freedom. For replicates, reject the null hypothesis if $F > 10.00$, the value of $F_{0.01}$ for $r-1 = 2-1 = 1$ and 10 degrees of freedom.

4. *Calculations*: Substituting $n = 3$, $r = 2$, $T_{1..} = 81.0$, $T_{2..} = 79.5$, $T_{3..} = 79.5$, $T_{.1.} = 70.0$, $T_{.2.} = 92.0$, $T_{.3.} = 78.0$, $T_{..1} = 119.5$, $T_{..2} = 120.5$, $T_{(A)} = 87.5$, $T_{(B)} = 86.5$, $T_{(C)} = 66.0$, $T_{...} = 240.0$, and $\sum \sum \sum y_{ij(k)l}^2 = 104.5$ into the formulas for the sums of squares, we get

$$C = \frac{(240)^2}{18} = 3200.0$$

$$SS(Tr) = \tfrac{1}{6}[(87.5)^2 + (86.5)^2 + (66.0)^2] - 3200.0 = 49.1$$

$$SSR = \tfrac{1}{6}[(81.0)^2 + (79.5)^2 + (79.5)^2] - 3200.0 = 0.2$$

$$SSC = \tfrac{1}{6}[(70.0)^2 + (92.0)^2 + (78.0)^2] - 3200.0 = 41.3$$

$$SS(Rep) = \tfrac{1}{9}[(119.5)^2 + (120.5)^2] - 3200.0 = 0.1$$

$$SST = (14.0)^2 + (16.5)^2 + \ldots + (11.5)^2 - 3200. = 104.5$$

$$SSE = 104.5 - 49.1 - 41.3 - 0.2 - 0.1 = 13.8$$

and the results are as shown in the following analysis-of-variance table:

Source of variation	Degrees of freedom	Sum of squares	Mean square	F
Treatments (methods)	2	49.1	24.6	17.6
Rows (operators)	2	0.2	0.1	0.1
Columns (fluxes)	2	41.3	20.6	14.7
Replicates	1	0.1	0.1	0.1
Error	10	13.8	1.4	
Total	17	104.5		

5. *Decisions*: For treatments (methods) and columns (fluxes), since $F = 17.6$ and 14.7 exceed 7.56, the corresponding null hypotheses must be rejected; for rows (operators), since $F = 0.1$ does not exceed 7.56, and for replicates, since $F = 0.1$ does not exceed 10.00, the corresponding null hypotheses cannot be rejected. In other words, we conclude that differences in methods and fluxes, but not differences in operators or replications, affect the solder strengths of the electrical leads. To go one step further, the Duncan multiple-range test of Section 12.4 gives the following *decision pattern* at the 0.01 level of significance:

	Method C	*Method B*	*Method A*
Mean	11.0	14.4	14.6

Thus, we conclude that Method *C* definitely yields weaker solder bonds than Methods *A* or *B*.

The elimination of *three* extraneous sources of variability can be accomplished by means of a design called a **Graeco-Latin square**. This design is a square array of *n* Latin letters and *n* Greek letters, with the Latin and Greek letters each forming a Latin square; furthermore, each Latin letter appears once and only once in conjunction with each Greek letter. The following is an example of a 4×4 Graeco-Latin square:

$A\alpha$	$B\beta$	$C\gamma$	$D\delta$
$B\delta$	$A\gamma$	$D\beta$	$C\alpha$
$C\beta$	$D\alpha$	$A\delta$	$B\gamma$
$D\gamma$	$C\delta$	$B\alpha$	$A\beta$

The construction of Graeco-Latin squares, also called **orthogonal Latin squares**, poses interesting mathematical problems, some of which are mentioned in the book by H. B. Mann listed in the Bibliography.

To give an example where the use of a Graeco-Latin square might be appropriate, suppose that in the soldering example an additional source of variability is the temperature of the solder. If three solder temperatures, denoted by α, β, and γ, are to be used together with the three methods, three operators (rows), and three solder fluxes (columns), a replicate of a suitable

Graeco-Latin-square experiment could be laid out as follows:

	Flux 1	*Flux 2*	*Flux 3*
Operator 1	$A\alpha$	$B\gamma$	$C\beta$
Operator 2	$C\gamma$	$A\beta$	$B\alpha$
Operator 3	$B\beta$	$C\alpha$	$A\gamma$

Thus, Method A would be used by operator 1 using solder flux 1 and temperature α, by operator 2 using solder flux 2 and temperature β, and by operator 3 using solder flux 3 and temperature γ. Similarly, Method B would be used by operator 1 using solder flux 2 and temperature γ, and so forth.

In a Graeco-Latin square each variable (represented by rows, columns, Latin letters, or Greek letters) is "distributed evenly" over the other variables. Thus, in comparing the means obtained for one variable, the effects of the other variables are all averaged out. The analysis of a Graeco-Latin square is similar to that of a Latin square, with the addition of an extra source of variability corresponding to the Greek letters.

There exists a wide variety of experimental designs, other than the ones discussed in this chapter, that are suitable for many diverse purposes. Among the more widely used designs are the **incomplete-block designs**, which are characterized by the feature that each treatment is not represented in each block. If the number of treatments under investigation in an experiment is large, it often happens that it is impossible to find homogeneous blocks such that all of the treatments can be accommodated in each block.

For instance, if n paints are to be compared by applying each paint to a sheet of metal and then baking the sheets in an oven, it may be impossible to put all of the sheets in the oven at one time. Consequently, it would be necessary to use an experimental design in which $k < n$ treatments (paints) are included in each block (oven run). One way of doing this is to assign the treatments to each oven run in such a way that each treatment occurs together with each other treatment in the same number of blocks. For example, for $n = 4$ and $k = 2$ we might use the following scheme:

Oven run	*Paints*
1	1 and 2
2	3 and 4
3	1 and 3
4	2 and 4
5	1 and 4
6	2 and 3

This kind of design is called a **balanced incomplete-block design**, and it has the important feature that *comparisons between any two treatments can be made with equal precision.*

Since balanced incomplete-block designs may require too many blocks, many other schemes have been developed. Most of these experimental designs arose to meet the specific needs of experimenters, notably in the field of agriculture. As we have pointed out earlier, much of the language of experimental design, including such terms as "treatments," "blocks," "plots," etc., has been borrowed from agriculture. Only in recent years have the more sophisticated designs been applied to industrial and engineering experimentation, and, with more widespread application, it is to be expected that many new designs will be developed to meet the requirements in these fields.

EXERCISES

12.38 Referring to the problem in which samples of tin plate are to be distributed among four laboratories (Section 12.1) suppose that we are concerned with systematic differences in tin-coating weight along the direction of rolling as well as across the rolling direction. To eliminate these two sources of variability, each of two sheets of tin plate is divided into 16 parts, representing four positions across and four positions along the rolling direction. Then, four samples from each sheet are sent to each of the Laboratories *A*, *B*, *C*, and *D*, as shown, and the resulting tin-coating weights are determined.

Replicate I

C 0.20	A 0.24	D 0.20	B 0.27
B 0.28	C 0.19	A 0.22	D 0.28
D 0.34	B 0.23	C 0.21	A 0.28
A 0.32	D 0.22	B 0.16	C 0.27

Rolling direction ↓

Replicate II

B 0.29	A 0.25	C 0.18	D 0.28
D 0.28	B 0.18	A 0.21	C 0.25
C 0.28	D 0.23	B 0.20	A 0.28
A 0.30	C 0.19	D 0.24	B 0.25

Determine from these data whether the laboratories were obtaining consistent results. Also determine whether there are actual tin-coating weight differences across and along the rolling direction. (Use the 0.05 level of significance.)

12.39 A Latin-square design was used to compare the bond strengths of gold semi-conductor lead wires bonded to the lead terminal by five different methods, *A*, *B*, *C*, *D*, and *E*. The bonds were made by five different operators, and the devices were encapsulated using five different plastics, with the following results, expressed as pounds force required to break the bond.

Operator

		O_1	O_2	O_3	O_4	O_5
	P_1	*A* 3.0	*B* 2.4	*C* 1.9	*D* 2.2	*E* 1.7
	P_2	*B* 2.1	*C* 2.7	*D* 2.3	*E* 2.5	*A* 3.1
Plastic	P_3	*C* 2.1	*D* 2.6	*E* 2.5	*A* 2.9	*B* 2.1
	P_4	*D* 2.0	*E* 2.5	*A* 3.2	*B* 2.5	*C* 2.2
	P_5	*E* 2.1	*A* 3.6	*B* 2.4	*C* 2.4	*D* 2.1

Analyze these results and apply the Duncan multiple-range test with $\alpha = 0.01$ to the mean breaking strengths of the five bonding methods.

12.40 A test was made to determine which of three different golf-ball designs, *A*, *B*, and *C* will give the greatest distance. Golf balls were driven by three different golf pros, P_1, P_2, and P_3, using three different drivers, D_1, D_2, and D_3. The experiment was performed on three different fairways, and the distances from the tee to the point where the ball came to rest were measured in yards, as follows:

First fairway

	D_1	D_2	D_3
P_1	*B* 265	*A* 311	*C* 249
P_2	*A* 350	*C* 284	*B* 330
P_3	*C* 258	*B* 305	*A* 351

Second fairway

	D_1	D_2	D_3
P_1	*B* 220	*A* 276	*C* 189
P_2	*A* 319	*C* 232	*B* 264
P_3	*C* 175	*B* 254	*A* 307

Third fairway

	D_1	D_2	D_3
P_1	B 159	A 205	C 142
P_2	A 262	C 168	B 246
P_3	C 198	B 237	A 283

Is any of the golf-ball designs superior to the others with respect to distance?

12.41 To study the effectiveness of five different kinds of front-seat passenger restraint systems in automobiles, A, B, C, D, and E, the following Graeco-Latin-square experiment was performed. The rows represent different automotive size classes (from subcompact to full size), the columns represent different barrier impact speeds, and the Greek letters α, β, γ, δ, and ϵ represent different impact angles. The experimental results are given in terms of an index of forces at critical points on the test dummy that relates to the probability of fatal injury.

$A\alpha$ 0.50	$B\beta$ 0.21	$C\gamma$ 0.43	$D\delta$ 0.35	$E\epsilon$ 0.46
$B\gamma$ 0.51	$C\delta$ 0.20	$D\epsilon$ 0.40	$E\alpha$ 0.25	$A\beta$ 0.39
$C\epsilon$ 0.45	$D\alpha$ 0.07	$E\beta$ 0.29	$A\gamma$ 0.20	$B\delta$ 0.31
$D\beta$ 0.39	$E\gamma$ 0.10	$A\delta$ 0.31	$B\epsilon$ 0.24	$C\alpha$ 0.27
$E\delta$ 0.43	$A\epsilon$ 0.17	$B\alpha$ 0.31	$C\beta$ 0.22	$D\gamma$ 0.32

Analyze this experiment.

12.42 A clothing manufacturer wishes to determine which of four different needle designs is best for his sewing machines. The sources of variability which must be eliminated to make this comparison are the actual sewing machine used, the operator, and the type of thread. Using the design shown (the rows represent the operators, the columns represent the machines, the Latin letters represent the needles, and the Greek letters stand for the types of thread), the manufacturer recorded the number of rejected garments at the end of each of two weeks, with the following results:

<table>
<tr><td colspan="4" align="center">*First week*</td></tr>
<tr><td>$D\gamma$
47</td><td>$B\alpha$
40</td><td>$A\beta$
23</td><td>$C\delta$
72</td></tr>
<tr><td>$C\beta$
74</td><td>$A\delta$
37</td><td>$B\gamma$
28</td><td>$D\alpha$
75</td></tr>
<tr><td>$A\alpha$
52</td><td>$C\gamma$
95</td><td>$D\delta$
57</td><td>$B\beta$
15</td></tr>
<tr><td>$B\delta$
10</td><td>$D\beta$
45</td><td>$C\alpha$
93</td><td>$A\gamma$
52</td></tr>
</table>

<table>
<tr><td colspan="4" align="center">*Second week*</td></tr>
<tr><td>$C\alpha$
105</td><td>$A\gamma$
38</td><td>$B\delta$
15</td><td>$D\beta$
60</td></tr>
<tr><td>$D\delta$
70</td><td>$B\beta$
20</td><td>$A\alpha$
60</td><td>$C\gamma$
85</td></tr>
<tr><td>$B\gamma$
13</td><td>$D\alpha$
82</td><td>$C\beta$
90</td><td>$A\delta$
28</td></tr>
<tr><td>$A\beta$
33</td><td>$C\delta$
75</td><td>$D\gamma$
53</td><td>$B\alpha$
31</td></tr>
</table>

Using the 0.05 level of significance, determine whether there is a difference in the effectiveness of the needles. Also, determine whether there are significant differences attributable to the operators, the machines, and the types of thread.

12.43 Give an example of a balanced incomplete-block design for which
(a) $n = 3$ and $k = 2$;   (c) $n = 6$ and $k = 4$.
(b) $n = 4$ and $k = 3$;
What is the minimum number of times each treatment must be replicated in each of these designs?

12.6 ANALYSIS OF COVARIANCE

The purpose of the methods of Sections 12.3 and 12.5 was to free the experimental error from variability due to identifiable and controllable extraneous causes. In this section, we shall introduce a method, called the **analysis of covariance**, which applies when such extraneous, or **concomitant**, variables cannot be held fixed, but can be measured, nevertheless. This would be the case, for example, if we wanted to compare the effectiveness of several industrial training programs and the results depended on the trainees' I.Q.'s; if we wanted to compare the durability of several kinds of leather soles and the results depended on the weight of the persons wearing the shoes; or if we wanted to compare the merits of several cleaning agents and the results depended on the original condition of the surfaces cleaned.

The method by which we analyze data of this kind is a combination of the linear regression method of Section 11.1 and the analysis of variance of

Section 12.2. The underlying model is given by

*Model equation
for analysis
of covariance*

$$y_{ij} = \mu + \alpha_i + \delta x_{ij} + \epsilon_{ij}$$

for $i = 1, 2, \ldots, k; j = 1, 2, \ldots, n$. As in the model on page 337, μ is the grand mean, α_i is the effect of the ith treatment, and the ϵ_{ij} are values of independent, normally distributed variables with zero means and the common variance σ^2; as in the model on page 295 where we called it β, δ is the slope of the linear regression equation.

In the analysis of such data, the values of the concomitant variable, the x_{ij}, are eliminated by regression methods, namely, by estimating δ by the method of least squares, and then an analysis of variance is performed on the adjusted y's, namely, on the quantities $y'_{ij} = y_{ij} - \hat{\delta} x_{ij}$. This procedure is referred to as an analysis of covariance, as it involves a partitioning of the total **sum of products**

$$SPT = \sum_{i=1}^{k} \sum_{j=1}^{n} (y_{ij} - \bar{y}.)(x_{ij} - \bar{x}.)$$

in the same way as an ordinary analysis of variance involves the partitioning of the total sum of squares. In actual practice, the calculations are performed as follows:

1. The total, treatment, and error sums of squares are calculated for the x's by means of the formulas for a one-way classification on page 340; they will be denoted by SST_x, $SS(Tr)_x$, and SSE_x.
2. The total, treatment, and error sums of squares are calculated for the y's by means of the formulas for a one-way classification on page 340; they will be denoted by SST_y, $SS(Tr)_y$, and SSE_y.
3. The total, treatment, and error sums of products are calculated by means of the formulas

*Sums of products—
analysis of
covariance*

$$SPT = \sum_{i=1}^{k} \sum_{j=1}^{n} x_{ij} \cdot y_{ij} - C$$

$$SP(Tr) = \frac{\sum_{i=1}^{k} T_{x_i} \cdot T_{y_i}}{n} - C$$

$$SPE = SPT - SP(Tr)$$

where the correction term, C, is given by

$$C = \frac{T_x \cdot T_y}{k \cdot n}$$

and where T_{x_i} is the total of the x's for the ith treatment, T_{y_i} is the total of the y's for the ith treatment, T_x is the total of all the x's, and T_y is the total of all the y's.

4. The total, error, and treatment sums of squares are calculated for the adjusted y's by means of the formulas

Adjusted sums of squares— analysis of covariance

$$SST_{y'} = SST_y - \frac{(SPT)^2}{SST_x}$$

$$SSE_{y'} = SSE_y - \frac{(SPE)^2}{SSE_x}$$

$$SS(Tr)_{y'} = SST_{y'} - SSE_{y'}$$

The results obtained in these calculations are conveniently summarized by means of the following kind of **analysis-of-covariance table**:

Source of variation	Sum of squares for x	Sum of squares for y	Sum of prod- ucts	Sum of squares for y'	Degrees of freedom	Mean square
Treatments	$SS(Tr)_x$	$SS(Tr)_y$	$SP(Tr)$	$SS(Tr)_{y'}$	$k-1$	$MS(Tr)_{y'}$ $= \frac{SS(Tr)_{y'}}{k-1}$
Error	SSE_x	SSE_y	SPE	$SSE_{y'}$	$nk-k-1$	$MSE_{y'}$ $= \frac{SSE_{y'}}{nk-k-1}$
Total	SST_x	SST_y	SPT	$SST_{y'}$	$nk-2$	

Note that each mean square is obtained by dividing the corresponding sum of squares by its degrees of freedom.

Finally, the null hypothesis $\alpha_1 = \alpha_2 = \ldots = \alpha_k = 0$ is tested against the alternative hypothesis that the α_i are not all equal to zero on the basis of the statistic

F ratio for adjusted treatments

$$F = \frac{MS(Tr)_{y'}}{MSE_{y'}}$$

It is rejected at the level of significance α if the value obtained for F exceeds F_α with $k-1$ and $nk-k-1$ degrees of freedom.

EXAMPLE   Suppose that a research worker has three different cleaning agents, A_1, A_2, and A_3, and he wishes to select the most efficient agent for cleaning a metallic surface. The cleanliness of a surface is measured by its reflectivity, expressed in arbitrary units as the ratio of the reflectivity observed to that of a standard mirror surface. Analysis of covariance must be used because the effect of a cleaning agent on reflectivity will depend on the original cleanliness, namely, the original reflectivity of the surface. The research worker obtained the following results:

A_1	Original reflectivity, x	0.50	0.55	0.60	0.35
	Final reflectivity, y	1.00	1.20	0.80	1.40
A_2	Original reflectivity, x	0.75	1.65	1.00	1.10
	Final reflectivity, y	0.75	0.60	0.55	0.50
A_3	Original reflectivity, x	0.60	0.90	0.80	0.70
	Final reflectivity, y	1.00	0.70	0.80	0.90

Perform an analysis of covariance to determine (at the 0.05 level of significance) whether there are differences in the reflectivity improvements produced by the three cleaning agents.

Solution

1. *Null hypothesis*: $\alpha_1 = \alpha_2 = \alpha_3 = 0$.
   Alternate hypothesis: The α's are not all equal to zero.
2. *Level of significance*: $\alpha = 0.05$
3. *Criterion*: Reject the null hypothesis if $F > 4.46$, the value of $F_{0.05}$ for $k - 1 = 3 - 1 = 2$ and $nk - k - 1 = 4 \times 3 - 3 - 1 = 8$ degrees of freedom.
4. *Calculations*: The totals are $T_{x_1} = 2.00$, $T_{x_2} = 4.50$, $T_{x_3} = 3.00$, $T_x = 9.50$, $T_{y_1} = 4.40$, $T_{y_2} = 2.40$, $T_{y_3} = 3.40$, and $T_y = 10.20$.
   For the x's, the correction term is $\dfrac{(9.50)^2}{3.4} = 7.52$, and the sums of squares are

$$SST_x = 0.50^2 + 0.55^2 + \ldots + 0.70^2 - 7.52 = 1.31$$

$$SS(Tr)_x = \frac{2.00^2 + 4.50^2 + 3.00^2}{4} - 7.52 = 0.79$$

$$SSE_x = 1.31 - 0.79 = 0.52$$

   For the y's, the correction term is $\dfrac{(10.20)^2}{3 \cdot 4} = 8.67$, and the sums of squares are

$$SST_y = 1.00^2 + 1.20^2 + \ldots + 0.90^2 - 8.67 = 0.79$$

$$SS(Tr)_y = \frac{4.40^2 + 2.40^2 + 3.40^2}{4} - 8.67 = 0.50$$

$$SSE_y = 0.79 - 0.50 = 0.29$$

For the sums of products, the correction term is $\frac{(9.50)(10.20)}{3 \cdot 4} = 8.08$, and we get

$$SPT = (0.50)(1.00) + (0.55)(1.20) + \ldots + (0.70)(0.90) - 8.08$$
$$= -0.80$$

$$SP(Tr) = \frac{(2.00)(4.40) + (4.50)(2.40) + (3.00)(3.40)}{4} - 8.08$$

$$= -0.63$$

$$SPE = -0.80 - (-0.63) = -0.17$$

Finally, for the adjusted y's we get

$$SST_{y'} = 0.79 - \frac{(-0.80)^2}{1.31} = 0.30$$

$$SSE_{y'} = 0.29 - \frac{(-0.17)^2}{0.52} = 0.23$$

$$SS(Tr)_{y'} = 0.30 - 0.23 = 0.07$$

All these results are summarized in the following analysis-of-covariance table:

Source of variation	Sum of squares for x	Sum of squares for y	Sum of products	Sum of squares for y'	Degrees of freedom	Mean squares
Treatments	0.79	0.50	−0.63	0.07	2	0.035
Error	0.52	0.29	−0.17	0.23	8	0.026
Total	1.31	0.79	−0.80	0.03	10	

5. *Decision*: Since $F = \frac{0.035}{0.026} = 1.34$ does not exceed 4.46, the null hypothesis cannot be rejected. In other words, one cannot conclude that any one of the cleaning agents is more effective than the others.

Analysis-of-covariance methods have not been widely used until recent years, due mainly to the rather extensive calculations that are required. Of course, with the widespread availability of computers and appropriate programs, this is no longer a problem. There are several ways in which the analysis-of-covariance method presented here can be generalized. First, there can be more than one concomitant variable; then, the method can be applied to more complicated kinds of designs, say, to a randomized block design, where the regression coefficient could even assume a different value for each block.

EXERCISES

12.44 To compare the life expectancy of a transistor under three storage conditions and account at the same time for a leakage current (collector to base), a laboratory technician obtained the following results, where the leakage current, x, is in microamperes, and the life times, y, are in hours:

Storage condition 1		Storage condition 2		Storage condition 3	
x	y	x	y	x	y
4.8	9,912	6.4	9,952	8.8	9,596
7.2	9,383	8.7	9,482	6.2	9,697
5.5	9,734	7.1	9,435	7.5	9,700
6.0	9,551	5.3	9,915	4.9	9,610
8.3	8,959	4.6	9,492	5.4	10,145
7.6	9,474	6.0	9,565	5.8	10,191
5.9	9,179	7.2	9,704	7.3	9,855
8.0	9,359	8.8	9,636	8.6	9,682
4.3	9,580	5.4	9,608	8.8	10,160
5.1	9,245	7.8	9,548	6.0	9,982

Perform an analysis of covariance, using the level of significance $\alpha = 0.05$. Also, estimate the value of the regression coefficient.

12.45 Four different railroad-track cross-section configurations were tested to determine which is most resistant to breakage under use conditions. Ten miles of each kind of track were laid in each of six locations, and the number of cracks and other fracture-related conditions (y) was measured over a two-year usage period. To compare these track designs adequately, however, it is necessary to correct for extent of usage (x), measured in terms of the average number of trains per day that ran over each section of track. Use the following experimental results to test (0.01 level of significance) whether the track designs were equally resistant to breakage and to estimate the effect of usage on breakage resistance.

Track design A		Track design B		Track design C		Track design D	
x	*y*	*x*	*y*	*x*	*y*	*x*	*y*
10.4	3	16.9	8	17.8	5	19.6	9
19.3	7	23.6	11	24.4	9	25.4	8
13.7	4	14.4	7	13.5	5	35.5	16
7.2	0	17.2	10	20.1	6	16.8	7
16.3	5	9.1	4	11.0	4	31.2	11

12.46 Three different instrument-panel configurations were tested by placing airline pilots in flight simulators and testing their reaction time to simulated flight emergencies. Each pilot was faced with ten emergency conditions in a randomized sequence, and the total time required to take corrective action for all ten conditions was measured, with the following results:

Instrument panel 1		Instrument panel 2		Instrument panel 3	
x	*y*	*x*	*y*	*x*	*y*
8.1	6.55	12.1	5.74	15.2	6.37
19.4	6.40	2.1	5.93	8.7	6.97
11.6	5.93	3.9	6.16	7.2	7.38
24.9	6.79	5.2	5.68	6.1	6.43
6.2	7.16	4.6	5.41	11.8	7.59
3.8	5.64	14.4	6.29	12.1	7.16
18.4	5.87	16.1	5.55	9.5	7.02
9.4	6.32	8.5	4.82	2.6	6.85

In this table, x is the number of years of experience of the pilot, and y is the total reaction time in seconds. Perform an analysis of covariance to test whether the instrument-panel configurations yield significantly different results ($\alpha = 0.05$). Also, perform a one-way analysis of variance (ignoring the covariate, x) and determine in that way what effect experience has on the results.

12.7 CHECK LIST OF KEY TERMS (with page references)

13

Factorial Experimentation

In Chapter 12 we were interested mainly in the effects of one variable, whose values we referred to as "treatments." Extraneous variables were accommodated by means of blocks, replicates, or the rows and columns of Latin squares and more complicated kinds of designs. In this chapter we shall be concerned with the individual and joint effects of several variables, and combinations of the values, or levels, of these variables will now play the roles of the different treatments. Extraneous variables, if any, will be handled as before.

In Sections 13.1 and 13.2 we deal with the analysis of experiments whose treatments can be regarded as combinations of the levels of two or more factors. In Section 13.3 we study the special case of factors having two levels, and in the remainder of the chapter we take up the analysis of experiments where there are too many combinations of all the factors to be included in the same block or experimental program.

To introduce the idea of a simple **two-factor (two-variable) experiment,** suppose that it is desired to determine the effects of flue temperature and oven width on the time required to make coke. The experimental conditions used are

Oven width (inches)	Flue temperature (degrees F)
4	1,600
4	1,900
8	1,600
8	1,900
12	1,600
12	1,900

and if several blocks (or replicates) were run, each consisting of these six "treatments," it would be possible to analyze the data as a two-way classification and test for significant differences among the six treatment means. In this instance, however, the experimenter is interested in knowing far more than that—he wishes to know whether variations in oven width or in flue temperature affect the coking time, and perhaps also whether any changes in coking time attributable to variations in oven width are the same at different temperatures.

It is possible to answer questions of this kind if the experimental conditions, the treatments, consist of appropriate combinations of the **levels** (or values) of the various **factors.** The factors in the preceding example are oven width and flue temperature; oven width has the *three levels* 4, 8, and 12 inches, while flue temperature has the *two levels* 1,600 and 1,900 degrees Fahrenheit. Note that the six treatments were chosen in such a way that each level of oven width is used once in conjunction with each level of flue temperature. In general, if two factors A and B are to be investigated at a levels and b levels, respectively, and if there are $a \cdot b$ experimental conditions (treatments) corresponding to all possible combinations of the levels of the two factors, the resulting experiment is referred to as a **complete $a \times b$ factorial experiment.** Note that if one or more of the $a \cdot b$ experimental conditions is omitted, the experiment can still be analyzed as a two-way classification, but it cannot readily be analyzed as a factorial experiment. It is customary to omit the word "complete," so that an *$a \times b$ **factorial experiment** is understood to contain experimental conditions corresponding to all possible combinations of the levels of the two factors.

In order to obtain an estimate of the experimental error in a two-factor experiment it is necessary to replicate, that is, to repeat the entire set of $a \cdot b$ experimental conditions, say, a total of r times, randomizing the order of applying the conditions in each replicate. If y_{ijk} is the observation in the kth replicate, taken at the ith level of factor A and the jth level of factor B, the model assumed for the analysis of this kind of experiment is usually written as

Model equation for two-factor experiment

$$y_{ijk} = \mu + \alpha_i + \beta_j + (\alpha\beta)_{ij} + \rho_k + \epsilon_{ijk}$$

for $i = 1, 2, \ldots, a, j = 1, 2, \ldots, b$, and $k = 1, 2, \ldots, r$. Here μ is the grand mean, α_i is the effect of the ith level of factor A, β_j is the effect of the jth level of factor B, $(\alpha\beta)_{ij}$ is the **interaction**, or joint effect, of the ith level of factor A and the jth level of factor B, and ρ_k is the effect of the kth replicate. As in the models used in Chapter 12 we shall assume that the ϵ_{ijk} are values of independent random variables having normal distributions with zero means and the common variance σ^2. Also, analogous to the restrictions imposed on the models on pages 337 and 349, we shall assume that

$$\sum_{i=1}^{a} \alpha_i = \sum_{j=1}^{b} \beta_j = \sum_{i=1}^{a} (\alpha\beta)_{ij} = \sum_{j=1}^{b} (\alpha\beta)_{ij} = \sum_{k=1}^{r} \rho_k = 0$$

It can be shown that these restrictions will assure unique estimates for the parameters μ, α_i, β_j, $(\alpha\beta)_{ij}$, and ρ_k.

To illustrate the model underlying a two-factor experiment, let us consider an experiment with two replicates in which factor A occurs at two levels, factor B occurs at two levels, and the replication effects are zero, that is, $\rho_1 = \rho_2 = 0$. In view of the restrictions on the parameters we also have

$$\alpha_2 = -\alpha_1, \beta_2 = -\beta_1 \quad \text{and} \quad (\alpha\beta)_{21} = (\alpha\beta)_{12} = -(\alpha\beta)_{22} = -(\alpha\beta)_{11}$$

and the population means corresponding to the four experimental conditions defined by the two levels of factor A and the two levels of factor B can be written as

$$\mu_{111} = \mu_{112} = \mu + \alpha_1 + \beta_1 + (\alpha\beta)_{11}$$
$$\mu_{121} = \mu_{122} = \mu + \alpha_1 - \beta_1 - (\alpha\beta)_{11}$$
$$\mu_{211} = \mu_{212} = \mu - \alpha_1 + \beta_1 - (\alpha\beta)_{11}$$
$$\mu_{221} = \mu_{222} = \mu - \alpha_1 - \beta_1 + (\alpha\beta)_{11}$$

Substituting for $\mu_{ij1} = \mu_{ij2}$ the mean of all observations obtained for the ith level of factor A and the jth level of factor B, we get four simultaneous linear equations which can be solved to yield estimates for the parameters μ, α_1, β_1, and $(\alpha\beta)_{11}$ (see Exercise 13.10 on page 398).

To continue our illustration, let us now suppose that $\mu = 10$. If all of the other effects equalled zero, each of the μ_{ijk} would equal 10, and the response surface would be the horizontal plane shown in Figure 13.1(a). If

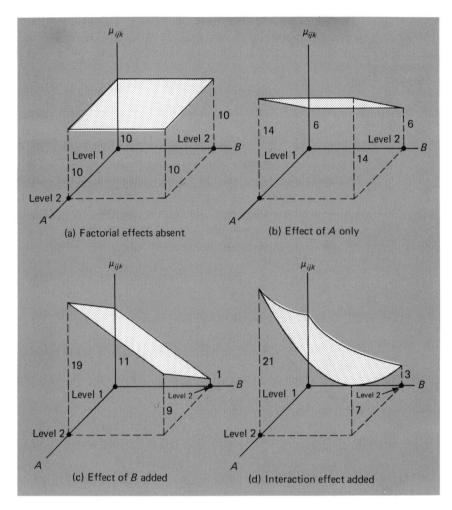

Figure 13.1 Factorial effects.

we now add an effect of factor A, with $\alpha_1 = -4$, the response surface be-comes the tilted plane shown in Figure 13.1(b), and if we add to this an effect of factor B, with $\beta_1 = 5$, we get the plane shown in Figure 13.1(c). Note that, so far, the effects of factors A and B are *additive*, that is, the change in the mean for either factor in going from level 1 to level 2 does not depend on the level of the other factor, and the response surface is a plane. If we now include an interaction, with $(\alpha\beta)_{11} = -2$, the plane becomes twisted as shown in Figure 13.1(d), the effects are no longer additive, and the response surface is no longer a plane. Note, also, that if the replication effects had not equalled zero, we would have obtained a different surface for each replicate;

for each replicate the surface of Figure 13.1(d) would have been shifted an appropriate number of units up or down.

Generalizing these ideas from the 2×2 factorial experiment, the analysis of an $a \times b$ factorial experiment is based on the following breakdown of the total sum of squares. First, we partition SST into components attributed to treatments, replicates (or blocks), and error, by means of the identity

$$\sum_{i=1}^{a} \sum_{j=1}^{b} \sum_{k=1}^{r} (y_{ijk} - \bar{y}...)^2 = r \sum_{i=1}^{a} \sum_{j=1}^{b} (\bar{y}_{ij.} - \bar{y}...)^2 + ab \sum_{k=1}^{r} (\bar{y}..._k - \bar{y}...)^2$$

$$+ \sum_{i=1}^{a} \sum_{j=1}^{b} \sum_{k=1}^{r} (y_{ijk} - \bar{y}_{ij.} - \bar{y}..._k + \bar{y}...)^2$$

Except for notation, this identity is equivalent to that of Theorem 12.2. The total sum of squares, on the left-hand side of the identity, has $abr - 1$ degrees of freedom. The terms on the right are, respectively, the treatment sum of squares having $ab - 1$ degrees of freedom, the replicate (block) sum of squares having $r - 1$ degrees of freedom, and the error sum of squares having $(ab - 1)(r - 1)$ degrees of freedom. (Note that the various degrees of freedom are the same as those of the analysis-of-variance table on page 351 if we substitute ab for a and r for b.)

So far there is nothing new about the analysis of the data; it is the analysis of a two-way classification, *but the distinguishing feature of a factorial experiment is that the treatment sum of squares can be further subdivided into components corresponding to the various factorial effects.* Thus, for a two-factor experiment we have the following partition, or breakdown, of the treatment sum of squares:

$$r \sum_{i=1}^{a} \sum_{j=1}^{b} (\bar{y}_{ij.} - \bar{y}...)^2 = rb \sum_{i=1}^{a} (\bar{y}_{i..} - \bar{y}...)^2 + ra \sum_{j=1}^{b} (\bar{y}_{.j.} - \bar{y}...)^2$$

$$+ r \sum_{i=1}^{a} \sum_{j=1}^{b} (\bar{y}_{ij.} - \bar{y}_{i..} - \bar{y}_{.j.} + \bar{y}...)^2$$

The first term on the right measures the variability of the means corresponding to the different levels of factor A, and we refer to it as the **factor A sum of squares**, SSA. Similarly, the second term is the sum of squares for factor B, SSB, and the third term is the **interaction sum of squares $SS(AB)$**, which measures the variability of the means $\bar{y}_{ij}$ that is not attributable to the individual (or separate) effects of factors A and B. The $ab - 1$ degrees of freedom for treatments are, accordingly, subdivided into $a - 1$ degrees of freedom for the effect of factor A, $b - 1$ for the effect of factor B, and

$$ab - 1 - (a - 1) - (b - 1) = (a - 1)(b - 1)$$

degrees of freedom for interaction.

EXAMPLE   Referring to the coking experiment described on page 380, suppose that three replicates of this experiment yield the following coking times (in hours):

Factor A Oven width	Factor B Flue temp.	Rep. 1	Rep. 2	Rep. 3	Total
4	1600	3.5	3.0	2.7	9.2
4	1900	2.2	2.3	2.4	6.9
8	1600	7.1	6.9	7.5	21.5
8	1900	5.2	4.6	6.8	16.6
12	1600	10.8	10.6	11.0	32.4
12	1900	7.6	7.1	7.3	22.0
	Total	36.4	34.5	37.7	108.6

Perform an analysis of variance based on this two-factor experiment and test for the significance of the factorial effects, using the 0.01 level of significance.

Solution   Following steps analogous to those used in the analysis of a two-way classification, we get
1. *Null hypotheses*:

$$\alpha_1 = \alpha_2 = \alpha_3 = 0, \qquad \beta_1 = \beta_2 = 0$$

$$(\alpha\beta)_{11} = (\alpha\beta)_{12} = (\alpha\beta)_{21} = (\alpha\beta)_{22} = (\alpha\beta)_{31} = (\alpha\beta)_{32} = 0$$

$$\rho_1 = \rho_2 = \rho_3 = 0$$

Alternative hypotheses: The α's are not all equal to zero; the β's are not all equal to zero; the $(\alpha\beta)$ terms are not all equal to zero; the ρ's are not all equal to zero.
2. *Levels of significance*: $\alpha = 0.01$ for all tests.
3. *Criteria*: For replications, reject the null hypothesis if $F > 7.56$, the value of $F_{0.01}$ for $r - 1 = 3 - 1 = 2$ and $(ab - 1)(r - 1) = (3 \times 2 - 1)(3 - 1) = 10$ degrees of freedom; for the main effect of factor A, reject the null hypothesis if $F > 7.56$, the value of $F_{0.01}$ for $a - 1 = 3 - 1 = 2$ and $(ab - 1)(r - 1) = (3 \times 2 - 1)(3 - 1) = 10$ degrees of freedom; for the main effect of factor B, reject if $F > 10.00$, the value of $F_{0.01}$ for $b - 1 = 2 - 1 = 1$ and $(ab - 1)(r - 1) = (3 \times 2 - 1)(3 - 1) = 10$ degrees of freedom; for the interaction effect, reject if $F > 7.56$, the value of $F_{0.01}$ for $(a - 1)(b - 1) = (3 - 1)(2 - 1) = 2$ and $(ab - 1)(r - 1) = (3 \times 2 - 1)(3 - 1) = 10$ degrees of freedom.
4. *Calculations*: Following the procedure used in analyzing a two-way classification, we first compute the correction term

$$C = \frac{(108.6)^2}{18} = 655.22$$

Then, the total sum of squares is given by

$$SST = (3.5)^2 + (2.2)^2 + \ldots + (7.3)^2 - 655.22 = 149.38$$

and the treatment and replicate (instead of blocks) sums of squares are given by

$$SS(Tr) = \tfrac{1}{3}[(9.2)^2 + (6.9)^2 + \ldots + (22.0)^2] - 655.22 = 146.05$$
$$SSR = \tfrac{1}{6}[(36.4)^2 + (34.5)^2 + (37.7)^2] - 655.22 = 0.86$$

Finally, by subtraction, we obtain

$$SSE = 149.38 - 146.05 - 0.86 = 2.47$$

Subdivision of the treatment sum of squares into components for factors A and B, and for interaction, can be facilitated by constructing the following kind of two-way table, where the entries are the totals in the right-hand column of the table giving the original data:

		Factor B Flue temperature		
		1600	*1900*	
Factor A Oven width	*4*	9.2	6.9	16.1
	8	21.5	16.6	38.1
	12	32.4	22.0	54.4
		63.1	45.5	108.6

Using formulas analogous to the ones with which we computed the sums of squares for various effects in Chapter 12, we now have for the two main effects

$$SSA = \frac{1}{b \cdot r} \sum_{i=1}^{a} T_{i \cdot \cdot}^2 - C$$

$$= \tfrac{1}{6}[(16.1)^2 + (38.1)^2 + (54.4)^2] - 655.22 = 123.14$$

$$SSB = \frac{1}{a \cdot r} \sum_{j=1}^{b} T_{\cdot j \cdot}^2 - C$$

$$= \tfrac{1}{9}[(63.1)^2 + (45.5)^2] - 655.22 = 17.21$$

and for the interaction

$$SS(AB) = SS(Tr) - SSA - SSB$$
$$= 146.05 - 123.14 - 17.21 = 5.70$$

Finally, dividing the various sums of squares by their degrees of freedom, and dividing the appropriate mean squares by the error mean square, we obtain the results shown in the following analysis-of-variance table:

Source of variation	Degrees of freedom	Sum of squares	Mean square	F
Replication	2	0.86	0.43	1.72
Main effects:				
A	2	123.14	61.57	246
B	1	17.21	17.21	68.8
Interaction	2	5.70	2.85	11.4
Error	10	2.47	0.25	
Total	17	149.38		

5. *Decisions*: For replications, since $F = 1.72$ does not exceed 7.56, the value of $F_{0.01}$ for 2 and 10 degrees of freedom, we cannot reject the null hypothesis. For the main effect of factor A, since $F = 246$ exceeds 7.56, and for the main effect of factor B, since $F = 68.8$ exceeds 10.00, the null hypotheses must be rejected. For the interaction effect, since $F = 11.4$ exceeds 7.56 we must reject the null hypothesis illustrated in Figure 13.2, showing the trend of mean coking times for changing oven width at each of the flue temperatures. It is apparent from this figure that the increase in coking time for changing oven width is *greater* at the lower flue temperature. In view of this interaction, great care must be exercised in stating the results of this experiment. For instance, it would be very misleading to state merely that the effect of increasing flue temperature from 1,600 to 1,900 degrees Fahrenheit is to lower the coking time by $\frac{63.1}{9} - \frac{45.5}{9} = 1.96$ hours. In fact, the coking time is lowered on the average by as little as 0.77 hour when the oven width is

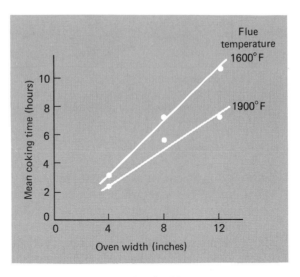

Figure 13.2 Results of coking experiment.

4 inches, and by as much as 3.47 hours when the oven width is 12 inches.

13.2 MULTIFACTOR EXPERIMENTS

Much industrial research and experimentation is conducted to discover the individual and joint effects of several factors on variables which are most relevant to phenomena under investigation. The experimental designs most often used are of the simple randomized-block or two-way classification type, but the distinguishing feature of most of them is the factorial arrangement of the treatments, or experimental conditions. As we observed in the preceding section, r sets of data pertaining to $a \cdot b$ experimental conditions can be analyzed as a factorial experiment in r replicates if the experimental conditions represent all possible combinations of the levels of two factors A and B. In this section, we shall extend the discussion to factorial experiments involving more than two factors, that is, to experiments where the experimental conditions represent all possible combinations of the levels of three or more factors.

To illustrate the analysis of a **multifactor experiment**, let us consider the following situation. A warm sulfuric pickling bath is used to remove oxides from the surface of a metal prior to plating, and it is desired to determine what factors in addition to the concentration of the sulfuric acid might affect the electrical conductivity of the bath. As it is felt that the salt concentration as well as the bath temperature might also affect the electrical conductivity, an experiment is planned to determine the individual and joint effects of these three variables on the electrical conductivity of the bath. In order to cover the ranges of concentrations and temperatures normally encountered, it is decided to use the following levels of the three factors:

Factor	Level 1	Level 2	Level 3	Level 4
A. Acid concentration (percent)	0	6	12	18
B. Salt concentration (percent)	0	10	20	
C. Bath temperature (degrees F)	80	100		

The resulting factorial experiment requires $4 \cdot 3 \cdot 2 = 24$ experimental conditions in each replicate, where each experimental condition is a pickling bath make up according to specifications. The order in which these pickling baths are made up should be random. Let us suppose that two replicates of the experiment have actually been completed, that is, the electrical conductivities of the various pickling baths have been measured, and that the results are as shown in the following table:

Results of Acid-Bath Experiment

Level of factor			Conductivity ($mhos/cm^3$)		
A	*B*	*C*	*Rep. 1*	*Rep. 2*	*Total*
1	1	1	0.99	0.93	1.92
1	1	2	1.15	0.99	2.14
1	2	1	0.97	0.91	1.88
1	2	2	0.87	0.86	1.73
1	3	1	0.95	0.86	1.81
1	3	2	0.91	0.85	1.76
2	1	1	1.00	1.17	2.17
2	1	2	1.12	1.13	2.25
2	2	1	0.99	1.04	2.03
2	2	2	0.96	0.98	1.94
2	3	1	0.97	0.95	1.92
2	3	2	0.94	0.99	1.93
3	1	1	1.24	1.22	2.46
3	1	2	1.12	1.15	2.27
3	2	1	1.15	0.95	2.10
3	2	2	1.11	0.95	2.06
3	3	1	1.03	1.01	2.04
3	3	2	1.12	0.96	2.08
4	1	1	1.24	1.20	2.44
4	1	2	1.32	1.24	2.56
4	2	1	1.14	1.10	2.24
4	2	2	1.20	1.19	2.39
4	3	1	1.02	1.01	2.03
4	3	2	1.02	1.00	2.02
		Total	25.53	24.64	50.17

The model we shall assume for the analysis of this experiment (or any similar three-factor experiment) is an immediate extension of the one used in Section 13.1. If y_{ijkl} is the conductivity measurement obtained at the ith level of acid concentration, the jth level of salt concentration, the kth level of bath temperature, in the lth replicate, we write

Model equation for three-factor experiment

$$y_{ijkl} = \mu + \alpha_i + \beta_j + \gamma_k + (\alpha\beta)_{ij} + (\alpha\gamma)_{ik} + (\beta\gamma)_{jk}$$
$$+ (\alpha\beta\gamma)_{ijk} + \rho_l + \epsilon_{ijkl}$$

for $i = 1, 2, \ldots, a, j = 1, 2, \ldots, b, k = 1, 2, \ldots, c,$ and $l = 1, 2, \ldots, r.$ We also assume that the sums of the **main effects** (α's, β's, and γ's) as well as the sum of the replication effects are equal to zero, that the sums of the **two-way interaction effects** summed on either subscript equal zero for any value of the other subscript, and that the sum of the **three-way interaction effects** summed on any one of the subscripts is zero for any values of the other two subscripts. As before, the ϵ_{ijkl} are assumed to be values of independent random variables having zero means and the common variance σ^2.

We begin the analysis of the data by treating the experiment as a two-way classification with $a \cdot b \cdot c$ treatments and r replicates (blocks), and using the shortcut formulas on page 350, we obtain

$$C = \frac{(50.17)^2}{48} = 52.4381$$

$$SST = (0.99)^2 + (1.15)^2 + \ldots + (1.00)^2 - 52.4381 = 0.6624$$

$$SS(Tr) = \tfrac{1}{2}[(1.92)^2 + (2.14)^2 + \ldots + (2.02)^2] - 52.4381 = 0.5712$$

$$SSR = \tfrac{1}{24}[(25.53)^2 + (24.64)^2] - 52.4281 = 0.0165$$

$$SSE = 0.6624 - 0.5712 - 0.0165 = 0.0747$$

The degrees of freedom for these sums of squares are, respectively, 47, 23, 1, and 23.

Next, we shall want to subdivide the treatment sum of squares into the three **main effect sums of squares** SSA, SSB, SSC, the three **two-way interaction sums of squares**, $SS(AB)$, $SS(AC)$, and $SS(BC)$, and the **three-way interaction sum of squares** $SS(ABC)$. To facilitate the calculation of these sums of squares we first construct the following three tables analogous to the one on page 385:

		B			
		1	*2*	*3*	
	1	4.06	3.61	3.57	11.24
	2	4.42	3.97	3.85	12.24
A	*3*	4.73	4.16	4.12	13.01
	4	5.00	4.63	4.05	13.68
		18.21	16.37	15.59	50.17

		C		
		1	*2*	
	1	5.61	5.63	11.24
	2	6.12	6.12	12.24
A	*3*	6.60	6.41	13.01
	4	6.71	6.97	13.68
		25.04	25.13	50.17

		B			
		1	*2*	*3*	
	1	8.99	8.25	7.80	25.04
C	*2*	9.22	8.12	7.79	25.13
		18.21	16.37	15.59	50.17

The entries of these tables are the totals of all measurements obtained at the respective levels of the two variables. Note the self-checking feature of these tables; the same marginal totals appear several times, thus providing a rapid and effective check on the calculations.

To calculate SSA, SSB, and $SS(AB)$ we refer to the first of the above tables and an identity analogous to the one on page 383. As a matter of fact, the calculations parallel those for calculating SSA, SSB, and $SS(AB)$ in the two-factor experiment. To take the place of the treatment sum of squares we first calculate

$$rc \sum_{i=1}^{a} \sum_{j=1}^{b} (\bar{y}_{ij..} - \bar{y}....)^2 = \frac{1}{r \cdot c} \sum_{i=1}^{a} \sum_{j=1}^{b} T_{ij..}^2 - C$$

$$= \tfrac{1}{4}[(4.06)^2 + (4.42)^2 + \ldots + (4.05)^2] - 52.4381$$

$$= 0.5301$$

and we then obtain

$$SSA = \frac{1}{bcr} \sum_{i=1}^{a} T_{i...}^2 - C$$

$$= \tfrac{1}{12}[(11.24)^2 + \ldots + (13.68)^2] - 52.4381$$

$$= 0.2750$$

$$SSB = \frac{1}{acr} \sum_{j=1}^{b} T^2_{\cdot j \cdot} - C$$

$$= \tfrac{1}{16}[(18.21)^2 + (16.37)^2 + (15.59)^2] - 52.4381$$

$$= 0.2262$$

and

$$SS(AB) = 0.5301 - 0.2750 - 0.2262$$

$$= 0.0289$$

Performing the same calculations for the second of the tables on page 390, we obtain, similarly,

$$SSC = 0.0002 \quad \text{and} \quad SS(AC) = 0.0085$$

and the analysis of the third table yields

$$SS(BC) = 0.0042$$

For the three-way interaction sum of squares we finally obtain by subtraction

$$SS(ABC) = SS(Tr) - SSA - SSB - SSC$$

$$- SS(AB) - SS(AC) - SS(BC)$$

$$= 0.5712 - 0.2750 - 0.2262 - 0.0002$$

$$- 0.0289 - 0.0085 - 0.0042$$

$$= 0.0282$$

Note that the degrees of freedom for each main effect is one less than the number of levels of the corresponding factor. The degrees of freedom for each interaction is the *product* of the degrees of freedom for those factors appearing in the interaction. Thus, the degrees of freedom for the three main effects are 3, 2, and 1 in this example, while the degrees of freedom for the two-way interactions are 6, 3, and 2, and the degrees of freedom for the three-way interaction is 6.

The following table shows the complete analysis of variance for the acid-bath experiment:

Source of variation	Degrees of freedom	Sum of squares	Mean square	F
Replicates	1	0.0165	0.0165	5.16
Main effects:				
A	3	0.2750	0.0917	28.66
B	2	0.2262	0.1131	35.34
C	1	0.0002	0.0002	< 1
Two-factor interactions:				
AB	6	0.0289	0.0048	1.50
AC	3	0.0085	0.0028	< 1
BC	2	0.0042	0.0021	< 1
Three-factor interaction:				
ABC	6	0.0282	0.0047	1.47
Error	23	0.0747	0.0032	
Total	47	0.6624		

Obtaining the appropriate values of $F_{0.05}$ and $F_{0.01}$ from Table 6, we find that the test for replicates is significant at the 0.05 level (perhaps the two replicates were performed under different atmospheric conditions or the thermometer used to measure bath temperatures went out of calibration, etc.), the tests for the factor A and factor B main effects are significant at the 0.01 level, while none of the other F's is significant at either level. We conclude from this analysis that variations in acid concentration and salt concentration affect the electrical conductivity, variations in bath temperature do not, and that there are no interactions. To go one step further, we might investigate the *magnitudes* of the effects by studying graphs of means like those shown in Figures 13.3 and 13.4. Here we find that the conductivity increases as acid is added and decreases as salt is added; using the methods of Chapter 11 we might even fit lines, curves, or surfaces to describe the response surface relating conductivity to the variables under consideration.

The general computational procedure for a multifactor experiment is similar to the method illustrated here for a $4 \times 3 \times 2$ factorial experiment. We first analyze the data as a two-way classification (or whatever other design is being used) and then we analyze the treatment sum of squares into the components attributed to the various main effects and interactions. In general, the sum of squares for each main effect is found by adding the squares of the totals corresponding to the different levels of that factor, dividing by the

Figure 13.3 Effect of acid concentration.

Figure 13.4 Effect of salt concentration.

number of observations comprising each of these totals, and then subtracting the correction term. The sum of squares for any interaction is found by adding the squares of all totals obtained by summing over those subscripts pertaining to factors not involved in the interaction, dividing by the number of observations comprising each of these totals, and then subtracting the correction term *and* all sums of squares corresponding to main effects and fewer-factor interactions involving the factors contained in that interaction.

EXERCISES

13.1 To determine optimum conditions for a plating bath, the effects of sulfone concentration and bath temperature on the reflectivity of the plated metal are studied in a 2×5 factorial experiment. The results of three replicates are as follows:

Concentration (grams/liter)	Temperature (degrees F)	Reflectivity		
		Rep. 1	*Rep. 2*	*Rep. 3*
5	75	35	39	36
5	100	31	37	36
5	125	30	31	33
5	150	28	20	23
5	175	19	18	22
10	75	38	46	41
10	100	36	44	39
10	125	39	32	38
10	150	35	47	40
10	175	30	38	31

Analyze these results and determine the bath condition or conditions that produce the highest reflectivity. Also construct a 0.95 confidence interval for the reflectivity of the plating bath corresponding to these optimum conditions.

13.2 A spoilage retarding ingredient is added in brewing beer. To determine the extent to which the taste of the beer is affected by the amount of this ingredient added to each batch, and how such taste changes might depend on the age of the beer, a 3×4 factorial experiment in two replications was designed. The taste of the beer was rated on a scale of 0, 1, 2, or 3 (3 being the most desirable) by a panel of trained experts, who reported the following mean ratings:

Amount of ingredient (grams per batch)	Aging period (weeks)	Mean ratings	
		Rep. 1	Rep. 2
2	2	2.1	1.6
2	4	2.6	1.9
2	6	2.9	2.4
3	2	1.4	1.7
3	4	1.9	2.2
3	6	2.3	2.7
4	2	0.5	0.9
4	4	1.2	0.8
4	6	1.7	1.4
5	2	1.0	1.6
5	4	2.2	1.3
5	6	2.3	2.1

(a) Analyze the data first as a two-way classification with 12 treatments and two blocks (replicates).
(b) Calculate the sums of squares corresponding to the main effects and interaction, and present the results in an analysis of variance table.
(c) Interpret the results of the experiment.

13.3 Suppose that in the experiment described on page 351 it is desired to determine also whether there is an interaction between the detergents and the washing machines, that is, whether one detergent might perform better in Machine 1, another might perform better in Machine 2, and so forth. Combining the data on page 351 with the replicate of the experiment given below, test for a significant interaction and discuss the results.

	Machine 1	Machine 2	Machine 3
Detergent A	39	42	58
Detergent B	44	46	48
Detergent C	34	47	45
Detergent D	47	45	57

13.4 Referring to Exercise 12.20 on page 357, it can be argued that the rocket-launcher experiment was poorly designed, as it was not replicated, and, thus, it is impossible to test whether there is an interaction between launchers and rocket fuels. Suppose that a second replicate were performed, with the following results:

	Fuel I	*Fuel II*	*Fuel III*	*Fuel IV*
Launcher X	66.8	51.1	40.1	49.2
Launcher Y	47.7	38.2	50.6	64.3

Combining the results of both replicates, perform an appropriate analysis of variance, and test for the presence of an interaction.

13.5 To test the design of a tennis racket, a subject plays tennis with a given racket for 1 hour and then rates the racket on a scale from 1 to 10. Two different sizes of racket face (standard and large) are tested, using three string tensions. Eighteen players, provided with rackets at random, yield the following results:

Racket face	*String tension*	*Rep. 1*	*Rep. 2*	*Rep. 3*
standard	low	5	6	5
standard	medium	6	6	7
standard	high	5	4	6
large	low	6	7	5
large	medium	8	9	7
large	high	3	5	2

Perform an appropriate analysis of variance and determine the conditions, if any, under which the large racket face is preferred.

13.6 The following tables give the weights (in grams) of food ingested by two different strains of rats after having been deprived of food for the stated number of hours and then having been given the stated dosage of a certain drug:

		Replicate 1		*Replicate 2*	
		Strain A	*Strain B*	*Strain A*	*Strain B*
Dosage 0.2 mg/kg	*1 hour*	8.77	7.59	9.07	6.02
	4 hours	11.82	9.21	9.16	7.05
	8 hours	14.65	15.35	16.08	12.01

		Replicate 1		Replicate 2	
		Strain A	Strain B	Strain A	Strain B
	1 hour	8.76	6.13	5.63	5.87
Dosage 0.4 mg/kg	4 hours	11.53	8.30	11.57	9.56
	8 hours	14.46	9.26	10.30	10.13
	1 hour	3.01	3.81	4.42	4.35
Dosage 0.8 mg/kg	4 hours	9.21	10.10	5.22	8.01
	8 hours	6.10	11.16	7.27	8.17

Perform an appropriate analysis of variance and interpret the results.

13.7 A study was conducted to measure the effect of three different meat tenderizers on the weight loss of steaks having the same initial (precooked) weights. The effects of cooking temperatures and cooking times also were measured by performing a $3 \times 2 \times 2$ factorial experiment in 3 replicates. The results are as follows:

Tenderizer	Cooking time (minutes)	Cooking temperature (°F)	Weight loss (ounces) Rep. 1	Rep. 2	Rep. 3
A	20	350	1.5	1.3	1.4
A	20	400	1.6	1.4	1.5
A	30	350	1.7	1.8	1.7
A	30	400	1.8	1.9	2.0
B	20	350	1.9	2.1	2.0
B	20	400	2.2	2.4	2.5
B	30	350	2.6	2.3	2.4
B	30	400	2.6	2.7	2.5
C	20	350	0.9	0.8	0.8
C	20	400	1.1	1.0	0.9
C	30	350	0.8	0.9	1.0
C	30	400	1.2	1.0	1.1

(a) Analyze this experiment first as a two-way classification with 12 treatments and three replicates (blocks).

(b) Complete the analysis by computing the sums of squares corresponding to the various main effects and interactions.

(c) Present the results in an analysis of variance table and interpret the experiment.

13.8 To study the effects of ingot location (A), slab position (B), specimen preparation (C), and twisting temperature (D) on the number of turns required to break a steel specimen by twisting, the following observations were recorded:

A	B	C	D	No. of turns Rep. 1	Rep. 2
top	1	turn	2,100°F	24	22
top	1	turn	2,200	25	28
top	1	turn	2,300	41	39
top	1	grind	2,100	18	18
top	1	grind	2,200	33	27
top	1	grind	2,300	35	41
top	2	turn	2,100	22	19
top	2	turn	2,200	26	31
top	2	turn	2,300	37	43
top	2	grind	2,100	23	7
top	2	grind	2,200	30	26
top	2	grind	2,300	34	30
mid	1	turn	2,100	26	19
mid	1	turn	2,200	30	31
mid	1	turn	2,300	39	42
mid	1	grind	2,100	19	19
mid	1	grind	2,200	31	31
mid	1	grind	2,300	26	35
mid	2	turn	2,100	30	26
mid	2	turn	2,200	31	34
mid	2	turn	2,300	39	42
mid	2	grind	2,100	22	20
mid	2	grind	2,200	32	26
mid	2	grind	2,300	38	22
bot	1	turn	2,100	18	21
bot	1	turn	2,200	35	32
bot	1	turn	2,300	34	37
bot	1	grind	2,100	21	19
bot	1	grind	2,200	20	29
bot	1	grind	2,300	44	31
bot	2	turn	2,100	23	22
bot	2	turn	2,200	31	26
bot	2	turn	2,300	38	41
bot	2	grind	2,100	18	19
bot	2	grind	2,200	31	24
bot	2	grind	2,300	35	41

Analyze this experiment.

13.9 A market test was performed to evaluate shelf position, label color, and package size for a canned food product, with the following results:

Shelf position	Label color	Package size (oz)	Sales (dollars) Day 1	Day 2	Day 3
low	red	10	70.10	68.00	69.50
low	red	12	72.25	71.90	74.40
low	red	16	78.05	74.85	82.60
low	red	24	61.50	62.10	59.15
low	green	10	65.75	62.35	68.60
low	green	12	69.45	71.05	75.45
low	green	16	75.15	70.70	71.25
low	green	24	64.80	60.85	59.90
medium	red	10	94.10	90.20	88.05
medium	red	12	104.85	99.55	96.80
medium	red	16	109.10	105.80	112.60
medium	red	24	59.90	62.50	54.75
medium	green	10	88.95	91.10	90.15
medium	green	12	100.60	94.05	101.35
medium	green	16	98.70	99.90	96.75
medium	green	24	62.50	53.85	59.40
high	red	10	92.60	88.80	85.50
high	red	12	100.55	102.15	99.10
high	red	16	111.95	108.25	109.45
high	red	24	61.40	65.20	59.70
high	green	10	97.35	98.70	92.60
high	green	12	120.65	115.45	108.65
high	green	16	118.10	116.35	121.90
high	green	24	70.30	65.05	71.40

Analyze this experiment.

13.10 Solve the four equations on page 381 for μ, α_1, β_1, and $(\alpha\beta)_{11}$ in terms of the population means μ_{ij1} corresponding to the four experimental conditions in the first replicate. Note that these equations serve as a guide for estimating the parameters in terms of the *sample means* corresponding to the various experimental conditions.

13.3 2^n FACTORIAL EXPERIMENTS

There are several reasons why factorial experiments are often performed with each factor taken at only two levels. Primarily, the number of experimental conditions in a factorial experiment increases multiplicatively with the number of levels of each factor; thus, if many factors are to be investigated simultaneously, it may be economically impossible to include more than two levels of each factor. Another important reason for treating **2^n factorial experiments** separately is that there exist computational shortcuts which apply only to this case. In fact, the remainder of this section will be devoted to these computational shortcuts, while other advantages, such as the ease of confounding higher-order interactions and the adaptability of 2^n factorials to experiments involving fractional replication, will be discussed in succeeding sections.

Before we introduce some of the special notation used in connection with 2^n factorial experiments, let us point out that such experiments do have some drawbacks. Since each factor is measured only at two levels, it is impossible to judge whether the effects produced by variations in a factor are linear or, perhaps, parabolic or exponential. For this reason 2^n factorial experiments are often used in "screening experiments," which are followed up by experiments involving fewer factors (ordinarily those found to be "significant" individually or jointly in the screening experiment) taken at more than two levels.

In the analysis of a 2^n factorial experiment it is convenient to denote the two levels of each factor by 0 and 1 (instead of 1 and 2). Thus, the models used for the analysis of this kind of experiment differ from those of Section 13.2 only inasmuch as we now have $i = 0$, 1 instead of $i = 1, 2, \ldots, a$, $j = 0$, 1 instead of $j = 1, 2, \ldots, b$, and so forth. For instance, for a 2^3 factorial experiment the model on page 389 becomes

Model equation for 2^3 factorial experiment

$$y_{ijkl} = \mu + \alpha_i + \beta_j + \gamma_k + (\alpha\beta)_{ij} + (\alpha\gamma)_{ik} + (\beta\gamma)_{jk}$$
$$+ (\alpha\beta\gamma)_{ijk} + \rho_l + \epsilon_{ijkl}$$

for $i = 0, 1, j = 0, 1, k = 0, 1$, and $l = 1, 2, \ldots, r$. The ϵ_{ijkl} are defined as before, and the parameters are now subject to the restrictions $\alpha_1 = -\alpha_0$, $\beta_1 = -\beta_0$, $\gamma_1 = -\gamma_0$, $(\alpha\beta)_{10} = (\alpha\beta)_{01} = -(\alpha\beta)_{11} = -(\alpha\beta)_{00}, \ldots$, and $\sum_{l=1}^{r} \rho_l = 0$. Note that besides the parameters for replicates we need only *one parameter of each kind:* that is, besides the parameters for replicates, we can express the entire model in terms of the parameters μ, α_0, β_0, γ_0, $(\alpha\beta)_{00}$, $(\alpha\gamma)_{00}$, $(\beta\gamma)_{00}$, and $(\alpha\beta\gamma)_{000}$.

A 2^n factorial experiment requires 2^n experimental conditions; since their number can be fairly large, it will be convenient to represent the experimental conditions by means of a special notation and list them in a so-called standard order. The notation consists of representing each experimental condition by the product of lowercase letters corresponding to the factors which are taken at level 1, called the "higher level." If a lowercase letter corresponding to a factor is missing, this means that the factor is taken at level 0, called the "lower level." Thus, in a three-factor experiment, ac represents the experimental condition where factors A and C are taken at the higher level and factor B is taken at the lower level, c represents the experimental condition where factor C is taken at the higher level and factors A and B are taken at the lower level, and so forth. The symbol "1" is used to denote the experimental condition in which all factors are taken at the lower level.

Although the experimental conditions are applied in a random order during the experiment itself, for the purpose of analyzing the results it is convenient to arrange them in a so-called **standard order**. For $n = 2$, this order is 1, a, b, ab, and for $n = 3$ it is the order shown in the following table:

Experimental condition	*Level of factor*		
	A	*B*	*C*
1	0	0	0
a	1	0	0
b	0	1	0
ab	1	1	0
c	0	0	1
ac	1	0	1
bc	0	1	1
abc	1	1	1

Note that the symbols for the first four experimental conditions are like those for a two-factor experiment, and that the second four are obtained by multiplying each of the first four symbols by c. Similarly, the arrangement for $n = 4$ on page 403 is obtained by first listing the eight symbols for a three-factor experiment and then repeating the set with each symbol multiplied by d.

Throughout this and the preceding chapter we referred to the total of all observations corresponding to a given experimental condition as a treatment total, and we represented these totals by means of symbols such as T_i., T_{ij}.., and so forth. Having introduced a special notation for the experimental conditions in a 2^n factorial experiment, we extend this notation by letting (1), (a), (b), (ab), (c), . . . , be the treatment totals corresponding to experimental conditions 1, a, b, ab, c, ... Thus, in a three-factor experiment

$$(1) = \sum_{l=1}^{r} y_{000l} \qquad (a) = \sum_{l=1}^{r} y_{100l}$$

. . . .

$$(bc) = \sum_{l=1}^{r} y_{011l} \qquad (abc) = \sum_{l=1}^{r} y_{111l}$$

The computational short-cuts referred to on page 399 consist, essentially, of expressing estimates of the various main effects and interactions as well as the corresponding sums of squares, in terms of *linear combinations* of the treatment totals. To illustrate what we mean, let us consider the quantity

$$-(1) + (a) - (b) + (ab) - (c) + (ac) - (bc) + (abc)$$

which is a linear combination, with coefficients $+1$ and -1, of the treatment totals corresponding to the eight experimental conditions. Referring to the model equation on page 399 and making use of the relationships among the parameters (but leaving all details to the reader in Exercises 13.16, 13.17, and 13.18 on pages 409) it can be shown that

$$-(1) + (a) - (b) + (ab) - (c) + (ac) - (bc) + (abc) = -8r\alpha_0 + \epsilon_A$$

where ϵ_A is a corresponding linear combination of sums of the ϵ_{ijkl}. From Theorem 7.1 on page 218 it follows that ϵ_A is a value of a random variable whose distribution has zero mean; as a matter of fact, it can be shown that ϵ_A is a value of a random variable having a normal distribution with zero mean and the variance $8r\sigma^2$. Referring to the above linear combination as the **effect total** $[A]$ for factor A, we find that $-[A]/8r$ provides an estimate of α_0, the main effect for factor A, and it can be shown that $[A]^2/8r$ actually equals SSA, the sum of squares for the main effect of factor A.

Similarly analyzing the linear combination

$$(1) + (a) - (b) - (ab) - (c) - (ac) + (bc) + (abc)$$

the reader will be asked to show in Exercise 13.18 on page 409 that it equals $8r(\beta\gamma)_{00} + \epsilon_{BC}$, where ϵ_{BC} is a corresponding linear combination of sums of the ϵ_{ijkl}. Referring to this linear combination of treatment totals as the effect total $[BC]$ for the two-way interaction of factors B and C, we find that $[BC]/8r$ provides an estimate of $(\beta\gamma)_{00}$, the effect of the BC interaction, and it can also be shown that $[BC]^2/8r$ equals $SS(BC)$, the sum of squares for the BC interaction. Proceeding in this fashion, we can present linear combinations of the treatment totals which yield estimates of the various other main effects and interactions, and whose squares, divided by $8r$, yield the corresponding sums of squares. These linear combinations, or effect totals, can easily be obtained with the use of the following **table of signs**:

(I)	(a)	(b)	(ab)	(c)	(ac)	(bc)	(abc)	Effect totals
1	1	1	1	1	1	1	1	$[I]$
−1	1	−1	1	−1	1	−1	1	$[A]$
−1	−1	1	1	−1	−1	1	1	$[B]$
1	−1	−1	1	1	−1	−1	1	$[AB]$
−1	−1	−1	−1	1	1	1	1	$[C]$
1	−1	1	−1	−1	1	−1	1	$[AC]$
1	1	−1	−1	−1	−1	1	1	$[BC]$
−1	1	1	−1	1	−1	−1	1	$[ABC]$

The entries of this table are the coefficients of the linear combinations of the treatment totals for the various main effects and interactions. As an aid for constructing similar tables for $n = 4$, $n = 5$, etc., note that for each main effect there is a "$+1$" when the factor is at the higher level and a "$−1$" when the factor is at the lower level. The signs for an interaction effect are obtained by multiplying the corresponding coefficients of *all* factors contained in the interaction. Thus, for $[AB]$ we multiply each sign for $[A]$ by the corresponding sign for $[B]$, getting

$$(-1)(-1) \quad (1)(-1) \quad (-1)(1) \quad (1)(1) \quad (-1)(-1) \quad (1)(-1) \quad (-1)(1) \quad (1)(1)$$

or

$$1 \qquad -1 \qquad -1 \qquad 1 \qquad 1 \qquad -1 \qquad -1 \qquad 1$$

Note also that in the above table $[I]$ stands for the grand total of all the observations, so that $[I]^2/8r$ gives the correction term for calculating SST, SSE, SSR, and $SS(Tr)$.

Although we have illustrated the above shortcut method for obtaining the various main-effect and interaction sums of squares with reference to a 2^3 factorial experiment, the only difference in a 2^n factorial experiment with $n > 3$ is that we require a more extensive table of signs and that the respective sums of squares are obtained by dividing the squares of the effect totals by $r \cdot 2^n$.

To illustrate this technique and introduce a further simplification, let us consider the following 2^4 factorial experiment, designed to determine the effects of certain variables on the reliability of a rotary stepping switch. The factors studied were as follows:

Factor	Low level	High level
A. Lubrication	dry	lubricated
B. Dust protection	unprotected	enclosed in dust cover
C. Spark suppression	no	yes
D. Current	0	0.5 amp

Each switch was operated continuously until a malfunction occurred, and the number of hours of operation was recorded. The whole experiment was performed twice, with the following results:

Experimental condition	Hours of operation Rep. 1	Rep. 2	Total
1	828	797	1,625
a	997	948	1,945
b	735	776	1,511
ab	807	1,003	1,810
c	994	949	1,943
ac	1,069	1,094	2,163
bc	989	1,215	2,204
abc	889	1,010	1,899
d	593	813	1,406
ad	773	1,026	1,799
bd	740	922	1,662
abd	936	1,138	2,074
cd	748	970	1,718
acd	1,202	1,182	2,384
bcd	1,103	966	2,069
abcd	985	1,154	2,139
Total	14,388	15,963	30,351

Analyzing these data first as a two-way classification with 16 treatments and two replications (blocks), we obtain

$$C = \frac{(30,351)^2}{32} = 28,786,975$$

$$SST = (828)^2 + (997)^2 + \ldots + (1,154)^2 - 28,786,975$$
$$= 744,876$$

$$SS(Tr) = \tfrac{1}{2}[(1,625)^2 + (1,945)^2 + \ldots + (2,139)^2] - 28,786,975$$
$$= 547,288$$

$$SSR = \tfrac{1}{16}[(14,388)^2 + (15,963)^2] - 28,786,975$$
$$= 77,520$$

$$SSE = 744,876 - 547,288 - 77,520 = 120,068$$

In order to subdivide the treatment sum of squares into SSA, SSB, . . . , and $SS(ABCD)$, we could construct a table of signs like the one on page 402, calculate the effect totals, and then divide the squares of the effect totals by $r \cdot 2^n = 2 \cdot 2^4 = 32$. For the A factor main effect we would thus obtain

$$[A] = -1,625 + 1,945 - 1,511 + 1,810 - 1,943 + 2,163 - 2,204$$
$$+ 1,899 - 1,406 + 1,799 - 1,662 + 2,074 - 1,718 + 2,384$$
$$- 2,069 + 2,139$$
$$= 2,075$$

and

$$SSA = \frac{(2,075)^2}{32} = 134,551$$

These calculations are quite tedious, but they can be simplified considerably by using a further shortcut, called the **Yates' method**. This method of calculating the effect totals is illustrated in the table shown below. The experimental conditions and the corresponding totals are listed in standard order. In the column marked (1), the upper half is obtained by adding successive pairs of treatment totals, and the lower half is obtained by subtracting successive pairs. Thus, in column (1) we obtained

$$1,625 + 1,945 = 3,570$$
$$1,511 + 1,810 = 3,321$$
$$\cdots\cdots\cdots\cdots\cdots$$
$$2,069 + 2,139 = 4,208$$
$$1,945 - 1,625 = 320$$
$$1,810 - 1,511 = 299$$
$$\cdots\cdots\cdots\cdots\cdots$$
$$2,139 - 2,069 = 70$$

Experimental condition	Treatment total	(1)	(2)	(3)	(4)	Identification	Sum of squares
1	1,625	3,570	6,891	15,100	30,351	[I]	28,786,975
a	1,945	3,321	8,209	15,251	2,075	[A]	134,551
b	1,511	4,106	6,941	534	385	[B]	4,632
ab	1,810	4,103	8,310	1,541	−1,123	[AB]	39,410
c	1,943	3,205	619	−252	2,687	[C]	225,624
ac	2,163	3,736	−85	637	−773	[AC]	18,673
bc	2,204	4,102	805	−546	−179	[BC]	1,001
abc	1,899	4,208	736	−577	−1,119	[ABC]	39,130
d	1,406	320	−249	1,318	151	[D]	713
ad	1,799	299	−3	1,369	1,007	[AD]	31,689
bd	1,662	220	531	−704	889	[BD]	24,698
abd	2,074	−305	106	−69	−31	[ABD]	30
cd	1,718	393	−21	246	51	[CD]	81
acd	2,384	412	−525	−425	635	[ACD]	12,601
bcd	2,069	666	19	−504	−671	[BCD]	14,070
abcd	2,139	70	−596	−615	−111	[ABCD]	385

Note that, in the lower half of column (1), the first total in each pair is subtracted from the second. Column (2) is then obtained by performing the identical operations on the entries of column (1), and columns (3) and (4) are obtained in the same manner from the entries in columns (2) and (3), respectively. Column (4), and in general column (n), gives the effect totals in standard order, as shown. Each sum of squares is then obtained as before, by squaring the corresponding effect total and then dividing the result by $r \cdot 2^n = 2 \cdot 2^4 = 32$.

Dividing the sums of squares by their degrees of freedom to obtain the mean squares, and dividing the various mean squares by the error mean square, we get the following analysis-of-variance table for the 2^4 factorial experiment:

Source of variation	Degrees of freedom	Sum of squares	Mean square	F
Replicates	1	77,520	77,520	9.68
Main effects:				
A	1	134,551	134,551	16.81
B	1	4,632	4,632	< 1
C	1	225,624	225,624	28.19
D	1	713	713	< 1
Two-factor interactions:				
AB	1	39,410	39,410	4.92
AC	1	18,673	18,673	2.33
AD	1	31,689	31,689	3.96
BC	1	1,001	1,001	< 1
BD	1	24,698	24,698	3.09
CD	1	81	81	< 1
Three-factor interactions:				
ABC	1	39,130	39,130	4.89
ABD	1	30	30	< 1
ACD	1	12,601	12,601	1.57
BCD	1	14,070	14,070	1.76
Four-factor interactions:				
ABCD	1	385	385	< 1
Error	15	120,068	8,005	
Total	31	744,876		

Since $F_{0.05} = 4.54$ and $F_{0.01} = 8.68$ for 1 and 15 degrees of freedom, we find that the replication effects as well as the effects of lubrication and spark suppression are significant at the 0.01 level, and that there are significant interactions at the 0.05 level between lubrication, dust protection, and spark suppression. The reader will be asked to interpret these results and estimate the magnitude of some of the effects in Exercise 13.19 on page 409.

EXERCISES

13.11 To determine the effect on taste of three different factors in manufacturing soft-drink cans, an experiment was performed where the taste of one soft drink was rated by a judge on a scale from 1 to 10. The results are as follows:

A *Lubricant*	*B* *Heat*	*C* *Resin*	*Ratings* *Rep. 1*	*Rep. 2*
fresh	unheated	*A*	6	8
fresh	unheated	*B*	8	7
fresh	heated	*A*	9	9
fresh	heated	*B*	1	2
aged	unheated	*A*	6	7
aged	unheated	*B*	6	8
aged	heated	*A*	9	8
aged	heated	*B*	2	3

(a) Analyze the results first as a two-way classification with 7 degrees of freedom for treatments and 1 degree of freedom for blocks (replicates).
(b) Use an appropriate table of signs to calculate the effect totals [*A*], [*B*], [*C*], [*AB*], [*AC*], [*BC*], [*ABC*].
(c) Using the results obtained in part (b) find the sums of squares corresponding to the main effects and interactions, and check their total against the treatment sum of squares obtained in part (a).
(d) Arrange the data with treatment combinations in standard order, and use the Yates method to find the effect totals. Compare with the results obtained in part (b).
(e) Construct an analysis of variance table and analyze the experiment.

13.12 An experiment was conducted to determine the effects of certain alloying elements on the ductility of a metal, and the following results were obtained:

Nickel	*Carbon*	*Manganese*	*Breaking strength (ft-lb)* *Rep. 1*	*Rep. 2*	*Rep. 3*
0.0%	0.3%	0.5%	36.7	39.6	38.2
0.0	0.3	1.0	47.5	43.5	45.9
0.0	0.6	0.5	40.6	36.8	36.0
0.0	0.6	1.0	41.1	45.8	46.4
4.0	0.3	0.5	37.8	32.7	31.6
4.0	0.3	1.0	34.2	37.2	36.5
4.0	0.6	0.5	39.5	41.7	39.1
4.0	0.6	1.0	46.4	43.7	49.4

Perform an appropriate analysis of variance and interpret the results.

13.13 A screening experiment was conducted to determine what factors are influential in controlling the final phosphorus content of steel produced in a converter. The levels of the factors studied and the experimental results are contained in the following table:

E Pouring temp. (°F)	D Lime ratio	C Oxygen	B Original phosphorus	A Original manganese	Final phosphorus Rep. 1	Final phosphorus Rep. 2
2,400	3	5%	0.15%	1%	0.003%	0.001%
2,400	3	5	0.15	3	0.004	0.009
2,400	3	5	0.30	1	0.002	0.008
2,400	3	5	0.30	3	0.015	0.007
2,400	3	15	0.15	1	0.002	0.005
2,400	3	15	0.15	3	0.011	0.006
2,400	3	15	0.30	1	0.004	0.001
2,400	3	15	0.30	3	0.002	0.004
2,400	4	5	0.15	1	0.000	0.003
2,400	4	5	0.15	3	0.008	0.002
2,400	4	5	0.30	1	0.003	0.007
2,400	4	5	0.30	3	0.005	0.012
2,400	4	15	0.15	1	0.010	0.006
2,400	4	15	0.15	3	0.006	0.001
2,400	4	15	0.30	1	0.006	0.014
2,400	4	15	0.30	3	0.011	0.015
2,600	3	5	0.15	1	0.003	0.007
2,600	3	5	0.15	3	0.007	0.004
2,600	3	5	0.30	1	0.011	0.005
2,600	3	5	0.30	3	0.010	0.017
2,600	3	15	0.15	1	0.004	0.008
2,600	3	15	0.15	3	0.019	0.013
2,600	3	15	0.30	1	0.004	0.008
2,600	3	15	0.30	3	0.017	0.023
2,600	4	5	0.15	1	0.007	0.004
2,600	4	5	0.15	3	0.015	0.009
2,600	4	5	0.30	1	0.004	0.011
2,600	4	5	0.30	3	0.010	0.006
2,600	4	15	0.15	1	0.017	0.011
2,600	4	15	0.15	3	0.005	0.010
2,600	4	15	0.30	1	0.014	0.009
2,600	4	15	0.30	3	0.016	0.011

Analyze the results of this experiment.

13.14 An experiment was conducted to determine the effects of the following factors on the gain of a semiconductor device:

Factor	Level 0	Level 1
A. Location of assembly	Laboratory	Production line
B. Partial pressure of controlling material	10^{-15}	10^{-4}
C. Relative humidity	1%	30%
D. Aging time	72 hours	144 hours

The results were as follows:

Experimental condition	Gain Rep. 1	Gain Rep. 2
1	39.0	43.2
a	31.8	43.7
b	47.0	51.4
ab	40.9	40.3
c	43.8	40.5
ac	29.3	52.9
bc	34.8	48.2
abc	45.6	58.2
d	40.1	41.9
ad	42.0	40.5
bd	54.9	53.0
abd	39.9	40.2
cd	43.1	40.2
acd	30.1	39.9
bcd	35.6	53.7
abcd	41.4	49.5

Perform an appropriate analysis of variance and interpret the results.

13.15 A study was performed to examine the effect of five different factors on the time required for odorants in natural gas to reach the surface once a leak has occurred. The results of the experiment were reported as follows:

A Odorant	B Odorant concentration mg/ml	C Soil moisture	D Flow rate ml/min	E Tempera-ture	Time (hours) Rep. 1	Time (hours) Rep. 2
A	4	8%	20	50°F	20	20
A	4	8	20	70	17	18
A	4	8	80	50	20	19
A	4	8	80	70	18	19
A	4	16	20	50	12	13
A	4	16	20	70	14	16
A	4	16	80	50	15	14
A	4	16	80	70	17	16
A	16	8	20	50	10	9
A	16	8	20	70	9	8
A	16	8	80	50	8	9
A	16	8	80	70	7	8
A	16	16	20	50	7	6
A	16	16	20	70	6	4
A	16	16	80	50	5	6
A	16	16	80	70	4	5

Continued on page 409

A Odorant	B Odorant concentration mg/ml	C Soil moisture	D Flow rate ml/min	E Tempera- ture	Time (hours) Rep. 1	Rep. 2
B	4	8	20	50	28	29
B	4	8	20	70	27	26
B	4	8	80	50	25	27
B	4	8	80	70	28	28
B	4	16	20	50	24	25
B	4	16	20	70	26	25
B	4	16	80	50	23	24
B	4	16	80	70	24	25
B	16	8	20	50	26	24
B	16	8	20	70	23	25
B	16	8	80	50	27	26
B	16	8	80	70	25	26
B	16	16	20	50	19	18
B	16	16	20	70	16	17
B	16	16	80	50	20	19
B	16	16	80	70	17	15

Analyze the results of this experiment.

13.16 Writing the treatment total (a) as the sum of the corresponding observations y_{100l} and substituting for these observations the expressions given by the model equation on page 399, it can be shown that

$$(a) = r[\mu + \alpha_1 + \beta_0 + \gamma_0 + (\alpha\beta)_{10} + (\alpha\gamma)_{10} + (\beta\gamma)_{00} + (\alpha\beta\gamma)_{100}]$$
$$+ \sum_{l=1}^{r} \epsilon_{100l}$$

Making use of the restrictions imposed on the parameters, rewrite this expression for (a) in terms of the parameters $\mu, \alpha_0, \beta_0, \gamma_0, (\alpha\beta)_{00}, (\alpha\gamma)_{00}, (\beta\gamma)_{00},$ and $(\alpha\beta\gamma)_{000}$.

13.17 Duplicating the work of Exercise 13.16, express (1), (b), (ab), (c), (ac), (bc), and (abc) in terms of the parameters $\mu, \alpha_0, \beta_0, \gamma_0, (\alpha\beta)_{00}, (\alpha\gamma)_{00}, (\beta\gamma)_{00},$ and $(\alpha\beta\gamma)_{000}$.

13.18 Using the results of Exercises 13.16 and 13.17, verify the expressions for $[A]$ and $[BC]$ obtained on page 401. Also express ϵ_A in terms of the quantities ϵ_{ijkl}.

13.19 Interpret the results of the analysis of variance given by the table on page 405, and estimate the magnitude of the significant effects.

13.20 A computational check on the sums of squares obtained for the various main effects and interactions is that their *sum* must equal the treatment sum of squares obtained by analyzing the data first as a two-way classification. Perform this check on the sums of squares given in the table on page 405.

13.21 If it is desired to find an expression for an effect total without constructing a complete table of signs, one can use the following method, illustrated by finding $[ABC]$ for a 2^4 factorial experiment. We take the expression $(a \pm 1)(b \pm 1)(c \pm 1)(d \pm 1)$ with a "+" if the corresponding letter does *not* appear in the symbol for the main effect or interaction for which we want to calculate an effect total, and a "−" if the corresponding letter does appear. Thus, for finding $[ABC]$ we write

$$(a - 1)(b - 1)(c - 1)(d + 1) = abcd + abc - abd - acd - bcd$$
$$- ab - ac + ad - bc + bd + cd$$
$$+ a + b + c - d - 1$$

and after arranging the terms in standard order and adding parentheses we finally obtain

$$[ABC] = -(1) + (a) + (b) - (ab) + (c) - (ac) - (bc) + (abc) - (d)$$
$$+ (ad) + (bd) - (abd) + (cd) - (acd) - (bcd) + (abcd)$$

(a) Use this method to express $[B]$, $[AC]$, and $[ABC]$ in terms of the treatment totals in a 2^3 factorial experiment.

(b) Use this method to express $[AC]$ and $[BCD]$ in terms of the treatment totals in a 2^4 factorial experiment.

13.4 CONFOUNDING IN A 2^n FACTORIAL EXPERIMENT

In some experiments it is impossible to run all the required experimental conditions in one block. For example, if a 2^3 factorial experiment involves eight combinations of paint pigments that are to be applied to a surface and baked in an oven which can accommodate only four specimens, it becomes necessary to divide the eight treatments into two blocks (oven runs) in each replicate. As we have pointed out earlier, if the block size is too small to accommodate all treatments, this requires special, so-called **incomplete-block designs**.

When the exeprimental conditions are distributed over several blocks, one or more of the effects may become confounded (inseparable) with possible block effects, that is, between-block differences. For example, if in the 2^3 factorial experiment referred to in the preceding paragraph, experimental conditions a, ab, ac, and abc are included in one oven run (Block 1) and experimental conditions 1, b, c, and bc are included in a second oven run (Block 2), then the "block effect," the difference between the two block totals, is given by

$$[(a) + (ab) + (ac) + (abc)] - [(1) + (b) + (c) + (bc)]$$

Referring to the table of signs on page 402, we observe that this quantity is, in fact, the effect total $[A]$, so that the estimate of the main effect of factor A is **confounded** with blocks. Note that all other factorial effects remain unconfounded; for each other effect total there are two $+1$ coefficients and two -1 coefficients in each block, so that the block effects cancel out. This kind of argument can also be used to decide what experimental conditions to put into each block to confound a given main effect or interaction. For instance, had we wanted to confound the ABC interaction with blocks in the above example, we could have put experimental conditions a, b, c, and abc, whose totals have $+1$ coefficients in $[ABC]$, into one block, and experimental conditions 1, ab, ac, and bc, whose totals have -1 coefficients, into another.

In general, confounding in a 2^n factorial experiment can be much more complicated than in the example just given. To avoid serious difficulties, we shall require that the number of blocks used is a power of 2, say 2^p. It turns out that the price paid for running a 2^n factorial experiment in 2^p blocks is that a total of $2^p - 1$ effects are confounded with blocks. To make it clear just which effects are confounded, and to indicate a method that can be used to confound only certain effects and no others, it is helpful to define the term "generalized interaction" as follows: the **generalized interaction** of two effects is the "product" of these effects, with like letters cancelled. Thus, the generalized interaction of AB and CD is $ABCD$, and the generalized interaction of ABC and BCD is $AB\!\!\!/C\!\!\!/B\!\!\!/C\!\!\!/D$, or AD. To confound a 2^n factorial experiment in 2^p blocks, the following method can be used: one selects any p effects for confounding, making sure that none is the generalized interaction of any of the others selected. Then, it can be shown that a further $2^p - (p + 1)$ effects are automatically confounded with blocks; together with the p effects originally chosen, this gives a total of $2^p - 1$ confounded effects in the experiment. The additional confounded effects are, in fact, the generalized interactions of the p effects originally chosen.

To illustrate the construction of a confounded design, let us divide a 2^4 factorial experiment into four blocks so that desired effects are confounded with blocks. In actual practice one ordinarily confounds only the higher-order interactions (in the hope that they are nonexistent anyhow). Since we have decided upon four blocks, we have $2^p = 4$ and $p = 2$, and we shall arbitrarily select two higher-order interactions for confounding. If we were to select $ABCD$ and BCD, then their generalized interaction, A, would also be confounded. Thus, to avoid confounding any main effects, and to confound as few two-factor interactions as possible, we shall select ABD and ACD, noting that the BC interaction is also confounded. (Observe that it is impossible to avoid confounding at least one main effect or two-factor interaction in this experiment.)

In order to assign the 16 experimental conditions to four blocks, we first distribute them into two blocks, so that the *ABD* interaction is confounded with blocks. Referring to an appropriate table of signs, we put all treatments whose totals have a "$+1$" in the row for [*ABD*] into one block, all those whose totals have a "-1" into a second block, and we get the following blocks:

First block:	a	b	ac	bc	d	abd	cd	abcd
Second block:	1	ab	c	abc	ad	bd	acd	bcd

Note that each experimental condition in the first block has an *odd number* of letters in common with *ABD*, while each experimental condition in the second block has an *even number* of letters in common with *ABD*. This **odd-even rule** provides an alternative way of distributing experimental conditions among two blocks to confound a given effect, and it has the advantage that it does not require the construction of a complete table of signs.

So far, we have confounded the *ABD* interaction by dividing the 16 experimental conditions into two blocks; now we shall confound the *ACD* interaction by dividing each of these blocks into two blocks of four conditions each. Using the odd-even rule just described (or a table of signs), we obtain the following four blocks:

Block 1:	a	bc	d	abcd
Block 2:	b	ac	abd	cd
Block 3:	ab	c	bd	acd
Block 4:	1	abc	ad	bcd

By comparing these blocks with a table of signs, or, equivalently, by applying the odd-even rule, the reader will be asked to verify in Exercise 13.28 on page 420 that the *BC* interaction is also confounded with blocks, while all other effects are left unconfounded.

The analysis of a confounded 2^n factorial experiment is similar to that of an unconfounded experiment, with the exception that the sums of squares for the confounded effects are not computed, and we compute a block sum of squares as if the experiment consisted of *br* blocks rather than *b* blocks in each of *r* replicates.

Referring to our example of a 2^4 factorial experiment with the *ABD*, *ACD*, and *BC* interactions confounded, and using two replicates, we have the following *dummy* analysis-of-variance table:

Source of variation	Degrees of freedom
Blocks	7
Main effects	4
Unconfounded two-factor interactions	5
Unconfounded three-factor interactions	2
Four-factor interaction	1
Intrablock error	12
Total	31

The sum of squares for blocks is obtained, as usual, by adding the squares of the eight block totals, dividing the result by 4 (the number of observations in each block), and subtracting the correction term. The total sum of squares and the sums of squares for the unconfounded factorial effects are obtained in the usual way, and the sum of squares for the **intrablock error**, a measure of the variability *within blocks*, is obtained by subtraction.

To illustrate the analysis of a confounded 2^n factorial experiment, let us suppose that each replicate of the stepping-switch experiment described in the preceding section was actually run in four blocks, because only four mountings were available for the 16 switches. (The order of running the blocks is assumed to have been randomized within each replicate, and the assignment of switches is assumed to have been randomized within each block.) Assuming also that the ABD, ACD, and BC interactions were confounded with the blocks as shown on page 412, we obtain the following block totals from the data on page 403:

	Block 1	Block 2	Block 3	Block 4
Replicate 1	3,564	3,488	3,743	3,593
Replicate 2	4,130	3,978	4,056	3,799

Thus, the sum of squares for blocks is given by

$$SS(Bl) = \frac{(3,564)^2 + (3,488)^2 + \ldots + (3,799)^2}{4} - 28,786,975$$
$$= 101,240$$

where the correction factor is the same as in the analysis on page 403.

Copying the total sum of squares and the sums of squares for the various unconfounded effects from the table on page 405, we obtain the analysis-of-variance table for the confounded factorial experiment shown below. In this analysis, the A and C main effects are again significant at the 0.01 level, but none of the other main effects or interactions is significant.

Source of variation	Degrees of freedom	Sum of squares	Mean square	F
Blocks	7	101,240	14,463	1.58
Main effects:				
A	1	134,551	134,551	14.68
B	1	4,632	4,632	< 1
C	1	225,624	225,624	24.62
D	1	713	713	< 1
Unconfounded two-factor interactions:				
AB	1	39,410	39,410	4.30
AC	1	18,673	18,673	2.04
AD	1	31,689	31,689	3.46
BD	1	24,698	24,698	2.69
CD	1	81	81	< 1
Unconfounded three-factor interactions:				
ABC	1	39,130	39,130	4.27
BCD	1	14,070	14,070	1.54
Four-factor interaction:				
ABCD	1	385	385	< 1
Intrablock error	12	109,980	9,165	
Total	31	744,876		

If there is replication, some of the lost information about the confounded effects can be recovered by a further breakdown of the above blocks sum of squares. This analysis consists of dividing the sum of squares for blocks into a component for each of the confounded effects, a component for replications, and a residual component called the "interblock error," which is a measure of the variability *between blocks*. Copying the sum of squares for replicates as well as those for the *BC*, *ABD*, and *ACD* interactions from the analysis-of-variance table on page 405, and copying the blocks sum of squares from the **intrablock analysis-of-variance** table above we obtain the following **interblock analysis-of-variance** table:

Source of variation	Degrees of freedom	Sum of squares	Mean square	F
Replicates	1	77,520	77,520	23.05
Confounded effects:				
BC	1	1,001	1,001	< 1
ABD	1	30	30	< 1
ACD	1	12,601	12,601	3.75
Interblock error	3	10,088	3,363	
Total (blocks)	7	101,240		

Note that the **interblock error** is obtained by subtraction and that the *F* ratios are obtained by dividing the mean squares for the confounded effects and the mean square for replicates by the mean square for the interblock error. Only the *F* test for replicates is significant (at the 0.05 level). The small number of degrees of freedom for the interblock error implies that the sensitivity of these significance tests is *very poor;* in fact, it rarely pays to make this kind of interblock analysis unless the number of replications is relatively large.

13.5 FRACTIONAL REPLICATION

In studies involving complex production lines, chemical processes such as may be encountered in the petroleum, plastics, or metals industries, physical-chemical processes such as may be encountered in the electronics or space-technology industries, and in many other engineering studies, the experimenter is often faced with a large and bewildering array of interrelated variables. The principles of factorial experimentation treated so far in this chapter help him to "sort out" these variables, to discover which of them have the greatest influence on the process under consideration, and what important interrelationships may exist.

There are, however, some serious limitations to the simultaneous study of a large number of factors. Even if each factor is assigned only two levels, one replicate of a six-factor experiment requires 64 observations; there are 128 observations in one replicate of a seven-factor experiment, and 1,024 observations in one replicate of a ten-factor experiment. The economic and practical limitations of these large numbers make it necessary to seek out ways in which the size of factorial experiments can be kept within manageable bounds. Of course, it must be emphasized that there is no substitute for careful preliminary planning which, coupled with engineering insight, can result in the elimination of many needless factors.

In spite of the most careful preliminary planning, however, it is often difficult to avoid having to include as many as six or ten (or more) factors in a single experiment. One way to reduce the size of such an experiment would be to break it up into several parts, each part involving the deliberate variation of one factor while all others are held fixed. This would have the undesirable consequence that we could not study any of the interactions. Even if we were to include half the factors in one part and the other half in another, such as replacing a ten-factor experiment (requiring 1,024 observations) by two five-factor experiments (each requiring 32 observations), any interaction between factors in the first part and factors in the second part would be irretrievably lost. It is possible to overcome some of these difficulties by observing that most of the time we are not interested in *all* the interactions. For instance, it is possible to perform only a fraction of a 2^n factorial and yet obtain most of the desired information, say, about the main effects and two-factor interactions (but not the higher interactions).

The principles involved in **fractional replication**, that is, in performing only a fraction of a complete 2^n factorial experiment, are similar to those used in confounding. To obtain a **half-replicate** one selects only one of the two blocks into which the experimental conditions have been divided by confounding one effect; to obtain a **quarter-replicate** one selects only one of the four blocks into which the experimental conditions have been divided by confounding two effects, and so forth. In contrast to a confounded experiment as discussed in Section 13.4, we find that in a fractional replicate *the effects are confounded, not with blocks, but with each other.*

To illustrate, suppose that only the experimental conditions *a*, *b*, *c*, and *abc* are included in a half-replicate of a 2^3 factorial experiment. (This is one block of a 2^3 factorial experiment with the *ABC* interaction confounded.) Considering the table of signs on page 402 with all columns except those corresponding to *a*, *b*, *c*, and *abc* crossed out, we find that the effect total for factor *A* is now given by

$$[A] = (a) - (b) - (c) + (abc)$$

If we write $(a) = \sum_{l=1}^{r} y_{100l}$, $(b) = \sum_{l=1}^{r} y_{010l}$, ..., and substitute for the y_{ijkl}

the expressions given by the model equation on page 399, we obtain

$$[A] = -4r[\alpha_0 - (\beta\gamma)_{00}] + \epsilon$$

where ϵ is the value of a random variable having zero mean (see Exercise 13.37 on page 422). We thus find that $[A]$ measures the main effect of factor A as well as the BC interaction, so that these two effects have become inseparable, or confounded. Note also that in the reduced table of signs (having columns only corresponding to a, b, c, and abc) the signs for $[A]$ and $[BC]$ are identical, and hence $[A] = [BC]$. In Exercise 13.38 on page 422, the reader will be asked to show that for the given fractional replicate the main effect for factor B is, similarly, confounded, or **aliased**, with the AC interaction, while the main effect for factor C is confounded, or aliased, with the AB interaction. The ABC interaction cannot be estimated.

With careful design, it is generally possible to confound all main effects and two-factor interactions *only* with interactions of higher order. This is illustrated in the following example, where we shall construct a half-replicate of a 2^5 factorial. First, we select an effect (usually a higher-order interaction) to split the experiment into two blocks, as in confounding. The effect chosen is called the **defining contrast**, and it cannot be estimated at all by the fractional replicate. Every other effect is aliased with another effect, namely, its generalized interaction (see page 411) with the defining contrast. Thus, if the defining contrast is $ABCDE$, the main effect for factor A has the four-factor interaction $BCDE$ as its alias, BC and ADE are an alias pair, and so forth. As we have seen, only the combined effect (the sum or difference of the aliased effects) can be estimated in the experiment. However, if it can be assumed that there are no higher-order interactions, one can attribute the effect of the alias pair BC and ADE entirely to the two-factor interaction BC, one can attribute the effect of the alias pair A and $BCDE$ entirely to the main effect of factor A, and so forth. A complete listing of the aliases in a half-replicate of a 2^5 factorial experiment having the defining contrast $ABCDE$ is as follows:

A and $BCDE$, B and $ACDE$, C and $ABDE$, D and $ABCE$
E and $ABCD$, AB and CDE, AC and BDE, AD and BCE
AE and BCD, BC and ADE, BD and ACE, BE and ACD
CD and ABE, CE and ABD, DE and ABC

Note that no main effect or two-factor interaction is aliased with another main effect or two-factor interaction.

The 16 experimental conditions to be included in the half-replicate are given by those in either of the two blocks obtained by confounding the defining contrast. Choosing "evens" in the odd-even rule, namely, those conditions which have an even number of letters in common with the defining contrast *ABCDE*, we obtain the following half-replicate:

$$
\begin{array}{llll}
1 & ad & ae & de \\
ab & bd & be & abde \\
ac & cd & ce & acde \\
bc & abcd & abce & bcde
\end{array}
$$

To go one step further, let us illustrate how to construct a quarter-replicate of the given 2^5 factorial. We shall do this by dividing the above half-replicate in half, confounding the three-factor interaction *ABC*. Again using "evens," we get the following eight experimental conditions:

$$
\begin{array}{llll}
1 & de & ab & abde \\
ac & acde & bc & bcde
\end{array}
$$

Since *DE*, the generalized interaction of the two confounded effects, is also confounded, we now have the three defining contrasts *ABCDE*, *ABC*, and *DE*. None of these effects can be estimated in the quarter-replicate, and each other effect is aliased with its three generalized interactions with the three defining contrasts. The complete aliasing is as follows:

Alias sets

A,	*BCDE,*	*BC,*	*ADE*
B,	*ACDE,*	*AC,*	*BDE*
C,	*ABDE,*	*AB,*	*CDE*
D,	*ABCE,*	*ABCD,*	*E*
AD,	*BCE,*	*BCD,*	*AE*
BD,	*ACE,*	*ACD,*	*BE*
CD,	*ABE,*	*ABD,*	*CE*

We have given this quarter-replicate merely as an illustration; it would hardly seem useful in actual practice because of the hopeless aliasing of main effects and two-factor interactions. Nevertheless, quarter-replicates of six- and seven-factor experiments (and even eighth-replicates of seven- and eight-factor experiments) can often provide much useful information.

The analysis of a fractional factorial is practically the same as that of a fully replicated factorial experiment. Given the fraction $1/2^p$ of a 2^n factorial, there are 2^{n-p} experimental conditions, and the Yates' method can be used as if the experiment were a 2^{n-p} factorial. Some care must be exercised to arrange the experimental conditions in a modified standard order as indicated in Exercise 13.35 on page 421.

When dealing with fractional factorials, there is the problem of obtaining an estimate of the experimental error. For example, in the half-replicate of a

2^5 factorial described on page 418, the breakdown of the total sum of squares having 15 degrees of freedom yields sums of squares for main effects (5 degrees of freedom), sums of squares for two-factor interactions (10 degrees of freedom), but no component (0 degrees of freedom) for the experimental error. In a situation like this, and in all other cases where the number of degrees of freedom for error is small, it is best to include a limited amount of replication. This may be accomplished by randomly selecting several experimental conditions, and making additional observations corresponding to these conditions. Furthermore, if it can be assumed that there are no higher-order interactions, the total of the sums of squares corresponding to higher-order interactions *which are not aliased with main effects or lower-order interactions* can be attributed to "error," and used in the denominator of the *F* test (see Exercise 13.36 below). In this way, use can be made of the "hidden replication" inherent in most large factorial experiments.

To summarize, fractional replication is useful whenever the number of factors to be included in an experiment is large and it is not economically feasible to include all possible experimental conditions. The reduction in size (and, therefore, in cost) of a fractionally replicated experiment is partially offset by the loss of information caused by aliasing, and by the difficulties inherent in estimating the experimental error. A more detailed discussion of fractional replication, including fractional replicates of 3^n experiments and a variety of other designs, can be found in the book by W. G. Cochran and G. M. Cox listed in the Bibliography.

EXERCISES

13.22 A 2^5 factorial experiment, having the factors A, B, C, D, and E, is to be run in several blocks. Show which treatments are assigned to each block if
    (a) there are to be two blocks, with the $ACDE$ interaction confounded;
    (b) there are to be four blocks, with the BDE and $ABCE$ interactions confounded. What other factorial effect or effects are also confounded?

13.23 List the experimental conditions included in each block if a 2^4 factorial experiment is to be confounded
    (a) in two blocks on $ABCD$;
    (b) in four blocks on ABC and BCD.

13.24 What is the largest number of blocks in which one can perform a 2^6 factorial experiment without confounding any main effects?

13.25 List the experimental conditions included in each block if a 2^6 factorial experiment is to be confounded in 8 blocks on $ABDE$, $BCDF$, and ABC. What other factorial effects are confounded?

13.26 Suppose that in Exercise 13.11 on page 406 the judge rates the soft drinks in sets of four, with a rest period in between, and that the experiment was actually performed so that each replicate consisted of two blocks with *ABC* confounded. Perform an intrablock analysis of the data.

13.27 Four new drugs are to be investigated to determine their effectiveness as tranquilizers, individually and in combination with each other. Each patient was given regular doses of one of the sixteen transquilizers formed from these drugs (including a placebo corresponding to the 0 level for each drug) and, after a two-week period, the effect of these tranquilizers on the emotional stability of each patient was judged (on a scale from 1 to 5) by five psychiatrists. To keep the staff work load within reasonable bounds, two hospitals were used in this experiment, and eight patients were selected by the staff of each hospital for each trial (replicate); thus, the experiment involved two replicates of two blocks each, with the *ABCD* interaction confounded. The results of the two 2-weeks trials were as follows:

FIRST HOSPITAL			SECOND HOSPITAL		
Treatment combination	*Mean rating*		*Treatment combination*	*Mean rating*	
	Trial 1	*Trial 2*		*Trial 1*	*Trial 2*
1	2.0	2.6	*a*	2.8	2.6
ab	3.8	3.4	*b*	3.6	2.0
ac	4.2	4.8	*c*	2.4	1.8
bc	4.8	4.0	*abc*	4.0	3.8
ad	1.8	2.4	*d*	1.8	2.2
bd	3.4	3.8	*abd*	1.6	2.0
cd	4.6	2.8	*acd*	3.6	2.4
abcd	4.2	4.6	*bcd*	3.4	3.8

If a high rating indicates satisfactory progress, which drug combination (or combinations) seems to be the most promising? Perform an intrablock analysis of variance to test for significant effects.

13.28 Verify that in the example on page 412 all treatment totals having +1 as a coefficient in [*BC*] are in Blocks 1 and 4, that all treatment totals having −1 as a coefficient in [*BC*] are in Blocks 2 and 3, and, hence, that the *BC* interaction is confounded with blocks.

13.29 Suppose that in Exercise 13.13 on page 407 only eight specimens could be tested in any one shift, and that the experiment was actually performed so that each replicate consists of four blocks with *ABC*, *ADE*, and *BCDE* confounded.
   (a) If for each factor levels 0 and 1 are, respectively, the lower and higher values, construct a table showing which experimental conditions go into each of the four blocks.
   (b) Perform an intrablock analysis of the experiment.

13.30 Show that the "odd-even" rule for assigning treatments to blocks, given on page 412, is equivalent to the method described in Exercise 13.21 on page 409.

13.31 Referring to Exercise 13.22, suppose that in each case the block containing treatment combination 1 was chosen as a fractional replicate of a 2^5 factorial.

(a) Show the alias pairs in the resulting half-replicate.

(b) Show the alias sets in the resulting quarter-replicate.

(In practice, random selection should be used to choose the block to be included in a fractional replicate.)

13.32 Referring to Exercise 13.23, list the alias sets if an experiment consists of

(a) a half-replicate of a 2^4 factorial experiment with $ABCD$ confounded;

(b) a quarter-replicate of a 2^4 factorial experiment with ABC and BCD confounded.

13.33 Construct a half-replicate of a 2^6 factorial having the defining contrast $ABCDEF$. List the 32 treatments in the block containing experimental condition 1, and show the alias pairs.

13.34 Design an experiment consisting of a quarter-replicate of a 2^7 factorial with $ABCDE$, $ABCFG$, and $DEFG$ confounded, if the experimental condition with each factor at the 0 level is to be included. Also exhibit all alias sets.

13.35 Referring to Exercise 13.33, we can define a **modified standard order** for the 32 treatment combinations in the half-replicate as follows. First, we list the 32 treatment combinations corresponding to the five factors A, B, C, D, and E in standard order. Then we append the letter f to 16 of these treatment combinations, so that the list contains the same ones as the block chosen for the half-replicate.

(a) Use this method to list the treatment combinations obtained in Exercise 13.33 in modified standard order.

(b) Generalize the above rule for arranging the treatment combinations in modified standard order so that it applies to a half-replicate of a 2^n factorial experiment. (For a $1/2^p$ fractional replicate of a 2^n factorial, a modified standard order can be obtained by noting that any block chosen contains a subset of $n - p$ letters which form a complete replicate of a 2^{n-p} factorial. The modified standard order is then obtained by using these letters only and later appending the remaining p letters to get the required treatment combinations.)

13.36 The following factors are to be studied in a half-replicate of a 2^6 factorial experiments (defining contrast $ABCDEF$), designed to evaluate several chemicals as insecticides.

Factor	Level 0	Level 1
A. BMC	0%	5%
B. Malathion	3%	6%
C. Tedion	1%	2%
D. Chlordane	2%	5%
E. Lindane	1%	4%
F. Pyrethrum	2%	4%

Each experimental unit consists of 10 insects, and the average lifetimes (in seconds) after application of the respective insecticides are as follows, in the random order in which they are obtained:

ce	181	*acdf*	162	*bd*	135	*abdf*	131
ae	172	1	182	*df*	171	*ab*	136
abef	140	*bf*	171	*acef*	159	*bcde*	105
bcdf	165	*cf*	176	*bc*	179	*abcdef*	109
acde	139	*be*	187	*ac*	165	*af*	176
ef	186	*abce*	131	*bcef*	181	*ad*	150
de	164	*abcf*	125	*cdef*	163	*abde*	115
abcd	112	*adef*	158	*bdef*	128	*cd*	166

(a) Write down the alias pairs.
(b) Arrange the results in modified standard order (see Exercise 13.35) and use the Yates' method to find the effect totals.
(c) Identify the effect totals as follows. In the last column of the Yates table, write down the 32 combinations $[I], [A], [B], [AB], \ldots, [ABCDE]$ in standard order. Each of these is aliased with another one; identify the alias pairs by using the main effect or lower-order interaction. (For example, $ABCDE$ is aliased with F—identify the effect total $[ABCDE]$ as that of the main effect F.) Note that 10 of the alias pairs are three-factor interactions aliased with three-factor interactions—label these "error."
(d) Obtain the mean squares and complete the analysis of variance. Note that the error mean square, having 10 degrees of freedom, is the average of the mean squares of the 10 effects labeled "error."

13.37 Verify the expression obtained for $[A]$ on page 417.

13.38 Duplicating the steps indicated on pages 416 and 417, show that in the given design the main effect for factor B is confounded with the AC interaction. [*Hint*: Express $[B]$ and $[AC]$ in terms of the parameters of the model.]

13.39 In Exercise 13.14 on page 407 suppose that only one-half of one of the replicates can be performed in the time available. Construct the half-replicate having the defining contrast $ABCD$. Using the appropriate data from Rep. 1, together with replicates of experimental conditions 1, *ab*, *bc*, and *abcd* taken from Rep. 2, perform an analysis of this half-replicate identifying those effects attributable to "error." [*Hint*: Each of the four pairs of replicated treatments provides one additional degree of freedom for error, and contributes $(y_1 - y_2)^2/2$ to the error sum of squares, where y_1 is the observation resulting from applying the treatment in the first replicate, and y_2 is the corresponding observation in the second replicate.]

13.40 In Exercise 13.15 on page 408 construct and analyze that half-replicate having the defining contract $ABCDE$, using the appropriate experimental results from Rep. 1.

13.6 CHECK LIST OF KEY TERMS (with page references)

14

Applications to Quality Assurance

Although there is the tendency to think of the subject of quality assurance as a recent development, there is nothing new about the basic idea of making a quality product characterized by a high degree of uniformity. For centuries skilled artisans have striven to make products distinctive through superior quality, and once a standard of quality was achieved, to eliminate insofar as possible all variability between products that were nominally alike.

The idea that statistics might be instrumental in assuring the quality of manufactured products goes back no further than the advent of mass production, and the widespread use of statistical methods in problems of quality assurance is even more recent than that. Many problems in the manufacture of a product are amenable to statistical treatment, and we have already studied some of them in earlier parts of this book. When we speak of (statistical) **quality assurance**, *we are referring essentially to the three special techniques which will be treated in this chapter: quality control will be discussed in Sections 14.1 through 14.3, the establishment of tolerance limits in Section 14.4, and acceptance sampling in Section 14.5. Note that the word "quality," when used technically as in this discussion, refers to some measurable or countable property of a product, such as the outside diameter of a ball bearing, the breaking strength of a yarn, the number of imperfections in a piece of cloth, the potency of a drug, and so forth.*

It may surprise some persons to learn that two apparently identical parts made under carefully controlled conditions, from the same batch of raw material, and only seconds apart by the same machine, can nevertheless be different in many respects. Indeed, any manufacturing process, however good, is characterized by a certain amount of variability which is of a random nature, and which cannot be completely eliminated.

When the variability present in a production process is confined to chance variation, the process is said to be in a state of **statistical control**. Such a state is usually attained by finding and eliminating trouble of the sort causing another kind of variation, called **assignable variation**, which may be due to poorly trained operators, poor-quality raw materials, faulty machine settings, worn parts, and the like. Since manufacturing processes are rarely free from trouble of this kind, it is important to have some systematic method of detecting serious deviations from a state of statistical control when, or if possible *before*, they actually occur. It is to this end that **control charts** are principally used.

In what follows, we shall differentiate between **control charts for measurements** and **control charts for attributes**, depending on whether the observations with which we are concerned are measurements or count data (say, the number of defectives in a sample of a given size). In either case, a control chart consists of a **central line** (see Figure 14.1) corresponding to the *average* quality at which the process is to perform, and lines corresponding to the **upper and lower control limits**. These limits are chosen so that values falling between them can be attributed to chance, while values falling beyond them are interpreted as indicating a lack of control. By plotting the results obtained from samples taken periodically at frequent intervals, it is possible to check

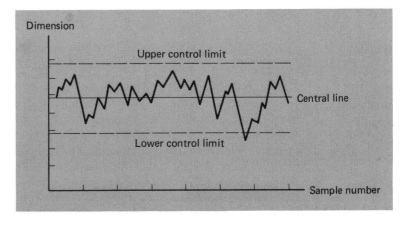

Figure 14.1 Control chart.

by means of such a chart whether the process is under control, or whether trouble of the sort indicated above has entered the process. When a sample point falls beyond the control limits, one looks for trouble, but even if the points fall between the control limits, a trend or some other systematic pattern may serve notice that action should be taken to avoid serious trouble.

The ability to "read" control charts and to determine from them just what corrective action should be taken is a matter of experience and highly developed judgment. A quality control engineer must not only understand the statistical foundation of his subject, but he must also be thoroughly acquainted with the processes themselves. The engineering and managerial aspects of quality control (and quality assurance in general), which nowadays includes incoming raw materials, outgoing products, and in-process control, constitute an extensive subject in themselves. In this chapter we shall present only the statistical aspects of the subject; more complete discussions of other aspects can be found in the book by D. J. Cowden listed in the Bibliography.

14.2 CONTROL CHARTS FOR MEASUREMENTS

When dealing with measurements, it is customary to exercise control over the average quality of a process as well as its variability. The first goal is accomplished by plotting the means of periodic samples on a so-called **control chart for means**, or more succinctly, an **$\bar{x}$ chart**. Variability is controlled by plotting the sample ranges or standard deviations, respectively, on an **R chart** or a **σ chart**, depending on which statistic is used to estimate the population standard deviation.

If the process mean and standard deviation, μ and σ, are known, and it is reasonable to treat the measurements as samples from a normal population, we can assert with probability $1 - \alpha$ that the mean of a random sample of size n will fall between

$$\mu - z_{\alpha/2}\frac{\sigma}{\sqrt{n}} \quad \text{and} \quad \mu + z_{\alpha/2}\frac{\sigma}{\sqrt{n}}$$

These two limits on $\bar{x}$ provide upper and lower control limits, and, under the given assumptions, they enable the quality-control engineer to determine whether or not to make an adjustment in the process.

In actual practice, μ and σ are usually unknown and it is necessary to estimate their values from a large sample (or samples) taken while the process is "in control." For this reason and because there may be no assurance that the measurements can be treated as samples from a normal population, the $(1 - \alpha)100\%$ confidence level associated with the control limits is only approximate, and such "probability limits" are seldom used in practice.

Instead, it is common industrial practice to use **three-sigma limits** obtained by substituting 3 for $z_{\alpha/2}$. With three-sigma limits one usually is highly confident that the process will not be declared out of control when, in fact, it is actually in control.

If there exists a long history of a process in good control, μ and σ can be estimated from past data practically without error. Thus, the central line of an $\bar{x}$ chart is given by μ, and the upper and lower three-sigma control limits are given by $\mu \pm A\sigma$, where $A = 3/\sqrt{n}$ and n is the size of each sample.† For convenience, values of A for $n = 2, 3, \ldots,$ and 15 are given in Table 13. The use of a constant sample size n simplifies the maintenance and interpretation of an $\bar{x}$ chart, but as the reader will observe in Exercise 14.4 on page 436, this restriction is not absolutely necessary.

In the more common case where the population parameters are *unknown*, it is necessary to estimate these parameters on the basis of preliminary samples. For this purpose, it is usually desirable to obtain the results of 20 or 25 consecutive samples taken when the process is in control. If k samples are used, each of size n, we shall denote the mean of the ith sample by $\bar{x}_i$, and the grand mean of the k sample means by $\bar{\bar{x}}$, that is,

Grand mean of sample means

$$\bar{\bar{x}} = \frac{1}{k} \sum_{i=1}^{k} \bar{x}_i$$

The process variability σ can be estimated either from the standard deviations or the ranges of the k samples. Since the sample size commonly used in connection with control charts for measurements is small, there is usually very little loss of efficiency in estimating σ from the sample ranges. (For an example where sample standard deviations are used in this connection, see Exercise 14.5 on page 437.) Denoting the range of the ith sample by R_i, we shall thus make use of the statistic

Mean sample range

$$\bar{R} = \frac{1}{k} \sum_{i=1}^{k} R_i$$

---

†Throughout this chapter we shall depart somewhat from the customary quality-control notation, in order to be consistent with the more widely accepted statistical notation used elsewhere in this book. (For instance, in quality control it is customary to denote the sample mean and standard deviation by $\bar{x}$ and σ, and the corresponding population parameters by $\bar{x}'$ and σ'.)

Since $\bar{x}$ provides an unbiased estimate of the population mean μ, the central line for the $\bar{x}$ chart is given by $\bar{\bar{x}}$. The statistic $\bar{R}$ does not provide an unbiased estimate of σ, but multiplying $\bar{R}$ by the constant A_2, we obtain an unbiased estimate of $3\sigma/\sqrt{n}$. The constant multiplier A_2, tabulated in Table 13 for various values of n, depends on the assumption that the measurements constitute a sample from a normal population. Thus, the central line and the upper and lower three-sigma control limits, *UCL* and *LCL*, for an $\bar{x}$ chart (with μ and σ estimated from past data) are given by

Control-chart values for an $\bar{x}$ chart

$$central\ line = \bar{\bar{x}}$$
$$UCL = \bar{\bar{x}} + A_2\bar{R}$$
$$LCL = \bar{\bar{x}} - A_2\bar{R}$$

An example of this kind of control chart for the mean is shown in Figure 14.2.

Figure 14.2 $\bar{x}$ chart.

In controlling a process, it may not be enough to monitor the population mean. Although an increase in process variability may become apparent from increased fluctuations of the $\bar{x}$'s, a more sensitive test of shifts in process variability is provided by a separate control chart, an R chart based on the sample ranges or a σ chart based on the sample standard deviations. An example of the latter may be found in Exercise 14.5 on page 437.

The central line and control limits of an R chart are based on the distribution of the range of samples of size n from a normal population. As we observed on page 231, the mean and the standard deviation of this sampling distribution are given by $d_2\sigma$ and $d_3\sigma$, respectively, when σ is known. Thus, three-sigma control limits for the range are given by $d_2\sigma \pm 3d_3\sigma$, and the complete set of control-chart values for an R chart (with σ known) is given by

Control-chart values for an R chart (σ known)

$$central\ line = d_2\sigma$$
$$UCL = D_2\sigma$$
$$LCL = D_1\sigma$$

Here $D_1 = d_2 - 3d_3$ and $D_2 = d_2 + 3d_3$, and values of these constants can be found in Table 13 for various values of n.

If σ is unknown, it is estimated from past data as previously described, and the control-chart values for an R chart (with σ unknown) are as follows:

Control-chart values for an R chart (σ unknown)

$$central\ line = \bar{R}$$
$$UCL = D_4\bar{R}$$
$$LCL = D_3\bar{R}$$

Here $D_3 = D_1/d_2$ and $D_4 = D_2/d_2$, and values of these constants can also be found in Table 13 for various values of n.

To illustrate the construction of an $\bar{x}$ chart and an R chart, suppose a manufacturer of a certain bearing knows from a preliminary record of 20 hourly samples of size 4 that for the diameters of these bearings $\bar{\bar{x}} = 0.9752$ and $\bar{R} = 0.0002$. Coding his data by means of the equation $\dfrac{x - 0.9750}{0.0001}$, that is, expressing each measurement as a deviation from 0.9750 in 0.0001 inch, he obtains

	$\bar{x}$ chart *(coded)*			R chart *(coded)*	
central line	$\bar{\bar{x}} = 2.0$		central line	$\bar{R} = 2.0$	
UCL	$\bar{\bar{x}} + A_2\bar{R} = 3.4$		UCL	$D_4\bar{R} = 4.6$	
LCL	$\bar{\bar{x}} - A_2\bar{R} = 0.6$		LCL	$D_3\bar{R} = 0$	

The values of $A_2 = 0.729$, $D_3 = 0$, and $D_4 = 2.282$ for samples of size 4 were obtained from Table 13. Graphically, these control charts are shown in Figures 14.2 and 14.3, where we have also indicated the results subsequently obtained in the following 20 samples:

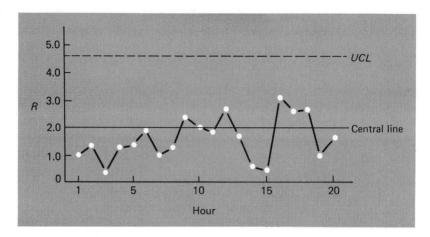

Figure 14.3 *R* chart.

Hour	Coded sample values				$\bar{x}$	R
1	1.7	2.2	1.9	1.2	1.75	1.0
2	0.8	1.5	2.1	0.9	1.32	1.3
3	1.0	1.4	1.0	1.3	1.18	0.4
4	0.4	−0.6	0.7	0.2	0.18	1.3
5	1.4	2.3	2.8	2.7	2.30	1.4
6	1.8	2.0	1.1	0.1	1.25	1.9
7	1.6	1.0	1.5	2.0	1.52	1.0
8	2.5	1.6	1.8	1.2	1.78	1.3
9	2.9	2.0	0.5	2.2	1.90	2.4
10	1.1	1.1	3.1	1.6	1.72	2.0
11	1.7	3 6	2 5	1.8	2.40	1.9
12	4.6	2.8	3.5	1.9	3.20	2.7
13	2.6	2.8	3.2	1.5	2.52	1.7
14	2.3	2.1	2.1	1.7	2.05	0.6
15	1.9	1.6	1.8	1.4	1.68	0.5
16	1.3	2.0	3.9	0.8	2.00	3.1
17	2.8	1.5	0.6	0.2	1.28	2.6
18	1.7	3.6	0.9	1.5	1.92	2.7
19	1.6	0.6	1.0	0.8	1.00	1.0
20	1.7	1.0	0.5	2.2	1.35	1.7

Inspection of Figure 14.2 shows that only one of the points falls outside of the control limits, but it also shows that there may nevertheless have been a downward shift in the process average. Figure 14.3 shows a definite downward shift in the process variability; note especially that most of the sample ranges fall below the central line of the *R* chart.

The reader may have observed the close connection between the use of control charts and the testing of hypotheses. A point on an $\bar{x}$ chart that is out of control corresponds to a sample for which the null hypothesis that $\mu = \mu_0$ is rejected. To be more precise, we should say that control-chart techniques provide sequential, *temporally ordered* sets of tests. We are interested not only in the positions of individual points, but also in possible trends or other patterns exhibited by the points representing successive samples. We might, thus, use the sign test (see page 272) to check whether there has been a shift in the product average in the above example, or a run test (see page 282) to determine whether there is a significant trend.

14.3 CONTROL CHARTS FOR ATTRIBUTES

Although more complete information can usually be gained from measurements made on a finished product, it is often quicker and cheaper to check the product against specifications on an "attribute" or "go, no-go" basis. For example, in checking the diameter and eccentricity of a ball bearing it is far simpler to determine whether it will pass through circular holes cut in a template than to make several measurements of the diameter with a micrometer. In this section we shall discuss two fundamental kinds of control charts used in connection with attribute sampling, the **fraction-defective chart**, also called a "*p* chart," and the **number-of-defects chart**, also called a "*c* chart." To clarify the distinction between "number defective" and "number of defects," note that a unit tested can have several **defects**, while, on the other hand, it is either defective or it is not. In many applications, a unit is referred to as **defective** if it has at least one defect.

Control limits for a fraction-defective chart are based on the sampling theory for proportions introduced in Section 9.1, and on the normal curve approximation to the binomial distribution. Thus, if a standard is given, that is, if the fraction defective should take on some preassigned value p, the central line is p and three-sigma control limits for the fraction defective in random samples of size n are given by

$$p \pm 3\sqrt{\frac{p(1-p)}{n}}$$

If no standard is given, which is more frequently the case in actual practice, p will have to be estimated from past data. If k samples are available, d_i is the number of defectives in the ith sample, and n_i is the number of observations in the ith sample, it is customary to estimate p as the proportion of defectives in the combined sample, namely, as

*Proportion
of defectives
in combined
sample*

$$\bar{p} = \frac{d_1 + d_2 + \ldots + d_k}{n_1 + n_2 + \ldots + n_k}$$

Substituting $\bar{p}$ for p in the above formulas, we obtain the following control-chart values for a fraction-defective chart based on the analysis of past data:

*Control-chart
values
for a fraction-
defective chart*

$$central~line = \bar{p}$$

$$UCL = \bar{p} + 3\sqrt{\frac{\bar{p}(1 - \bar{p})}{n}}$$

$$LCL = \bar{p} - 3\sqrt{\frac{\bar{p}(1 - \bar{p})}{n}}$$

Note that if p is small, as is often the case in practice, substitution in the formula for the lower control limit might yield a negative number. When this occurs, it is customary to regard the lower control limit as if it were zero and, in effect, to use only the upper control limit. Another complication that can arise if p is small is that the binomial distribution may not be adequately approximated by the normal distribution. Generally speaking, the use of the above control limits for p charts is unrealistic whenever n and p are such that the underlying binomial (or hypergeometric) distribution cannot be approximated by a normal curve (see page 112). In such cases it is best to use an upper control limit obtained directly from a table of binomial probabilities or, perhaps, use the Poisson approximation to the binomial distribution.

Figure 14.4 *p* chart.

As an illustration of a p chart, suppose that it is desired to control the output of a certain integrated circuit production line to maintain a "yield" of 60 percent, that is, a proportion defective of 40 percent. To this end, daily samples of 100 units are checked to electrical specifications, with the following results:

Date	Number of defectives	Date	Number of defectives	Date	Number of defectives
3-12	24	3-26	44	4- 9	23
3-13	38	3-27	52	4-10	31
3-16	62	3-30	45	4-13	26
3-17	34	3-31	30	4-14	32
3-18	26	4- 1	34	4-15	35
3-19	36	4- 2	33	4-16	15
3-20	38	4- 3	22	4-17	24
3-23	52	4- 6	34	4-20	38
3-24	33	4- 7	43	4-21	21
3-25	44	4- 8	28	4-22	16

Since the standard is given as $p = 0.40$, the control-chart values are

$$\text{central line} = 0.40$$

$$UCL = 0.40 + 3\sqrt{\frac{(0.40)(0.60)}{100}} = 0.55$$

$$LCL = 0.40 - 3\sqrt{\frac{(0.40)(0.60)}{100}} = 0.25$$

The corresponding control chart with points for the 30 sample fractions defective is shown in Figure 14.4, and it exhibits some interesting characteristics. Note that there is only one point out of control on the high side, but there are seven points out of control on the low side. Most of these seven "low points" occurred after April 1, and there appears to be a general downward trend. In fact, there is an unbroken run of eleven points below the central line after April 7. It would appear from this chart that the yield is not yet stabilized, and that the process is potentially capable of maintaining a yield well above the nominal 60% value.

Equivalent to the p chart for the fraction defective is the control chart for the number of defectives. Instead of plotting the fraction defective in a sample of size n, one plots the number of defectives, and the control-chart values for this kind of chart are obtained by multiplying the above values for the central line and the control limits by n. Thus, if p is estimated by $\bar{p}$, the control-chart values for a **number-of-defectives** chart are as follows:

Control-chart
values
for a number-
of-defectives chart

$$central\ line = n\bar{p}$$
$$UCL = n\bar{p} + 3\sqrt{n\bar{p}(1 - \bar{p})}$$
$$LCL = n\bar{p} - 3\sqrt{n\bar{p}(1 - \bar{p})}$$

There are situations where it is necessary to control the number of defects in a unit of product, rather than the fraction defective or the number of defectives. For example, in the production of carpeting it is important to control the number of defects per hundred yards; in the production of newsprint one may wish to control the number of defects per roll. These situations are similar to the one described in Section 3.7, which led to the Poisson distribution. Thus, if c is the number of defects per manufactured unit, c is taken to be a value of a random variable having the Poisson distribution.

It follows that the center line for a number-of-defects chart is the parameter λ of the corresponding Poisson distribution, and that three-sigma control limits can be based on the fact that the standard deviation of this distribution is $\sqrt{\lambda}$. If λ is unknown, that is, if no standard is given, its value is usually estimated from at least 20 values of c observed from past data. If k is the number of units of product available for estimating λ, and if c_i is the number of defects in the ith unit, then λ is estimated by

Mean number
of defects

$$\bar{c} = \frac{1}{k} \sum_{i=1}^{k} c_i$$

and the control-chart values for the *c chart* are

Control-chart
values
for a number-
of-defects chart

$$central\ line = \bar{c}$$
$$UCL = \bar{c} + 3\sqrt{\bar{c}}$$
$$LCL = \bar{c} - 3\sqrt{\bar{c}}$$

To illustrate this kind of control chart, suppose that it is known from past experience that on the average an aircraft assembly made by a certain company has $\bar{c} = 4$ missing rivets. The corresponding control chart for the number of missing rivets is shown in Figure 14.5, on which we have also plotted the results of inspections which revealed 4, 6, 5, 1, 2, 3, 5, 7, 1, 2, 2, 4, 6, 5, 3, 2, 4, 1, 8, 4, 5, 6, 3, 4, and 2 missing rivets in 25 assemblies.

Figure 14.5 *c* chart.

EXERCISES

14.1 A plastics manufacturer extrudes blanks for use in the manufacture of eyeglass temples. Specifications require that the thickness of these blanks have $\mu = 0.150$ inch and $\sigma = 0.002$ inch.

(a) Use the specifications to calculate a central line and three-sigma control limits for an $\bar{x}$ chart with $n = 5$.

(b) Use the specifications to calculate a central line and three-sigma control limits for an R chart with $n = 5$.

(c) Plot the following means and ranges, obtained in 20 successive random samples of size 5, on charts based on the control-chart constants obtained in parts (a) and (b), and discuss the process.

Sample	$\bar{x}$	R	Sample	$\bar{x}$	R
1	0.152	0.004	11	0.149	0.003
2	0.147	0.006	12	0.153	0.004
3	0.153	0.004	13	0.150	0.005
4	0.153	0.002	14	0.152	0.001
5	0.151	0.003	15	0.149	0.003
6	0.148	0.002	16	0.146	0.002
7	0.149	0.006	17	0.154	0.004
8	0.144	0.001	18	0.152	0.005
9	0.149	0.003	19	0.151	0.002
10	0.152	0.005	20	0.149	0.004

14.2 Calculate $\bar{\bar{x}}$ and $\bar{R}$ for the data of part (c) of Exercise 14.1, and use these values to construct the central lines and three-sigma control limits for new $\bar{x}$ and R charts to be used in the control of the thickness of the extruded plastic blanks.

14.3 The following data give the means and ranges of 25 samples, each consisting of four compression test results on steel forgings, in thousands of pounds per square inch:

Sample	1	2	3	4	5	6	7	8
$\bar{x}$	45.4	48.1	46.2	45.7	41.9	49.4	52.6	54.5
R	2.7	3.1	5.0	1.6	2.2	5.7	6.5	3.6

Sample	9	10	11	12	13	14	15	16
$\bar{x}$	45.1	47.6	42.8	41.4	43.7	49.2	51.1	42.8
R	2.5	1.0	3.9	5.6	2.7	3.1	1.5	2.2

Sample	17	18	19	20	21	22	23	24	25
$\bar{x}$	51.1	52.4	47.9	48.6	53.3	49.7	48.2	51.6	52.3
R	1.4	4.3	2.2	2.7	3.0	1.1	2.1	1.6	2.4

(a) Use these data to find the central line and control limits for an $\bar{x}$ chart.
(b) Use these data to find the central line and control limits for an R chart.
(c) Plot the given data on $\bar{x}$ and R charts based on the control-chart constants computed in parts (a) and (b), and interpret the results.
(d) Using runs above and below the central line (similar to runs above and below the median discussed on page 283), test at a level of significance of 0.05 whether there is a trend in the $\bar{x}$ values.
(e) Would it be reasonable to use the control limits found in this exercise in connection with subsequent compression test measurements from the same process? Why?

14.4 Reverse-current readings (in nanoamperes) are made on a sample of 10 transistors every half hour. Since some of the units may prove to be "shorts" or "opens," it is not always possible to obtain 10 readings. The following table shows the number of readings made at the end of each half-hour interval during an 8-hour shift, and the mean reverse currents obtained.

Sample	1	2	3	4	5	6	7	8
n	10	6	9	8	8	10	7	9
$\bar{x}$	12.5	11.1	10.2	11.6	21.9	12.3	9.7	15.6

Sample	9	10	11	12	13	14	15	16
n	7	8	10	9	7	8	9	10
$\bar{x}$	16.7	9.8	11.6	17.2	10.1	9.5	13.1	14.2

(a) Find the central line for an $\bar{x}$ chart by taking the weighted mean of the 16 $\bar{x}$'s, weighting each value with the size of the corresponding sample.

Figure 14.6 Exercise 14.4.

(b) Construct a table showing the central line found in part (a) and three-sigma control limits corresponding to $n = 6, 7, 8, 9$, and 10. Use $R = 4.0$, a value based on prior data.

(c) Plot the data on a control chart like the one of Figure 14.6 and interpret the results.

14.5 If the sample standard deviations instead of the sample ranges are used to estimate σ, the control limits for the resulting $\bar{x}$ chart are given by $\bar{\bar{x}} \pm A_1 \bar{s}$, where $\bar{s}$ is the mean of the sample standard deviations obtained from given data, and A_1 can be found in Table 13. Note that in connection with problems of quality control the sample standard deviation is defined using the divisor n instead of $n - 1$. The corresponding R chart is replaced by a σ chart, having the central line $c_2 \bar{s}$ and the upper and lower control limits $B_3 \bar{s}$ and $B_4 \bar{s}$, where B_3 and B_4 can be obtained from Table 13.

(a) Construct an $\bar{x}$ chart and a σ chart for 20 samples of size 3 which had $\bar{x}$ equal to 21.2, 19.4, 20.4, 20.4, 20.4, 19.0, 20.3, 21.1, 21.6, 22.1, 24.4, 23.9, 24.9, 24.1, 21.8, 19.5, 20.3, 22.5, 23.4, 23.3, and s equal to 2.0, 0.8, 1.1, 0.9, 1.0, 0.3, 1.3, 2.0, 0.8, 1.0, 1.5, 1.0, 1.5, 0.8, 1.3, 2.9, 4.3, 1.2, 0.3, 3.1.

(b) Would it be reasonable to use these control limits for subsequent data? Why?

14.6 In order to establish control charts for a boring process, 30 samples of five measurements of the inside diameters are taken, and the results are $\bar{\bar{x}} = 1.317$ inches and $\bar{s} = 0.002$ inch. Using the method of Exercise 14.5, construct an $\bar{x}$ chart for $n = 5$, and on it plot the following means obtained in 25 successive samples: 1.328, 1.330, 1.321, 1.325, 1.332, 1.340, 1.327, 1.321, 1.324, 1.325, 1.329, 1.326, 1.330, 1.324, 1.328, 1.322, 1.326, 1.327, 1.329, 1.325, 1.324, 1.329, 1.330, 1.321, and 1.329. Discuss the results.

14.7 Suppose that in the example of Exercise 14.6 it is desired to establish control also over the variability of the process. Using the method of Exercise 14.5 and the values of $\bar{\bar{x}}$ and $\bar{s}$ given in Exercise 14.6, calculate the central line and control limits for a σ chart with $n = 5$.

14.8 Thirty-five successive samples of 100 castings each, taken from a production line, contained, respectively, 3, 3, 5, 3, 5, 0, 3, 2, 3, 5, 6, 5, 9, 1, 2, 4, 5, 2, 0, 10, 3, 6, 3, 2, 5, 6, 3, 3, 2, 5, 1, 0, 7, 4, and 3 defectives. If the fraction defective is to be maintained at 0.02, construct a p chart for these data and state whether or not this standard is being met.

14.9 The data of Exercise 14.8 may be looked upon as evidence that the standard of 2% defectives is being exceeded.
(a) Use the data of Exercise 14.8 to construct new control limits for the fraction defective.
(b) Using the control limits found in part (a), continue the control of the process by plotting the following data on the number of defectives obtained in 20 subsequent samples of size $n = 100$: 2, 4, 2, 4, 7, 5, 3, 2, 2, 3, 5, 6, 4, 5, 8, 0, 5, 5, 4, and 2.

14.10 The specifications for a certain mass-produced valve prescribe a testing procedure according to which each valve can be classified as satisfactory or unsatisfactory (defective). Past experience has shown that the process can perform so that $\bar{p} = 0.03$. Construct a three-sigma control chart for the number of defectives obtained in samples of size 100, and on it plot the following numbers of defectives obtained in such samples randomly selected from 30 successive half-day's production: 3, 4, 2, 1, 5, 2, 1, 2, 3, 1, 3, 2, 2, 2, 1, 1, 2, 0, 4, 3, 1, 0, 2, 4, 0, 1, 5, 7, 3, and 2.

14.11 The standard for a process producing tin plate in a continuous strip is five defects in the form of pinholes or visual blemishes per hundred feet. Based on the following set of 25 observations, giving the number of defects per hundred feet, can it be concluded that the process is in control to this standard?

Inspection number	1	2	3	4	5	6	7	8	9	10	11	12
Number of defects	3	2	2	4	4	4	6	4	1	7	5	5

Inspection number	13	14	15	16	17	18	19	20	21	22	23	24	25
Number of defects	4	6	6	9	5	2	6	5	11	6	6	8	2

14.12 A process for the manufacture of 4 feet by 8 feet woodgrained panels has performed in the past with an average of 2.7 imperfections per 100 panels. Construct a chart to be used in the inspection of the panels and discuss the control if 25 successive 100-panel lots contained, respectively, 4, 1, 0, 3, 5, 3, 5, 4, 1, 4, 0, 1, 4, 2, 3, 7, 4, 2, 1, 3, 0, 2, 6, 1, and 3 imperfections.

14.4 TOLERANCE LIMITS

Inherent in every phase of industrial quality control is the problem of comparing some quality characteristic or measurement of a finished product against given specifications. Sometimes the specifications, or **tolerance limits,** are so stated by the customer or by the design engineer that any appreciable departure will make the product unusable. There remains, however, the problem of producing the part so that an acceptably high proportion of units will fall within tolerance limits specified for the given quality characteristic. Also, if a product is made without prior specifications, or if modifications are made, it is desirable to know within what limits the process can hold a quality characteristic a reasonably high percentage of the time. We, thus, speak of "natural" tolerance limits; that is, we let the process establish its own limits which, according to experience, can be met in actual practice.

If reliable information is available about the distribution underlying the measurement in question, it is a relatively simple matter to find natural tolerance limits. For instance, if long experience with a product enables us to assume that a certain dimension is normally distributed with the mean μ and the standard deviation σ, it is easy to construct limits between which we can expect to find any given proportion P of the population. For $P = 0.90$ we have the tolerance limits $\mu \pm 1.645\sigma$, and for $P = 0.95$ we have $\mu \pm 1.96\sigma$, as can easily be verified from a table of normal-curve areas.

In most practical situations the true values of μ and σ are not known, and tolerance limits must be based on the mean $\bar{x}$ and the standard deviation s of a random sample. Whereas $\mu \pm 1.96\sigma$ are limits including 95% of a normal population, the same cannot be said for the limits $\bar{x} \pm 1.96s$. These limits are values of random variables and they may or may not include a given proportion of the population. Nevertheless, it is possible to determine a constant K so that *one can assert with* $(1 - \alpha)100\%$ *confidence that the proportion of the population contained between* $\bar{x} - Ks$ *and* $\bar{x} + Ks$ *is at least* P. Such values of K for random samples from normal populations are given in Table 14 for $P = 0.90, 0.95$, and 0.99, 95% and 99% levels of confidence, and selected values of n from 2 to 1,000.

To illustrate this technique, suppose that a manufacturer takes a sample of size $n = 100$ from a very large lot of mass-produced compression springs and that he obtains $\bar{x} = 1.507$ and $s = 0.004$ inch for the free lengths of the springs. Choosing the 99% level of confidence and a minimum proportion of $P = 0.95$, he obtains the tolerance limits $1.507 \pm (2.355)(0.004)$; in other words, the manufacturer can assert with 99% confidence that at least 95% of the springs in the entire lot have free lengths from 1.497 to 1.517 inches. Note that in problems like these the minimum proportion P as well as the degree of confidence $1 - \alpha$ must be specified; also note that the lower tolerance limit is rounded *down* and the upper tolerance limit is rounded *up*.

To avoid confusion, let us also point out that there is an essential difference between confidence limits and tolerance limits. Whereas confidence limits are used to estimate a parameter of a population, tolerance limits are used to indicate between what limits one can find a certain proportion of a population. This distinction is emphasized by the fact that when n becomes large the length of a confidence interval approaches zero, while the tolerance limits will approach the corresponding values for the population. Thus, for large n, K approaches 1.96 in the columns for $P = 0.95$ in Table 14.

EXERCISES

14.13 To check the strength of carbon steel for use in chain links, the yield stress of a random sample of 25 pieces was measured, yielding a mean and a standard deviation of 52,800 psi and 4,600 psi, respectively. Establish tolerance limits with $\alpha = 0.05$ and $P = 0.99$, and express *in words* what these tolerance limits mean.

14.14 In a study designed to determine the number of turns required for an artillery-shell fuse to arm, 75 fuses, rotated on a turntable, averaged 38.7 turns with a standard deviation of 4.3 turns. Establish tolerance limits for which one can assert with 99% confidence that *at least 95%* of the fuses will arm within these limits.

14.15 In a random sample of 40 piston rings chosen from a production line, the mean edge width was 0.1063 inch, and the standard deviation was 0.0004 inch.
(a) Between what limits can it be said with 95% confidence that at least 90% of the edge widths of piston rings produced by this production line will lie?
(b) Find 95% confidence limits for the true mean edge width, and explain the difference between these limits and the tolerance limits found in part (a).

14.16 *Nonparametric tolerance limits* can be based on the extreme values in a random sample of size n taken from any continuous population. The following equation relates the quantities n, P, and α, where P is the minimum proportion of the population contained between the smallest and largest observations with $(1 - \alpha)100\%$ confidence:

$$nP^{n-1} - (n - 1)P^n = \alpha$$

An approximate solution for n is given by

$$n \simeq \frac{1}{2} + \frac{1 + P}{1 - P} \cdot \frac{\chi_\alpha^2}{4}$$

where χ_α^2 is the value of chi square for 4 degrees of freedom that corresponds to a right-hand tail area α.
(a) How large a sample is required to be 95% certain that at least 90% of the population will be included between the extreme values of the sample?
(b) With 95% confidence, at least what proportion of the population can be expected to be included between the extreme values of a sample of size 100?

14.5 ACCEPTANCE SAMPLING

Manufactured goods are shipped to the purchaser in lots ranging in size from only a few to many thousands of individual items. Ideally, each lot should not contain any defectives, but practically speaking it is rarely possible to meet this goal. Recognizing the fact that some defective goods are bound to be delivered, even if each lot were to be inspected 100%, most consumers require that evidence, based on careful inspection, be given that the proportion of defectives in each lot is not excessive.

A frequently used and highly effective method for providing such evidence is that of sampling inspection, where items are selected from each lot prior to shipment (or prior to acceptance by the consumer), and a decision is made on the basis of this sample whether to accept or reject the lot. Acceptance of a lot ordinarily implies that it can be shipped (or be accepted by the consumer), even though it may contain some defective items. Arrangements between the producer and the consumer may allow for some form of credit to be given for defectives subsequently discovered by the consumer. Rejection of a lot need not mean that it is to be scrapped; a rejected lot may be subjected to closer inspection with the aim of eliminating all defective items.

Since the cost of inspection is rarely negligible (sometimes it is nearly as high as or higher than the cost of production), it is seldom desirable to inspect each item in a lot. Thus, acceptance inspection usually involves **sampling;** more specifically, a random sample is selected from each lot and the lot is accepted if the number of defectives found in the sample does not exceed a given **acceptance number**. This procedure is equivalent to a test of the null hypothesis that the proportion defective p in the lot equals some specified value p_0 against the alternative that it equals p_1, where $p_1 > p_0$. In acceptance sampling the value p_0 is called the **acceptable quality level** or **AQL**, and p_1 is called the **lot tolerance percent defective** or **LTPD**. The probability of a Type I error, α, can be interpreted as an upper limit to the proportion of "good" lots (lots with $p \leq p_0$) that will be rejected, and in this context it is called the **producer's risk**. The probability of a Type II error, β, gives an upper bound to the proportion of "bad" lots (lots with $p \geq p_1$) that will be accepted, and it is called the **consumer's risk**.

A **single-sampling plan** is simply a specification of the sample size and the acceptance number to be used, and its choice is usually based on a specified AQL and (or) $LTPD$ in association with given producer's and (or) consumer's risks. A given sampling plan is best described by its operating characteristic or OC curve, which gives the probability of acceptance for each value that can be assumed by the lot proportion defective p. Thus, the OC curve describes the degree of protection offered by the sampling plan against incoming lots of various qualities. If a sample of size n is taken from a lot containing N units, and if the acceptance number is c, the probability of accepting a lot containing the proportion of defectives p (the lot contains Np defectives) can be calculated by using the hypergeometric distribution, as follows:

Probability of lot acceptance

$$L(p) = \sum_{x=0}^{c} h(x; n, Np, N)$$

To illustrate the calculation of points on an OC curve, suppose that the lot size is $N = 100$, the sample size is $n = 10$, and the acceptance number is $c = 1$. We have

$$L(p) = \frac{\binom{100p}{0}\binom{100(1-p)}{10} + \binom{100p}{1}\binom{100(1-p)}{9}}{\binom{100}{10}}$$

Since calculations involving the hypergeometric distribution are fairly tedious, especially when n and N are large, it is customary in acceptance sampling to approximate the hypergeometric distribution with the binomial distribution, as on page 64. For instance, for $p = 0.10$ the exact value is $L(0.10) = 0.739$, as the reader will be asked to verify in Exercise 14.21 on page 450, while Table 1 with $n = 10$ and $p = 0.10$ yields the binomial approximation

$$L(0.10) \simeq B(1; 10. 0.10) = 0.736$$

A sketch of the OC curve for the sampling plan $n = 10$, $c = 1$ can be made rapidly with the aid of Table 1, and the result is shown in Figure 14.7. From this curve it can be seen that the producer's risk is approximately 0.05 when the AQL is 0.04, and the consumer's risk is approximately 0.10 when the $LTPD$ is 0.34.

A sampling plan can also be described by means of its **average outgoing quality** or AOQ curve. This curve describes the degree of protection offered by the sampling plan by showing the average quality of outgoing lots corresponding to each quality level of incoming lots (that is, lots prior to inspection). If incoming lots are of good quality, that is, if their proportion defective is smaller than the AQL, very few lots will be rejected and the average outgoing quality or AOQ will be good. If the incoming lots are of poor quality, that is, if their proportion defective is larger than the $LTPD$, most of them will be rejected. If all rejected lots are inspected 100% and all defective units are replaced by good units prior to acceptance of the lot, then the average out-

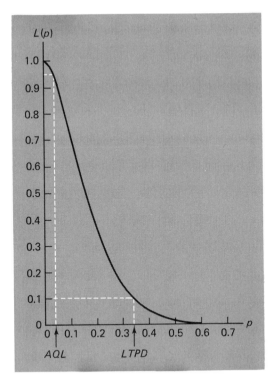

Figure 14.7 *OC* curve.

going quality will be good even though the average incoming quality is poor. It is when the average incoming quality lies between the *AQL* and the *LTPD* that the poorest quality of lots will be shipped. In general, there will be a maximum *AOQ* over all values of incoming quality p, and this value is called the **average outgoing quality limit**, or *AOQL*.

It is not difficult to derive a formula for finding the *AOQ* corresponding to a given incoming quality p under the assumption that all defectives in rejected lots are replaced by acceptable items prior to their final acceptance. If the incoming quality is p, the probability that a lot will be accepted is $L(p)$, and each such lot contains the proportion p of defectives. The proportion $1 - L(p)$ of lots which are eventually rejected contain no defectives, and it follows that the *AOQ* is given by $p \cdot L(p) + 0 \cdot [1 - L(p)]$, or

Average outgoing quality

$$AOQ = p \cdot L(p)$$

More common practice is to remove or replace defectives found in accepted as well as rejected lots, but the modification thus required in the AOQ is usually minor and it is customary to use the above formula regardless of the inspection procedure.

The AOQ curve for the sampling plan $n = 10$, $c = 1$ is shown in Figure 14.8, and it is apparent from this figure that the $AOQL$ is approximately 0.081.

Figure 14.8 *AOQ* curve.

Sometimes, smaller samples (and hence, reductions in cost) can be achieved without sacrifice in the degree of protection by the use of so-called **double** or **multiple sampling**. A double-sampling plan involves the selection of a random sample of size n_1 from a lot; if the sample contains c_1 or fewer defectives, the lot is accepted; if it contains c'_1 or more defectives ($c'_1 > c_1$) the lot is rejected; otherwise, a second sample of size n_2 is taken from the lot, and the lot is accepted unless the total number of defectives in the combined sample of size $n_1 + n_2$ exceeds c_2. A multiple-sampling plan is similar in nature to a double-sampling plan, but it involves more than two stages.

An illustration of a multiple-sampling plan is given in the following table:

Sample	Sample size	Combined samples		
		Size	Acceptance number	Rejection number
First	20	20		3
Second	20	40	1	4
Third	20	60	3	5
Fourth	20	80	3	6
Fifth	20	100	5	7
Sixth	20	120	6	8
Seventh	20	140	7	8

This table summarizes a procedure that can be described as follows. In the first step, the lot is rejected if there are 3 or more defectives; otherwise, sampling continues. In the second step, the lot is accepted if the combined sample contains at most 1 defective, it is rejected if there are 4 or more defectives; otherwise, sampling continues. This goes on, if necessary, until in the final step the lot is accepted if there are at most 7 defectives in the combined sample of size 140, and otherwise it is rejected.

By an appropriate choice of the sample sizes and the acceptance and rejection numbers it is possible to match the *OC* curve of a double- or multiple-sampling plan closely to that of an equivalent single-sampling plan. Thus, the degree of protection offered by a double- or multiple-sampling plan can be made essentially the same as that offered by an equivalent single-sampling plan.

The advantage of double or multiple sampling is that there is a high probability that a very good lot will be accepted or a very poor lot will be rejected on the basis of the first sample (or, at least, an early sample), thus reducing the required amount of inspection. On the other hand, if the lot quality is "intermediate," the total sample size required may actually be larger than that of the equivalent single-sampling plan.

To observe how a double- or multiple-sampling plan can conserve sampling effort, and to introduce the idea of the **average sample number (*ASN*) curve**, we consider the following double-sampling plan:

Sample	Sample size	Combined samples		
		Size	Acceptance number	Rejection number
First	15	15	1	5
Second	30	45	5	6

If the incoming lot quality is $p = 0.05$, the probability that a second sample will be required is the same as the probability that there will be 2, 3, or 4 defectives in a sample of size 15. According to Table 1, this probability is equal to

$$B(4; 15, 0.05) - B(1; 15, 0.05) = 0.170$$

Thus, on the average, it will require a sample of size $15 + (0.170)(30) = 20.1$ to decide whether to accept or reject an incoming lot of quality $p = 0.05$. Similar calculations enable us to find the average sample size required to inspect a lot having any given incoming quality p. A graph showing the relation between the average sample size (also called the **average sample number**) and the incoming lot quality is called an *ASN* curve; the *ASN* curve for the double-sampling plan described above is shown in Figure 14.9.

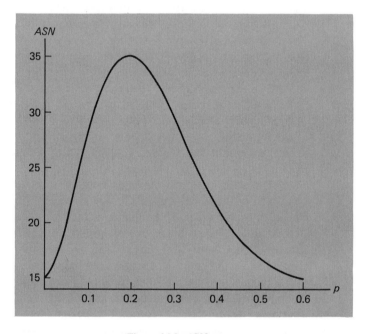

Figure 14.9 *ASN* curve.

Several standard sampling plans have been published to facilitate the use of acceptance sampling (see the book by I. W. Burr listed in the Bibliography). Among the most widely used standard plans are those contained in **Military Standard *105D*** (see the Bibliography). These plans stress the maintenance of a specified *AQL*, and they are designed to encourage the producer to offer only good products to the consumer. To accomplish this there are three general levels of inspection corresponding to different consumer's risks.

(Inspection level II is normally chosen; level I uses smaller sample sizes and level III uses larger sample sizes than level II.) There are also three types of inspection: normal, tightened, and reduced. The type of inspection depends on whether the average proportion defectives for prior samples has been above or below the AQL, and it may be changed during the course of inspection. Under tightened inspection the producer's risk is increased and the consumer's risk is (slightly) decreased; under reduced inspection the consumer's risk is increased and the producer's risk is (slightly) decreased. Tables are available for single, double, and multiple sampling, and a brief portion of these tables is included in this book in Tables 15 and 16.

The procedure in using MIL-STD-105D for single sampling is first to find the sample size code letter corresponding to the lot size and the desired inspection level and type. Then, using the sample size code letter thus obtained and the appropriate AQL, one finds the sample size and the acceptance number from the master table. A portion of the table for finding sample size code letters is given in Table 15, and a portion of the master table for normal inspection is included in Table 16.

To illustrate the use of MIL-STD-105D, suppose that incoming lots contain 2,000 items, and inspection level II is to be used in conjunction with normal inspection and an AQL of 0.025, or 2.5 percent. From Table 15 we find that the sample size code letter is K. Then, entering Table 16 in the row labeled K, we find that the sample size to be used is 125. Using the column labeled 2.5, we find that the acceptance number is 7 and the rejection number is 8. Thus, if a single sample of size 125, selected at random from a lot of 2,000 items, contains 7 or fewer defectives the lot is to be accepted; if the sample contains 8 or more defectives the lot is to be rejected.

The concept of multiple sampling is carried to its extreme in **sequential sampling**. A sampling procedure is said to be sequential if, after each observation, one of the following decisions is made: accept whatever hypothesis is being tested, reject the hypothesis, or take another observation. Although sequential procedures are used also in connection with other kinds of problems, we shall discuss this kind of sampling only in connection with acceptance sampling, where we shall thus decide after the inspection of each successive item whether to accept a lot, reject it, or continue sampling.

The construction of a sequential sampling plan consists of finding two sequences of numbers a_n and r_n, where n is the number of observations, so that the lot is accepted as soon as the number of defectives is less than or equal to a_n for some n, the lot is rejected as soon as the number of defectives is greater than or equal to r_n for some n, and sampling continues so long as the number of defectives in a sample of size n falls between a_n and r_n. If an acceptance plan is to have p_0 and p_1 as its AQL and $LTPD$, the producer's risk α, and the consumer's risk β, it can be shown (see the book by A. Wald in the Bibliography) that the required values of a_n and r_n can be computed by means of the formulas

$$a_n = \frac{\log \dfrac{\beta}{1-\alpha} + n \cdot \log \dfrac{1-p_0}{1-p_1}}{\log \dfrac{p_1}{p_0} - \log \dfrac{1-p_1}{1-p_0}}$$

$$r_n = \frac{\log \dfrac{1-\beta}{\alpha} + n \cdot \log \dfrac{1-p_0}{1-p_1}}{\log \dfrac{p_1}{p_0} - \log \dfrac{1-p_1}{1-p_0}}$$

If a_n is not an integer, it is replaced by the largest integer less than a_n; if r_n is not an integer, it is replaced by the smallest integer greater than r_n.

To illustrate this procedure, let $p_0 = 0.05$, $p_1 = 0.20$, $\alpha = 0.05$, and $\beta = 0.10$. Substituting these values into the above formulas for a_n and r_n, we obtain

$$a_n = -1.45 + 0.11n$$

$$r_n = 1.86 + 0.11n$$

and, letting $n = 1, 2, 3, \ldots$, and 25, we get the acceptance and rejection numbers shown in the second and fourth columns of the following table:

Number of items inspected n	Acceptance number a_n	Number of defectives d_n	Rejection number r_n
1	–	0	–
2	–	0	–
3	–	0	3
4	–	0	3
5	–	0	3
6	–	0	3
7	–	0	3
8	–	1	3
9	–	1	3
10	–	1	3
11	–	1	4
12	–	1	4
13	–	1	4
14	0	1	4
15	0	1	4
16	0	1	4
17	0	2	4
18	0	2	4
19	0	3	4
20	0	3	5
21	0	3	5
22	0	4	5
23	1	5	5
24	1		5
25	1		5

In the third column we have indicated the results obtained in an inspection where the 8th, 17th, 19th, 22nd, and 23rd items are defective, and where the inspection terminates with rejection of the lot after inspection of the 23rd item.

The tabular procedure illustrated can be replaced by an equivalent graphical procedure for carrying out sequential sampling inspection. Plotting the values of a_n and r_n obtained from the equations on page 448 (without rounding), we obtain two *straight lines* like those of Figure 14.10. Sampling terminates with rejection or acceptance if the number of defectives observed falls above the line for r_n or below the line for a_n, respectively.

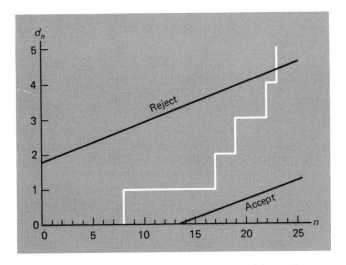

Figure 14.10 Graphical procedure for sequential sampling.

The main advantage of sequential sampling is that it can materially reduce the required amount of inspection. Studies have shown that the average decrease in sample size is often near 50 % when compared with the sample size of equivalent single-sampling plans. The major disadvantage is that, in a sequential sampling plan, there is no upper limit to the number of items that might have to be inspected in order to reach a decision concerning the acceptance or rejection of a lot. In fact, the sample size is a random variable and its value will occasionally be very large. For this reason it is customary to *truncate* sequential sampling procedures by selecting a number N such that a decision must be made to accept or reject a lot prior to or at $n = N$.

EXERCISES

14.17 A sampling plan, calling for a sample of size $n = 50$, has the acceptance number $c = 3$. Assuming that the lot size is very large, calculate the probability of accepting a lot of incoming quality 15 % defective and the probability of rejecting a lot of incoming quality 4% defective

(a) by calculating the corresponding binomial probabilities;

(b) by using the Poisson approximation to the binomial distribution.

14.18 A single-sampling plan calls for a sample of size 150.

(a) Use the normal approximation to find the acceptance number c if the *AQL* is to be 2.5% and the producer's risk is to be $\alpha = 0.05$.

(b) Using the acceptance number obtained in part (a), determine the consumer's risk if the *LTPD* is to be 6%.

14.19 Referring to the sampling plan of Exercise 14.17, use the Poisson approximation to calculate $L(p)$ for $p = 0.01, 0.02, \ldots, 0.19$, and 0.20. Sketch the *OC* curve for this sampling plan and read off the consumer's and producer's risks corresponding to an *AQL* of 4% and an *LTPD* of 14%.

14.20 A single-sampling plan has $n = 100$ and $c = 5$.

(a) Find the *AQL* if the producer's risk is to be 0.025.

(b) Find the *LTPD* if the consumer's risk is to be 0.05. [*Hint*: Use the normal approximation and set up equations leading to quadratic equations in p_0 and p_1, respectively.]

14.21 With reference to the example of page 442, verify that

(a) $L(0.10) = 0.739$, using the exact (hypergeometric) formula;

(b) $L(0.10) \simeq 0.736$, using the approximate (binomial) formula;

(c) the *OC* curve is as given in Figure 14.7;

(d) the *AOQ* curve is as given in Figure 14.8.

14.22 Calculate $L(p)$ for selected values of p and sketch the *OC* curve for the single-sampling plan $n = 120$, $c = 3$. [*Hint*: Assume a large lot size, and use the normal approximation to the binomial distribution.]

14.23 Sketch the *AOQ* curve for the sampling plan of Exercise 14.22, and estimate the *AOQL*.

14.24 Using the values of $L(p)$ obtained in Exercise 14.19, sketch the *AOQ* curve for the given sampling plan and estimate the *AOQL*.

14.25 Referring to the double-sampling plan on page 445, use Table 1 and the normal approximation (when necessary) to calculate the producer's risk α when the *AQL* is $p_0 = 0.10$.

14.26 Calculate the average sample size required for selected values of p and sketch the *ASN* curve for the following double-sampling plan:

		Combined samples		
Sample	*Sample size*	*Size*	*Acceptance number*	*Rejection number*
First	10	10	0	3
Second	35	45	3	4

14.27 A lot of 200 items is to be inspected at an AQL of 6.5%. If MIL-STD-105D is to be used, with normal inspection at general inspection level II, what single-sampling plan is required?

14.28 An incoming lot of 5,000 items is to be inspected using MIL-STD-105D, with normal inspection at general inspection level II and an AQL of 2.5%. What single-sampling plan should be used?

14.29 Find formulas for the acceptance and rejection numbers for the sequential sampling plan having an AQL of 0.10, an $LTPD$ of 0.30, a producer's risk of 0.05, and a consumer's risk of 0.10. If a sample should contain defectives on the third, fifth, seventh, and eighth trials, would this plan accept or reject the corresponding lot prior to the tenth trial? If so, on which trial?

14.30 A sequential sampling plan is to have $p_0 = 0.01$, $p_1 = 0.10$, $\alpha = 0.05$, and $\beta = 0.20$.
   (a) Determine the acceptance and rejection numbers for $n = 1, 2, \ldots$, and 50.
   (b) Use random numbers and the acceptance and rejection numbers obtained in part (a) to simulate the inspection of a very large lot containing 20% defectives.

14.6 CHECK LIST OF KEY TERMS (*with page references*)

15

Applications to Reliability

and Life Testing

The task of designing and supervising the manufacture of a product has been made increasingly difficult by rapid strides in the sophistication of modern products and the severity of the environmental conditions under which they must perform. No longer can an engineer be satisfied if the operation of a product is technically feasible, or if it can be made to "work" under optimum conditions. In addition to such considerations as cost and ease of manufacture, increasing attention must now be paid to size and weight, ease of maintenance, and reliability. The magnitude of the problem of maintainability and reliability is illustrated by surveys which have uncovered the fact that frequently a high percentage of space-age electronic equipment has been in inoperative condition. Military surveys have further shown that maintenance and repair expenses for electronic equipment often exceed the original cost of procurement, even during the first year of operation.

In Section 15.1 we define the concept of reliability; we discuss and apply special probability distributions to the calculation of reliabilities in Sections 15.2 and 15.3. Some theory and applications relating to testing products for useful lifetime are introduced in Sections 15.4 and 15.5.

The problem of assuring and maintaining reliability has many facets, including original equipment design, control of quality during production, acceptance inspection, field trials, life testing, and design modifications. To complicate matters further, reliability competes directly or indirectly with a host of other engineering considerations, chiefly cost, complexity, size and weight, and maintainability. In spite of its complicated engineering aspects, it is possible to give a relatively simple mathematical definition of reliability. To motivate this definition, we call the reader's attention to the fact that a product may function satisfactorily under one set of conditions but not under other conditions, and that satisfactory performance for one purpose does not assure adequate performance for another purpose. For example, a vacuum tube which is perfectly satisfactory for use in a home radio may be entirely unsatisfactory for use in the airborne guidance system of a missile. Accordingly, we shall define the **reliability** of a product as *the probability that it will function within specified limits for at least a specified period of time under specified environmental conditions.* Thus, the reliability of a "standard equipment" automobile tire is close to unity for 10,000 miles of normal operation on a passenger car, but it is virtually zero for use at the Indianapolis "500."

Since reliability has been defined as a probability, the theoretical treatment of this subject is based essentially on the material introduced in the early chapters of this book. Thus, the rules of probability introduced in Chapter 2 can be applied directly to the calculation of the reliability of a complex system, if the reliabilities of the individual components are known. (Estimates of the reliabilities of the individual components are usually obtained from statistical life tests, such as those discussed in Sections 15.4 and 15.5.)

Many systems can be considered to be series or parallel systems, or a combination of both. A **series system** is one in which all components are so interrelated that the entire system will fail if any one of its components fails; a **parallel system** is one that will fail only if all of its components fail.

Let us first discuss a system of n components connected in series, and let us suppose that the components are independent, namely, that the performance of any one part does not affect the reliability of the others. Under these conditions, the probability that the system will function is given by the special rule of multiplication for probabilities, and we have

Product law of reliabilities

$$R_s = \prod_{i=1}^{n} R_i$$

where R_i is the reliability of the ith component and R_s is the reliability of the series system. This simple **product law of reliabilities**, applicable to series

453

systems of independent components, vividly demonstrates the effect of increased complexity on reliability.

EXAMPLE A system consists of five independent components in series, each having a reliability of 0.970. What is the reliability of the system? What happens to the system reliability if its complexity is increased so that it contains 10 similar components?

> *Solution* The reliability of the five-component series system is
>
> $$(0.970)^5 = 0.859$$
>
> Increasing system complexity to 10 components will decrease the system reliability to
>
> $$(0.970)^{10} = 0.738$$
>
> Looking at the effect of increasing complexity in another way, we find that each of the components in the 10-component system would require a reliability of 0.985, instead of 0.970, for the 10-component system to have a reliability equal to that of the original 5-component system.

One way to increase the reliability of a system is to replace certain components by several similar components connected in parallel. If a system consists of n independent components connected in parallel, it will fail to function only if all n components fail. Thus, if $F_i = 1 - R_i$ is the "unreliability" of the ith component, we can again apply the special rule of multiplication for probabilities to obtain

$$F_p = \prod_{i=1}^{n} F_i$$

where F_p is the unreliability of the parallel system, and $R_p = 1 - F_p$ is the reliability of the parallel system. Thus, for parallel systems we have a **product law of unreliabilities** analogous to the product law of reliabilities for series systems. Writing this law in another way, we get

Product law of unreliabilities

$$R_p = 1 - \prod_{i=1}^{n} (1 - R_i)$$

for the reliability of a parallel system.

EXAMPLE The two basic formulas for the reliability of series and parallel systems can be used in combination to calculate the reliability of a system having both series and parallel parts. To illustrate such a calculation, consider the system

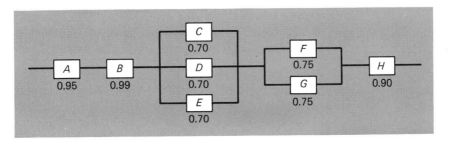

Figure 15.1 Systems reliability.

diagrammed in Figure 15.1, which consists of eight components having the reliabilities shown in that figure. Find the reliability of this system.

> **Solution**   The parallel assembly C, D, E can be replaced by an equivalent component C' having the reliability $1 - (1 - 0.70)^3 = 0.973$, without affecting the overall reliability of the system. Similarly, the parallel assembly F, G can be replaced by a single component F' having the reliability $1 - (1 - 0.75)^2 = 0.9375$. The resulting series system A, B, C', F', H, equivalent to the original system, has the reliability
>
> $$(0.95)(0.99)(0.973)(0.9375)(0.90) = 0.772.$$

15.2 FAILURE-TIME DISTRIBUTIONS

According to the definition of reliability given in the preceding section, the reliability of a system or a component will often depend on the length of time it has been in service. Thus, of fundamental importance in reliability studies is the **failure-time distribution**, that is, the distribution of the time to failure of a component under given environmental conditions. A useful way to characterize this distribution is by means of its associated **instantaneous failure rate**. To develop this concept, first let $f(t)$ be the probability density of the time to failure of a given component, that is, the probability that the component will fail between times t and $t + \Delta t$ is given by $f(t) \cdot \Delta t$. Then, the probability that the component will fail on the interval from 0 to t is given by

$$F(t) = \int_0^t f(x)\, dx$$

and the **reliability function**, expressing the probability that it survives to time t, is given by

$$R(t) = 1 - F(t)$$

Thus, the probability that the component will fail in the interval from t to $t + \Delta t$ is $F(t + \Delta t) - F(t)$, and the conditional probability of failure in this interval, *given that the component survived to time t*, is expressed by

$$\frac{F(t + \Delta t) - F(t)}{R(t)}$$

Dividing by Δt, we find that the average rate of failure in the interval from t to $t + \Delta t$, given that the component survived to time t, is

$$\frac{F(t + \Delta t) - F(t)}{\Delta t} \cdot \frac{1}{R(t)}$$

Taking the limit as $\Delta t \to 0$, we then get the instantaneous failure rate, or simply the **failure rate**

$$Z(t) = \frac{F'(t)}{R(t)}$$

where $F'(t)$ is the derivative of $F(t)$ with respect to t. Finally, observing that $f(t) = F'(t)$ (see page 102) we get the relation

General equation for failure-rate function

$$Z(t) = \frac{f(t)}{R(t)} = \frac{f(t)}{1 - F(t)}$$

which expresses the failure rate in terms of the failure-time distribution.

A failure-rate curve that is typical of many manufactured items is shown in Figure 15.2. The curve is conveniently divided into three parts. The first part is characterized by a decreasing failure rate and it represents the period

Figure 15.2 Typical failure-rate curve.

during which poorly manufactured items are weeded out. (It is common in the electronics industry to "burn in" components prior to actual use in order to eliminate any early failures.) The second part, which is often characterized by a constant failure rate, is normally regarded as the period of useful life during which only chance failures occur. The third part is characterized by an increasing failure rate, and it is the period during which components fail primarily because they are worn out. Note that the same general failure-rate curve is typical of human mortality, where the first part represents infant mortality, and the third part corresponds to old-age mortality.

Let us now derive an important relationship expressing the failure-time density in terms of the failure-rate function. Making use of the fact that $R(t) = 1 - F(t)$ and, hence, that $F'(t) = -R'(t)$, we can write

$$Z(t) = -\frac{R'(t)}{R(t)} = -\frac{d[\ln R(t)]}{dt}$$

Solving this differential equation for $R(t)$, we obtain

$$R(t) = e^{-\int_0^t Z(x)\,dx}$$

and, making use of the relation $f(t) = Z(t) \cdot R(t)$, we finally get

General equation for failure-time distribution

$$f(t) = Z(t) \cdot e^{-\int_0^t Z(x)\,dx}$$

As illustrated in Figure 15.2, it is often assumed that the failure rate is constant during the period of useful life of a component. Denoting this constant failure rate by α, where $\alpha > 0$, and substituting α for $Z(t)$ in the formula for $f(t)$, we obtain

$$f(t) = \alpha \cdot e^{-\alpha t} \qquad t > 0$$

Thus, we have an **exponential failure-time distribution** when it can be assumed that the failure rate is constant. For this reason, the assumption of constant failure rates is sometimes also called the "exponential assumption." Interpreting the time to failure as a waiting time, we can use the results of Section 4.7 to conclude that the occurrence of failures is a Poisson process, if a component which fails is immediately replaced with a new one having the same constant failure rate α. As we observed on page 120, the mean waiting time between successive failures is $1/\alpha$, or the reciprocal of the failure rate. Thus, the constant $1/\alpha$ is often referred to as the **mean time between failures**, and it is abbreviated **MTBF**.

There are situations in which the assumption of a constant failure rate is not realistic, and in many of these situations one assumes instead that the failure rate function increases or decreases "smoothly" with time. In other words, it is assumed that there are no discontinuities or turning points. This assumption would be consistent with the initial and last stages of the failure-rate curve shown in Figure 15.2.

A useful function that is often used to approximate such failure-rate curves is given by

$$Z(t) = \alpha \beta t^{\beta-1} \qquad t > 0$$

where α and β are positive constants. Note the generality of this function: If $\beta < 1$ the failure rate *decreases* with time, if $\beta > 1$ it *increases* with time, and if $\beta = 1$ the failure rate equals α. Note that the assumption of a constant failure rate, the exponential assumption, is thus included as a special case.

If we substitute the above expression for $Z(t)$ into the formula for $f(t)$ on page 457, we obtain

$$f(t) = \alpha \beta t^{\beta-1} e^{-\alpha t^\beta} \qquad t > 0$$

where α and β are positive constants. This density, or distribution, is the Weibull distribution, introduced in Section 4.9, and we shall discuss its application to problems of life testing in Section 15.5.

15.3   THE EXPONENTIAL MODEL IN RELIABILITY

If we make the exponential assumption about the distribution of failure times, some very useful results can be derived concerning the *MTBF*, the mean time between failure, of series and parallel systems. In order to use the product laws of Section 15.1, we shall first have to obtain a relation expressing the reliability of a component in terms of its service time t. Making use of the fact that

$$R(t) = 1 - F(t) = 1 - \int_0^t f(x)\, dx$$

we obtain

$$R(t) = 1 - \int_0^t \alpha e^{-\alpha x}\, dx = e^{-\alpha t}$$

for the reliability function of the exponential model. Thus, if a component has a failure rate of 0.05 per thousand hours, the probability that it will survive at least 10,000 hours of operation is $e^{-(0.05)10} = 0.607$.

Suppose now that a system consists of n components connected in *series*, and that these components have the respective failure rates $\alpha_1, \alpha_2, \ldots,$ and

α_n. The product law of reliabilities can be written as

Product law of reliabilities— exponential case

$$R_s(t) = \prod_{i=1}^{n} e^{-\alpha_i t} = e^{-\left(\sum_{i=1}^{n} \alpha_i\right)t}$$

and it can be seen that the reliability function of the series system also satisfies the exponential assumption. The failure rate of the entire series system is readily identified as $\sum_{i=1}^{n} \alpha_i$, the *sum* of the failure rates of its components. Since the *MTBF* is the reciprocal of the failure rate when each component which fails is replaced immediately with another having the identical failure rate, we obtain the formula

Mean time between failures—series system

$$\mu_s = \frac{1}{\dfrac{1}{\mu_1} + \dfrac{1}{\mu_2} + \ldots + \dfrac{1}{\mu_n}}$$

expressing the *MTBF* μ_s of a series system in terms of the *MTBF*'s μ_i of its components. In the special case where all n components have the same failure rate α and hence the same *MTBF* μ, the system failure rate is $n\alpha$, and the system *MTBF* is $1/n\alpha = \mu/n$.

For parallel systems the results are not quite so simple. If a system consists of n components in parallel, having the respective failure rates α_1, $\alpha_2, \ldots, \alpha_n$, the system "unreliability" to time t is given by

$$F_p(t) = \prod_{i=1}^{n} (1 - e^{-\alpha_i t})$$

Thus, the failure-time distribution of a parallel system is not exponential even when each of its components satisfies the exponential assumption. The system failure-rate function can be obtained by means of the formula $Z_p(t) = F'_p(t)/R_p(t)$, but the result is fairly complicated. Note, however, that the system failure rate is not constant, but depends on t, the "age" of the system.

The mean time to failure of a parallel system is also difficult to obtain in general, but in the special case where all components have the same failure rate α, an interesting and useful result can be obtained. In this special case the system reliability function becomes

$$R_p(t) = 1 - (1 - e^{-\alpha t})^n$$

$$= \binom{n}{1} e^{-\alpha t} - \binom{n}{2} e^{-2\alpha t} + \ldots + (-1)^{n-1} e^{-n\alpha t}$$

after using the binomial theorem to expand $(1 - e^{-\alpha t})^n$. Then, making use of the fact that $f_p(t) = -R'_p(t)$, we obtain

$$f_p(t) = \alpha \binom{n}{1} e^{-\alpha t} - 2\alpha \binom{n}{2} e^{-2\alpha t} + \ldots + (-1)^{n-1} n\alpha e^{-n\alpha t}$$

and the mean of the failure-time distribution is given by

$$\mu_p = \int_0^\infty t \cdot f_p(t) \, dt$$

$$= \alpha \binom{n}{1} \int_0^\infty t e^{-\alpha t} \, dt - 2\alpha \binom{n}{2} \int_0^\infty t e^{-2\alpha t} \, dt + \ldots$$

$$+ (-1)^{n-1} n\alpha \int_0^\infty t e^{-n\alpha t} \, dt$$

$$= \frac{1}{\alpha} \binom{n}{1} - \frac{1}{2\alpha} \binom{n}{2} + \ldots + (-1)^{n-1} \frac{1}{n\alpha}$$

It can be proved by induction that this expression is equivalent to

Mean time between failures— parallel system

$$\mu_p = \frac{1}{\alpha} \left(1 + \frac{1}{2} + \ldots + \frac{1}{n} \right)$$

Thus, if a parallel system consists of n components having the identical failure rate α, the mean time between failures of the system equals $\left(1 + \frac{1}{2} + \ldots + \frac{1}{n} \right)$ times the common *MTBF* of its components, provided each defective component is replaced whenever the whole parallel system fails. Thus, if we use two parallel components rather than one, the mean time to failure of the pair exceeds that of the single component by 50 percent, rather than doubling it. In general, the above formula for μ_p expresses a rather severe law of diminishing returns for parallel redundancy.

The formulas derived in this section can be used in system design. To illustrate, we again consider the system diagrammed in Figure 15.1 on page 455. Assuming the exponential model and that the reliabilities are given for 10 hours of operation, we can calculate the failure rate of component A by solving the equation $0.95 = e^{-10\alpha}$ for α, and we obtain $\alpha = 5.1 \cdot 10^{-3}$ failures per hour, or 5.1 failures per thousand hours. The failure rates of all eight components (in failures per thousand hours) are as shown in the following table:

Component	A	B	C	D	E	F	G	H
Failure rate	5.1	1.0	35.7	35.7	35.7	28.8	28.8	10.5

To compute the mean time to failure for the entire system, we first obtain the mean failure times for the parallel assemblies C, D, E, and F, G, respectively. For C, D, E we have $\mu_{CDE} = \frac{1}{35.7}\left(1 + \frac{1}{2} + \frac{1}{3}\right) = 0.051$ thousand hours, or 51 hours; for F, G we have $\mu_{FG} = \frac{1}{28.8}\left(1 + \frac{1}{2}\right) = 0.052$ thousand hours, or 52 hours. Although the two parallel assemblies do not have constant failure rates, we shall approximate their respective failure rates by $1/0.051 = 19.6$ and $1/0.052 = 19.2$ failures per thousand hours, and treat the entire system as a series system. Thus, the system failure rate is given approximately by $5.1 + 1.0 + 19.6 + 19.2 + 10.5 = 55.4$ failurs per thousand hours, and the mean time to failure of the system is approximately $1/55.4 = 0.018$ or 18 hours.

EXERCISES

15.1 An old-fashioned string of Christmas-tree lights has eight bulbs connected in series. What would have to be the reliability of each bulb if there is to be a 95% chance of the string's lighting after a year's storage?

15.2 A system consists of seven identical components connected in parallel. What must be the reliability of each component if the overall reliability of the system is to be 0.90?

15.3 A system consists of six components connected as in Figure 15.3. Find the overall reliability of the system, given that the reliabilities of A, B, C, D, E, and F are, respectively, 0.95, 0.80, 0.90, 0.99, 0.90, and 0.85.

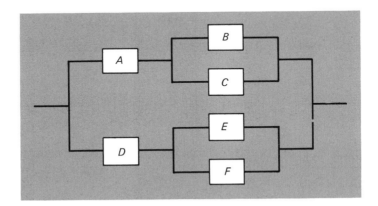

Figure 15.3 Exercise 15.3.

15.4 Suppose that the flight of an aircraft is regarded as a system having the three main components A (aircraft), B (pilot), and C (airport). Suppose, furthermore, that component B can be regarded as a parallel subsystem consisting of B_1 (captain), B_2 (first officer), and B_3 (flight engineer); and C is a parallel subsystem consisting of C_1 (scheduled airport) and C_2 (alternate airport). Under given flight conditions, the reliabilities of components A, B_1, B_2, B_3, C_1, and C_2 (defined as the probabilities that they can contribute to the successful completion of the scheduled flight) are, respectively, 0.9999, 0.9995, 0.999, 0.20, 0.95, and 0.85.

(a) What is the reliability of the system?

(b) What is the effect on system reliability of having a flight engineer who is also a trained pilot, so that the reliability of B_3 is increased from 0.20 to 0.99?

(c) If the flight crew did not have a first officer, what then would be the effect of increasing the reliability of B_3 from 0.20 to 0.99?

(d) What is the effect of adding a second alternate landing point, C_3, with reliability 0.80?

15.5 In some reliability problems we are concerned only with initial failures, treating a component as if (for all practical purposes) it never fails, once it has survived past a certain time $t = \alpha$. In a problem like this, it may be reasonable to use the failure rate

$$Z(t) = \begin{cases} \beta\left(1 - \dfrac{t}{\alpha}\right) & \text{for } 0 < t < \alpha \\ 0 & \text{elsewhere} \end{cases}$$

(a) Find expressions for $f(t)$ and $F(t)$.

(b) Show that the probability of an initial failure is given by

$$1 - e^{-\alpha\beta/2}$$

15.6 As has been indicated in the text, one often distinguishes between initial failures, random failures during the useful life of the product, and wear-out failures. Thus, suppose that for a given product the probability of an initial failure (a failure prior to time $t = \alpha$) is θ_1, the probability of a wear-out failure (a failure beyond time $t = \beta$) is θ_2, and that for the interval $\alpha \leq t \leq \beta$ the failure-time density is given by

$$f(t) = \frac{1 - \theta_1 - \theta_2}{\beta - \alpha}$$

(a) Find an expression for $F(t)$ for the interval $\alpha \leq t \leq \beta$.

(b) Show that for the interval $\alpha \leq t \leq \beta$ the failure rate is given by

$$Z(t) = \frac{1 - \theta_1 - \theta_2}{(\beta - \alpha)(1 - \theta_1) - (1 - \theta_1 - \theta_2)(t - \alpha)}$$

(c) Suppose that the failure of a color television set is considered to be an initial failure if it occurs during the first 100 hours of usage, and a wear-out failure if it occurs after 15,000 hours. Assuming that the model given in this exercise holds and that θ_1 and θ_2 equal 0.05 and 0.75, respectively, sketch the graph of the failure-rate function from $t = 100$ to $t = 15,000$ hours.

15.7 An integrated-circuit chip has a constant failure rate of 0.02 per thousand hours.
   (a) What is the probability that it will operate satisfactorily for at least 20,000 hours?
   (b) What is the 5,000-hour reliability of a component consisting of four such chips connected in series?

15.8 A certain component has an exponential life distribution with a failure rate if $\alpha = 0.0045$ failures per hour.
   (a) What is the probability that the component will fail during the first 250 hours it is in operation?
   (b) What is the probability that two such components will both survive the first 100 hours of operation?

15.9 A system containing several identical components in parallel is to have a failure rate of at most $5 \cdot 10^{-5}$ per hour. What is the least number of components that must be used if each has a constant failure rate of $2.0 \cdot 10^{-4}$ per hour?

15.10 A system consists of six different components connected in series. Find the *MTBF* of the system if the six components have exponential time-to-failure distributions with failure rates of 1.8, 2.4, 2.0, 1.3, 3.0, and 1.5 per 1,000 hours, respectively.

15.11 A certain part has an exponential life distribution with a mean life, that is, an *MTBF*, of 1,000 hours.
   (a) What is the probability that such a part will last at least 500 hours?
   (b) What is the probability that among three such parts at least one will fail during the first 1,000 hours?
   (c) What is the probability that among four such parts exactly two will fail during the first 600 hours?

15.12 If a component has the Weibull failure-time distribution with the parameters $\alpha = 0.005$ per hour and $\beta = 0.80$, find the probability that it will operate successfully for at least 5,000 hours.

15.13 Throughout Sections 15.2 and 15.3 we assumed that the products with which we were concerned were in continuous operation. Consequently, the models discussed in these sections do not apply if we want to investigate the ability of light bulbs to withstand successive voltage overloads, the performance of switches which are repeatedly turned on and off, etc. In each of these cases failure can occur on the xth trial ($x = 1, 2, 3, \ldots$), and it is often assumed that the probability of failure on the xth trial equals some constant p, provided the item has not failed prior to that trial.
   (a) Show that the probability of failure on the xth trial is given by

$$f(x) = p(1 - p)^{x-1}$$

   for $x = 1, 2, 3, \ldots$. This probability distribution is the *geometric distribution* introduced in Section 3.8.

(b) Find $F(x)$ for the probability distribution obtained in part (a).

(c) What is the probability that a switch will survive 2,000 cycles of operation, if the above model holds and the constant probability of failure as the result of any one of the switching cycles is $p = 6 \times 10^{-4}$?

15.4 THE EXPONENTIAL MODEL IN LIFE TESTING

An effective and widely used method of handling problems of reliability is that of **life testing**. For the purpose of such tests, a random sample of n components is selected from a lot, put on test under specified environmental conditions, and the times to failure of the individual components are observed. If each component that fails is immediately replaced by a new one, the resulting life test is called a **replacement test**; otherwise, the life test is called a **nonreplacement test**. Whenever the mean lifetime of the components is so large that it is not practical, or economically feasible, to test each component to failure, the life test may be **truncated**, that is, it may be terminated after the first r failures have occurred ($r \leq n$), or after a fixed period of time has elapsed.

A special method that is often used when early results are required in connection with very-high-reliability components is that of **accelerated life testing**. In an accelerated life test the components are put on test under environmental conditions that are far more severe than those normally encountered in practice. This causes the components to fail more quickly, and it can drastically reduce both the time required for the test and the number of components that must be placed on test. Accelerated life testing can be used to compare two or more types of components for the purpose of obtaining a rapid assessment of which is the most reliable. Sometimes, preliminary experimentation is carried out to determine the relationship between the proportion of failures that can be expected under nominal conditions and under various levels of accelerated environmental conditions. The methods of Sections 11.4 and 13.2 can be applied in this connection to determine "derating curves," relating the reliability of the component to the severity of the environmental conditions under which it is to operate.

In the remainder of this section we shall assume that the exponential model holds, namely, that the failure-time distribution of each component is given by

$$f(t) = \alpha \cdot e^{-\alpha t} \qquad t > 0, \alpha > 0$$

In what follows, we shall assume that n components are put on test, life testing is discontinued after a fixed number, $r(r \leq n)$, of components have failed, and that the observed failure times are $t_1 \leq t_2 \leq \ldots \leq t_r$. We shall be concerned with estimating and testing hypotheses about the mean life of the component, namely, $\mu = 1/\alpha$.

Using theory developed in the article by B. Epstein mentioned in the Bibliography, it can be shown that unbiased estimates of the mean life of the component are given by

Estimate of mean life

$$\hat{\mu} = \frac{T_r}{r}$$

where T_r is the accumulated life on test until the rth failure occurs, and hence

Accumulated life to r failures— nonreplacement test

$$T_r = \sum_{i=1}^{r} t_i + (n - r)t_r$$

for nonreplacement tests and

Accumulated life to r failures— replacement test

$$T_r = nt_r$$

if the test is with replacement. Note that if the test is without replacement and $r = n$, $\hat{\mu}$ is simply the mean of the observed times to failure.

To make inferences concerning the mean life μ of the component, we use the fact that $2T_r/\mu$ is a value of a random variable having the chi-square distribution with $2r$ degrees of freedom (see reference to B. Epstein in the Bibliography). With the appropriate expression substituted for T_r, this is true regardless of whether the test is conducted with or without replacement. Thus, in either case a two-sided $(1 - \alpha)100\%$ confidence interval for μ is given by

Confidence interval for mean life

$$\frac{2T_r}{\chi_2^2} < \mu < \frac{2T_r}{\chi_1^2}$$

where χ_1^2 and χ_2^2 cut off left- and right-hand tails of area $\alpha/2$ under the chi-square distribution with $2r$ degrees of freedom. (See Exercise 15.19 on page 473.)

Tests of the null hypothesis that $\mu = \mu_0$ can also be based on the sampling distribution of $2T_r/\mu$, using the appropriate expression for T_r depending on whether the test is with or without replacement. Thus, if the alternative hypothesis is $\mu > \mu_0$, we reject the null hypothesis at the level of significance α when $2T_r/\mu_0$ exceeds χ_α^2, or

Critical region for testing $H_0: \mu = \mu_0$

$$T_r > \tfrac{1}{2}\mu_0\chi_\alpha^2$$

where χ_α^2, to be determined for $2r$ degrees of freedom, is as defined on page 181. In Exercises 15.15 and 15.17 on page 472 the reader will be asked to construct and perform similar tests corresponding to the alternative hypotheses $\mu < \mu_0$ and $\mu \neq \mu_0$.

An alternate life-testing procedure consists of discontinuing the test after a fixed accumulated amount of lifetime T has elapsed, and treating the observed number of failures k as the value of a random variable. In the important special case where n items are tested with replacement for a length of time t^*, we have $T = nt^*$. Regardless of whether the test is with or without replacement, an *approximate* $1 - \alpha$ confidence interval for the mean life of the component is given by

$$\frac{2T}{\chi_4^2} < \mu < \frac{2T}{\chi_3^2}$$

Here χ_4^2 cuts off a right-hand tail of area $\alpha/2$ under the chi-square distribution with $2k + 2$ degrees of freedom, while χ_3^2 cuts off a left-hand tail of area $\alpha/2$ under the chi-square distribution with $2k$ degrees of freedom.

EXAMPLE Suppose that 50 units are placed on life test (without replacement) and that the test is to be truncated after $r = 10$ of them have failed. We shall suppose, furthermore, that the first 10 failure times are 65, 110, 380, 420, 505, 580, 650, 840, 910, and 950 hours. Estimate the mean life of the component, its failure rate, and calculate a 0.90 confidence interval for μ.

Solution Since $n = 50$, $r = 10$,

$$T_{10} = (65 + 110 + \ldots + 950) + (50 - 10)950$$
$$= 43{,}410 \text{ hours}$$

and we estimate the mean life of the component as $\hat{\mu} = \dfrac{43{,}410}{10} = 4{,}341$ hours. The failure rate α is estimated by $1/\hat{\mu} = 0.00023$ failure per hour, or 0.23 failure per thousand hours. Also, a 0.90 confidence interval for μ is given by

$$\frac{2(43{,}410)}{31.410} < \mu < \frac{2(43{,}410)}{10.851}$$

or

$$2{,}764 < \mu < 8{,}001$$

EXAMPLE Using the data of the preceding example, test whether the failure rate is 0.40 failure per thousand hours against the alternative that the failure rate is less. Use the 0.05 level of significance.

> **Solution**
>
> 1. *Null hypothesis*: $\mu = \dfrac{1{,}000}{0.40} = 2{,}500$ hours
>
>   *Alternative hypothesis*: $\mu > 2{,}500$ hours
> 2. *Level of significance*: $\alpha = 0.05$
> 3. *Criterion*: Reject the null hypothesis if $T_r > \frac{1}{2}\mu_0\chi^2_{0.05}$ where $\chi^2_{0.05} = 31.410$ is the value of $\chi^2_{0.05}$ having 20 degrees of freedom.
> 4. *Calculations*: Substituting $r = 10$ and $\mu_0 = 2{,}500$, we find the critical value for this test to be
>
> $$\frac{1}{2}\mu_0\chi^2_{0.05} = \frac{1}{2}(2{,}500)(31.410) = 39{,}263$$
>
> 5. *Decision*: Since $T_{10} = 43{,}410$ exceeds the critical value, we must reject the null hypothesis, concluding that the mean life-time exceeds 2,500 hours, or, equivalently, that the failure rate is less than 0.40 failure per thousand hours.

15.5 THE WEIBULL MODEL IN LIFE TESTING

Although life testing of components during the period of useful life is generally based on the exponential model, we have already pointed out that the failure rate of a component may not be constant throughout a period under investigation. In some instances the period of initial failure may be so long that the component's main use is during this period, and in other instances the main purpose of life testing may be that of determining the time to wear-out failure rather than chance failure. In such cases the exponential model generally does not apply, and it is necessary to substitute a more general assumption for that of a constant failure rate.

As we observed on page 458, the Weibull distribution adequately describes the failure times of components when their failure rate either increases or decreases with time. It has the parameters α and β, its formula is given by

***Weibull
distribution***

$$f(t) = \alpha\beta t^{\beta-1}e^{-\alpha t^\beta} \qquad t > 0,\ \alpha > 0,\ \beta > 0$$

and it follows (see Exercise 15.24 on page 473) that the reliability function associated with the **Weibull failure-time distribution** is given by

Weibull reliability function

$$R(t) = e^{-\alpha t^\beta}$$

We already showed on page 458 that the failure rate leading to the Weibull distribution is given by

Weibull failure-rate function

$$Z(t) = \alpha \beta t^{\beta - 1}$$

The range of shapes a graph of the Weibull density can take on is very broad, depending primarily on the value of the parameter β. As illustrated in Figure 15.4, the Weibull curve is asymptotic to both axes and highly skewed to the right for values of β less than 1; it is identical to that of the exponential density for $\beta = 1$, and it is "bell-shaped" but skewed for values of β greater than 1.

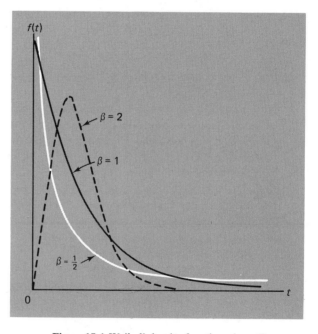

Figure 15.4 Weibull density functions ($\alpha = 1$).

The mean of the Weibull distribution having the parameters α and β may be obtained by evaluating the integral

$$\mu = \int_0^\infty t \cdot \alpha \beta t^{\beta-1} e^{-\alpha t^\beta} \, dt$$

Making the change of variable $u = \alpha t^\beta$, we get

$$\mu = \alpha^{-1/\beta} \int_0^\infty u^{1/\beta} e^{-u} \, du$$

Recognizing the integral as $\Gamma\left(1 + \dfrac{1}{\beta}\right)$, we find that the mean time to failure for the Weibull model is

Mean time to failure— Weibull model

$$\mu = \alpha^{-1/\beta} \Gamma\left(1 + \frac{1}{\beta}\right)$$

The reader will be asked to show in Exercise 15.25 on page 473 that the variance of this distribution is given by

Variance of Weibull model

$$\sigma^2 = \alpha^{-2/\beta} \left\{ \Gamma\left(1 + \frac{2}{\beta}\right) - \left[\Gamma\left(1 + \frac{1}{\beta}\right)\right]^2 \right\}$$

Estimates of the parameters α and β of the Weibull distribution are somewhat difficult to obtain. Although there exist analytical methods for estimating these parameters, they involve the solution of a system of transcendental equations, and they will not be presented here. Instead, a more rapid and commonly used method, based on a graphical technique, will be described. This method is based on the fact that the reliability function of the Weibull distribution can be transformed into a linear function of $\ln t$ by means of a double-logarithmic transformation. Taking the natural logarithm of $R(t)$, we obtain

$$\ln R(t) = -\alpha t^\beta \quad \text{or} \quad \ln \frac{1}{R(t)} = \alpha t^\beta$$

Again taking logarithms, we have

$$\ln \ln \frac{1}{R(t)} = \ln \alpha + \beta \cdot \ln t$$

and it can be seen that the right-hand side is linear in ln t.

To estimate α and β we require estimates of $R(t)$ for various values of t, and the usual procedure is to place n units on life test and observe their failure times. If the ith unit fails at time t_i, we estimate $F(t_i) = 1 - R(t_i)$ by the same method as on page 145, namely, we use the estimator

$$\widehat{F(t_i)} = \frac{i - \frac{1}{2}}{n}$$

Before going any further, it is customary to check whether it is actually reasonable to use the Weibull model. To this end, we plot points having the coordinates t_i and $\widehat{F(t_i)}$ on special graph paper having its scales transformed so that the divisions on the horizontal axis are proportional to ln t and those on the vertical scale are proportional to $\ln \ln \dfrac{1}{1 - F(t)}$. If the points fall reasonably close to a straight line, it can be assumed that the underlying failure-time distribution is of the Weibull type. The parameters α and β of this distribution can then be estimated by applying the least-squares method of Chapter 11 to fit a straight line to the transformed data.

As an illustration of this procedure, suppose that a sample of 100 components is put on life test for 500 hours and that the times to failure of the 12 components that failed during the test are as follows: 6, 21, 50, 84, 95, 130, 205, 260, 270, 370, 440, and 480 hours. The points $\left(t_i, \dfrac{i - \frac{1}{2}}{n} \right)$ are plotted in Figure 15.5 on the transformed scales previously described, and it can be seen that they fall fairly close to a straight line.

Figure 15.5 Weibull failure-time distribution.

The parameters α and β are estimated by applying the method of least squares to the transformed points (x_i, y_i), where

$$x_i = \ln t_i$$

$$y_i = \ln \ln \frac{1}{1 - \widehat{F(t_i)}}$$

Thus, in this illustration, we obtain

$\widehat{F(t_i)}$	t_i	x_i	y_i
0.005	6	1.79	−5.30
0.015	21	3.04	−4.20
0.025	50	3.91	−3.68
0.035	84	4.43	−3.33
0.045	95	4.55	−3.08
0.055	130	4.87	−2.87
0.065	205	5.32	−2.70
0.075	260	5.56	−2.55
0.085	270	5.60	−2.42
0.095	370	5.91	−2.30
0.105	440	6.09	−2.20
0.115	480	6.17	−2.10

and, by the methods of Section 11.1, the regression line becomes

$$y = -6.44 + 0.71x$$

Note that in calculating the values of y_i for the above table, it is convenient to use the approximation $\ln \ln \dfrac{1}{1-z} \simeq \ln z$ (see Exercise 15.26 on page 473) for small values of $\widehat{F(t_i)}$.

Thus, the parameter β of the underlying Weibull distribution is estimated as $\hat{\beta} = 0.71$, and α is estimated as $\hat{\alpha} = e^{-6.44} = 0.0016$. It follows that the mean time to failure is estimated as

$$\hat{\mu} = (0.0016)^{-1/0.71}\Gamma\left(1 + \frac{1}{0.71}\right)$$

which equals approximately 11,000 hours. Also, values of the failure-rate function may be obtained by substituting for t into

$$\widehat{Z(t)} = (0.0016)(0.71)t^{-0.29}$$

Since $\hat{\beta} < 1$ the failure rate is decreasing with time. After 1 hour ($t = 1$), units are failing at the rate of $(0.0016)(0.71) = 0.00114$ unit per hour, and after 1,000 hours the failure rate has decreased to $(0.00114)(1000)^{-0.29} = 0.00015$ unit per hour.

EXERCISES

15.14 Suppose that 50 units are put on life test, each unit that fails is immediately replaced, and the test is discontinued after 8 units have failed. If the eighth failure occurred at 760 hours,
  (a) construct a 95% confidence interval for the mean life of such units;
  (b) test at the 0.05 level of significance whether or not the mean life is less than 10,000 hours.

15.15 In a life test with replacement, 35 space heaters were put into continuous operation, and the first five failures occurred after 250, 380, 610, 980, and 1,250 hours.
  (a) Assuming the exponential model, construct a 99% confidence interval for the mean life of this kind of space heater.
  (b) To check the manufacturer's claim that the mean life of these heaters is at least 5,000 hours, test the null hypothesis $\mu = 5,000$ against an appropriate alternative, so that the burden of proof is put on the manufacturer. Use $\alpha = 0.05$.

15.16 Fifteen assemblies are put on accelerated life test without replacement, and the test is truncated after four failures. If the first four failures occurred at 16.5, 19.2, 20.8, and 37.3 hours,
  (a) find a 90% confidence interval for the failure rate of such assemblies under these accelerated conditions;
  (b) test the null hypothesis that the failure rate is 0.004 failure per hour against the alternative that it is less than 0.004, using the 0.01 level of significance.

15.17 To investigate the average time to failure of a certain weld subjected to continuous vibration, seven welded pieces were subjected to specified frequencies and amplitudes of vibration and their times to failure were 211, 350, 384, 510, 539, 620, and 715 thousand cycles.
  (a) Assuming the exponential model, construct a 95% confidence interval for the mean life (in thousands of cycles) of such a weld under the given vibration conditions.
  (b) Assuming the exponential model, test the null hypothesis that the mean life of the weld under the given vibration conditions is 500,000 cycles against the two-sided alternative $\mu \neq 500,000$. Use the level of significance 0.10.

15.18 In life testing we are sometimes interested in establishing tolerance limits for the life of a component (see Section 14.5); in particular, we may be interested in a one-sided tolerance limit t^*, for which we can assert with a $(1 - \alpha)100\%$ confidence that at least $100 \cdot P$ percent of the components have a life exceeding t^*. Using the exponential model, it can be shown that a good approximation is given by

$$t^* = \frac{-2T_r(\ln P)}{\chi_\alpha^2}$$

where T_r is as defined on page 465 and the value of χ_α^2 is to be obtained from Table 5 with $2r$ degrees of freedom.

(a) Using the data of Exercise 15.15, establish a lower tolerance limit for which one can assert with 95% confidence that it is exceeded by at least 80% of the lifetimes of the heaters.

(b) Using the data of Exercise 15.17, establish a lower tolerance limit for which one can assert with 99% confidence that it is exceeded by at least 90% of the lifetimes of the given welds.

15.19 Using the fact that $2T_r/\mu$ is a value of a random variable having the chi-square distribution with $2r$ degrees of freedom, derive the confidence interval for μ given on page 465.

15.20 One hundred devices are put on life test and the times to failure (in hours) of the first 10 that fail are as follows: 7.0, 14.1, 18.9, 31.6, 52.8, 80.0, 164.5, 355.4, 451.0, and 795.1. Assuming a Weibull failure-time distribution, estimate the parameters α and β as well as the failure rate at 1,000 hours. How does this value of the failure rate compare with the value we would obtain if we assumed the exponential model?

15.21 A sample of 300 high-reliability capacitors was placed on life test for 2,000 hours and there were no failures; the test was then terminated. Find a 0.95 *lower* confidence limit for the mean life of the capacitors.

15.22 To investigate the performance of a logic circuit for a small electronic calculator, a laboratory puts 75 of the circuits on life test (without replacement) under specified environmental conditions, and the first 10 failures are observed after 28, 46, 50, 63, 81, 101, 116, 137, 159, and 175 hours. Using the Weibull model, estimate the mean life of such a circuit. How does this value compare with the mean life that would have been obtained under the exponential assumption?

15.23 Using the estimates of the parameters of the Weibull model obtained in Exercise 15.22, estimate the probability that this kind of circuit will perform satisfactorily for at least 100 hours.

15.24 Show that the reliability function associated with the Weibull failure-time distribution is given by

$$R(t) = e^{-\alpha t^\beta}$$

15.25 Derive the formula for the variance of the Weibull distribution given on page 469.

15.26 Show that $\ln \ln \dfrac{1}{1-z}$ can be approximated by $\ln z$ for small values of z. [*Hint*: Note that $\dfrac{1}{1-z} = 1 + z + z^2 + z^3 \dots$ for $|z| < 1$, and then use the Maclaurin's series for $\ln(1+x)$.]

15.6 *CHECK LIST OF KEY TERMS* (*with page references*)

Accelerated life testing, 464

Exponential failure-time distribution, 457

Failure rate, 456

Failure-rate function, 456

Failure-time distribution, 455

Instantaneous failure rate, 455

Life testing, 464

Mean time between failures (MTBF), *457*

Nonreplacement test, 464

Parallel system, 453

Product law of reliabilities, 453

Product law of unreliabilities, 454

Reliability, 453

Reliability function, 455

Replacement test, 464

Series system, 453

Truncated test, 464

Weibull failure-time distribution, 468

Bibliography

1. Engineering Statistics

Bowker, A. H., and Lieberman, G. J., *Engineering Statistics*, 2nd ed. Englewood Cliffs, N.J.: Prentice-Hall, Inc., 1972.

Brownlee, K. A., *Statistical Theory and Methodology in Science and Engineering*, 2nd ed. New York: John Wiley & Sons, Inc., 1965.

Gibra, I. N., *Probability and Statistical Inference for Scientists and Engineers*. Englewood Cliffs, N.J.: Prentice-Hall, Inc., 1973.

Guttman, I., Wilks, S. S., and Huntor, J. S., *Introductory Engineering Statistics*, 2nd ed. New York: John Wiley & Sons, Inc., 1971.

Hald, A., *Statistical Theory with Engineering Applications*. New York: John Wiley & Sons, Inc., 1952.

Johnson, N. L., and Leone, F. C., *Statistics and Experimental Design in Engineering and the Physical Sciences*, 2nd ed., Vols. I and II. New York: John Wiley & Sons, Inc., 1977.

Kirkpatrick, E. G., *Introductory Statistics and Probability for Engineering, Science, and Technology*. Englewood Cliffs, N.J.: Prentice-Hall, Inc., 1974.

Volk, W., *Engineering Statistics with a Programmable Calculator*. New York: McGraw-Hill Book Company, 1982.

Walpole, R. E., and Myers, R. H., *Probability and Statistics for Engineers and Scientists*, 2nd ed. New York: Macmillan Publishing Company, 1978.

2. Theoretical Statistics

Brunk, H. D., *An Introduction to Mathematical Statistics*, 3rd ed. New York: John Wiley & Sons, Inc., 1975.

Freund, J. E., and Walpole, R., *Mathematical Statistics*, 3rd ed. Englewood Cliffs, N.J.: Prentice-Hall, Inc., 1980.

Fry, T. C., *Probability and Its Engineering Uses*, 2nd ed. Princeton, N.J.: D. Van Nostrand Company, 1965.

Hoel, P., *Introduction to Mathematical Statistics*, 4th ed. New York: John Wiley & Sons, Inc., 1971.

Hogg, R. V., and Craig, A. T., *Introduction to Mathematical Statistics*, 4th ed. New York: Macmillan Publishing Company, 1978.

Kendall, M. G., and Stuart, A., *The Advanced Theory of Statistics*, Vol. 1, 4th ed., Vol. 2, 4th ed., Vol. 3, 3rd ed. New York: Hafner Press, 1977, 1979, and 1976.

Mann, H. B., *Analysis and Design of Experiments*. New York: Dover Press, 1949.

Mood, A. M., Graybill, F. A., and Boes, D. C., *Introduction to the Theory of Statistics*, 3rd ed. New York: McGraw-Hill Book Company, 1973.

Wilks, S. S., *Mathematical Statistics*. New York: John Wiley & Sons, Inc., 1962.

3. Experimental Design and Analysis of Variance

Cochran, W. G., and Cox, G. M., *Experimental Designs*, 2nd ed. New York: John Wiley & Sons, Inc., 1957.

Davies, O. L., Ed., *The Design and Analysis of Industrial Experiments*, 2nd rev. ed. New York: Hafner Press, 1967.

Federer, W. T. *Experimental Designs, Theory and Application*. New York: Macmillan Publishing Company, 1955.

Fisher, R. A., *The Design of Experiments*, 9th ed. New York: Hafner Press, 1971.

Hicks, C. R., *Fundamental Concepts in the Design of Experiments*, 2nd ed. New York: Holt, Rinehart and Winston, 1973.

Lipson, C., and Sheth, N.J., *Statistical Design and Analysis of Engineering Experiments*. New York: McGraw-Hill Book Company, 1973.

Scheffe, H., *The Analysis of Variance*. New York: John Wiley & Sons, Inc., 1959.

Snedecor, G. W., and Cochran, W. G., *Statistical Methods*, 6th ed. Ames, Iowa: Iowa University Press, 1967.

4. Special Topics

Barlow, R. E., and Proschan, F., *Mathematical Theory of Reliability*. New York: John Wiley & Sons, Inc., 1965.

Barlow, R. E., and Proschan, F., *Statistical Theory of Reliability and Life Testing: Probability Models*. New York: Holt, Rinehart and Winston, 1975.

Burr, I. W., *Engineering Statistics and Quality Control*. New York: McGraw-Hill Book Company, 1953.

Cowden, D. J., *Statistical Methods in Quality Control*. Englewood Cliffs, N.J.: Prentice-Hall, Inc., 1957.

Draper, N. R., and Smith, H. Jr., *Applied Regression Analysis*, 2nd ed. New York: John Wiley & Sons, Inc., 1981.

Duncan, A. J., *Quality Control and Industrial Statistics*, 4th ed. Homewood, Ill.: Richard D. Irwin, Inc., 1974.

Epstein, B., "Statistical Life Test Acceptance Procedures," *Technometrics*, Vol. 2, November, 1960.

Grant, E. L., and Leavenworth, R., *Statistical Quality Control*, 4th ed. New York: McGraw-Hill Book Company, 1972.

Kalbfleisch, J. D., and Prentice, R. L., *The Statistical Analysis of Failure Time Data*. New York: John Wiley & Sons, Inc., 1980.

Lawless, J. F., *Statistical Models and Methods for Lifetime Data*. New York: John Wiley & Sons, Inc., 1981.

Miller, R. G., Jr., *Survival Analysis*. New York: John Wiley & Sons, Inc., 1981.

Montgomery, D., and Peck, E., *Introduction to Linear Regression Analysis*. New York: John Wiley & Sons, Inc., 1982.

Nelson, W., *Applied Life Data Analysis*. New York: John Wiley & Sons, Inc., 1981.

Noether, G. E., *Elements of Nonparametric Statistics*. New York: John Wiley & Sons, Inc., 1967.

Wald, A., *Sequential Analysis*. New York: John Wiley & Sons, Inc., 1947.

Zelen, M., *Statistical Theory of Reliability*. Madison, Wis.: University of Wisconsin Press, 1963.

5. General Reference Works and Tables

Kendall, M. G., and Buckland, W. R., *A Dictionary of Statistical Terms*, 3rd ed. New York: Hafner Press, 1971.

Military Standard 105D. Washington, D.C.: U.S. Government Printing Office, 1963.

National Bureau of Standards Handbook 91: Experimental Statistics. Washington, D.C.: U.S. Government Printing Office, 1963.

National Bureau of Standards: Tables of the Binomial Distribution. Washington, D.C.: U.S. Government Printing Office, 1950.

Owen, D. B., *Handbook of Statistical Tables*. Reading, Mass.: Addison-Wesley Publishing Co., Inc., 1962.

Pearson, E. S., and Hartley, H. O., *Biometrika Tables for Statisticians*, 3rd ed. Cambridge: Cambridge University Press, 1966.

RAND Corporation, *A Million Random Digits with 100,000 Normal Deviates*. New York: Macmillan Publishing Company, third printing 1966.

Romig, H. G., *50–100 Binomial Tables*. New York: John Wiley & Sons, Inc., 1953.

Statistical Tables

TABLE 1

Binomial Distribution Function

$$B(x;n,p) = \sum_{k=0}^{x} \binom{n}{k} p^k (1-p)^{n-k}$$

n	x	0.05	0.10	0.15	0.20	0.25	0.30	0.35	0.40	0.45	0.50
2	2	0.9025	0.8100	0.7225	0.6400	0.5625	0.4900	0.4225	0.3600	0.3025	0.2500
	1	0.9975	0.9900	0.9775	0.9600	0.9375	0.9100	0.8755	0.8400	0.7975	0.7500
3	0	0.8574	0.7290	0.6141	0.5120	0.4219	0.3430	0.2746	0.2160	0.1664	0.1250
	1	0.9928	0.9720	0.9392	0.8960	0.8438	0.7840	0.7182	0.6480	0.5748	0.5000
	2	0.9999	0.9990	0.9966	0.9920	0.9844	0.9730	0.9571	0.9360	0.9089	0.8750
4	0	0.8145	0.6561	0.5220	0.4096	0.3164	0.2401	0.1785	0.1296	0.0915	0.0625
	1	0.9860	0.9477	0.8905	0.8192	0.7383	0.6517	0.5630	0.4752	0.3910	0.3125
	2	0.9995	0.9963	0.9880	0.9728	0.9492	0.9163	0.8735	0.8208	0.7585	0.6875
	3	1.0000	0.9999	0.9995	0.9984	0.9961	0.9919	0.9850	0.9744	0.9590	0.9375
5	0	0.7738	0.5905	0.4437	0.3277	0.2373	0.1681	0.1160	0.0778	0.0503	0.0312
	1	0.9774	0.9185	0.8352	0.7373	0.6328	0.5282	0.4284	0.3370	0.2562	0.1875
	2	0.9988	0.9914	0.9734	0.9421	0.8965	0.8369	0.7648	0.6826	0.5931	0.5000
	3	1.0000	0.9995	0.0078	0.9933	0.9844	0.9692	0.9460	0.9130	0.8688	0.8125
	4	1.0000	1.0000	0.9999	0.9997	0.9990	0.9976	0.9947	0.9898	0.9815	0.9688
6	0	0.7351	0.5314	0.3771	0.2621	0.1780	0.1176	0.0754	0.0467	0.0277	0.0156
	1	0.9672	0.8857	0.7765	0.6554	0.5339	0.4202	0.3191	0.2333	0.1636	0.1094
	2	0.9978	0.9842	0.9527	0.9011	0.8306	0.7443	0.6471	0.5443	0.4415	0.3438
	3	0.9999	0.9987	0.9941	0.9830	0.9624	0.9295	0.8826	0.8208	0.7447	0.6562
	4	1.0000	0.9999	0.9996	0.9984	0.9954	0.9891	0.9777	0.9590	0.9308	0.8906
	5	1.0000	1.0000	1.0000	0.9999	0.9998	0.9993	0.9982	0.9959	0.9917	0.9844
7	0	0.6983	0.4783	0.3206	0.2097	0.1335	0.0824	0.0490	0.0280	0.0152	0.0078
	1	0.9556	0.8503	0.7166	0.5767	0.4449	0.3294	0.2338	0.1586	0.1024	0.0625
	2	0.9962	0.9743	0.9262	0.8520	0.7564	0.6471	0.5323	0.4199	0.3164	0.2266
	3	0.9998	0.9973	0.9879	0.9667	0.9294	0.8740	0.8002	0.7102	0.6083	0.5000
	4	1.0000	0.9998	0.9988	0.9953	0.9871	0.9712	0.9444	0.9037	0.8471	0.7734
	5	1.0000	1.0000	0.9999	0.9996	0.9987	0.9962	0.9910	0.9812	0.9643	0.9375
	6	1.0000	1.0000	1.0000	1.0000	0.9999	0.9998	0.9994	0.9984	0.9963	0.9922
8	0	0.6634	0.4305	0.2725	0.1678	0.1001	0.0576	0.0319	0.0168	0.0084	0.0039
	1	0.9428	0.8131	0.6572	0.5033	0.3671	0.2553	0.1691	0.1064	0.0632	0.0352
	2	0.9942	0.9619	0.8948	0.7969	0.6785	0.5518	0.4278	0.3154	0.2201	0.1445
	3	0.9996	0.9950	0.9786	0.9437	0.8862	0.8059	0.7064	0.5941	0.4770	0.3633
	4	1.0000	0.9996	0.9971	0.9896	0.9727	0.9420	0.8939	0.8263	0.7396	0.6367
	5	1.0000	1.0000	0.9998	0.9988	0.9958	0.9887	0.9747	0.9502	0.9115	0.8555
	6	1.0000	1.0000	1.0000	0.9999	0.9996	0.9987	0.9964	0.9915	0.9819	0.9648
	7	1.0000	1.0000	1.0000	1.0000	1.0000	0.9999	0.9998	0.9993	0.9983	0.9961
9	0	0.6302	0.3874	0.2316	0.1342	0.0751	0.0404	0.0207	0.0101	0.0046	0.0020
	1	0.9288	0.7748	0.5995	0.4362	0.3003	0.1960	0.1211	0.0705	0.0385	0.0195
	2	0.9916	0.9470	0.8591	0.7382	0.6007	0.4628	0.3373	0.2318	0.1495	0.0898
	3	0.9994	0.9917	0.9661	0.9144	0.8343	0.7297	0.6089	0.4826	0.3614	0.2539
	4	1.0000	0.9991	0.9944	0.9804	0.9511	0.9012	0.8283	0.7334	0.6214	0.5000
	5	1.0000	0.9999	0.9994	0.9969	0.9900	0.9747	0.9464	0.9006	0.8342	0.7461
	6	1.0000	1.0000	1.0000	0.9997	0.9987	0.9957	0.9888	0.9750	0.9502	0.9102
	7	1.0000	1.0000	1.0000	1.0000	0.9999	0.9996	0.9986	0.9962	0.9909	0.9805
	8	1.0000	1.0000	1.0000	1.0000	1.0000	1.0000	0.9999	0.9997	0.9992	0.9980

TABLE 1

Binomial Distribution Function (Continued)

							p				
n	x	0.05	0.10	0.15	0.20	0.25	0.30	0.35	0.40	0.45	0.50
10	0	0.5987	0.3487	0.1969	0.1074	0.0563	0.0282	0.0135	0.0060	0.0025	0.0010
	1	0.9139	0.7361	0.5443	0.3758	0.2440	0.1493	0.0860	0.0464	0.0232	0.0107
	2	0.9885	0.9298	0.8202	0.6778	0.5256	0.3828	0.2616	0.1673	0.0996	0.0547
	3	0.9990	0.9872	0.9500	0.8791	0.7759	0.6496	0.5138	0.3823	0.2660	0.1719
	4	0.9999	0.9984	0.9901	0.9672	0.9219	0.8497	0.7515	0.6331	0.5044	0.3770
	5	1.0000	0.9999	0.9986	0.9936	0.9803	0.9527	0.9051	0.8338	0.7384	0.6230
	6	1.0000	1.0000	0.9999	0.9991	0.9965	0.9894	0.9740	0.9452	0.8980	0.8281
	7	1.0000	1.0000	1.0000	0.9999	0.9996	0.9984	0.9952	0.9877	0.9726	0.9453
	8	1.0000	1.0000	1.0000	1.0000	1.0000	0.9999	0.9995	0.9983	0.9955	0.9893
	9	1.0000	1.0000	1.0000	1.0000	1.0000	1.0000	1.0000	0.9999	0.9997	0.9990
11	0	0.5688	0.3138	0.1673	0.0859	0.0422	0.0198	0.0088	0.0036	0.0014	0.0005
	1	0.8981	0.6974	0.4922	0.3221	0.1971	0.1130	0.0606	0.0302	0.0139	0.0059
	2	0.9848	0.9104	0.7788	0.6174	0.4552	0.3127	0.2001	0.1189	0.0652	0.0327
	3	0.9984	0.9815	0.9306	0.8389	0.7133	0.5696	0.4256	0.2963	0.1911	0.1133
	4	0.9999	0.9972	0.9841	0.9496	0.8854	0.7897	0.6683	0.5328	0.3971	0.2744
	5	1.0000	0.9997	0.9973	0.9883	0.9657	0.9218	0.8513	0.7535	0.6331	0.5000
	6	1.0000	1.0000	0.9997	0.9980	0.9924	0.9784	0.9499	0.9006	0.8262	0.7256
	7	1.0000	1.0000	1.0000	0.9998	0.9988	0.9957	0.9878	0.9707	0.9390	0.8867
	8	1.0000	1.0000	1.0000	1.0000	0.9999	0.9994	0.9980	0.9941	0.9852	0.9673
	9	1.0000	1.0000	1.0000	1.0000	1.0000	1.0000	0.9998	0.9993	0.9978	0.9941
	10	1.0000	1.0000	1.0000	1.0000	1.0000	1.0000	1.0000	1.0000	0.9998	0.9995
12	0	0.5404	0.2824	0.1422	0.0687	0.0317	0.0138	0.0057	0.0022	0.0008	0.0002
	1	0.8816	0.6590	0.4435	0.2749	0.1584	0.0850	0.0424	0.0196	0.0083	0.0032
	2	0.9804	0.8891	0.7358	0.5583	0.3907	0.2528	0.1513	0.0834	0.0421	0.0193
	3	0.9978	0.9744	0.9078	0.7946	0.6488	0.4925	0.3467	0.2253	0.1345	0.0730
	4	0.9998	0.9957	0.9761	0.9274	0.8424	0.7237	0.5833	0.4382	0.3044	0.1938
	5	1.0000	0.9995	0.9954	0.9806	0.9456	0.8822	0.7873	0.6652	0.5269	0.3872
	6	1.0000	0.9999	0.9993	0.9961	0.9857	0.9614	0.9154	0.8418	0.7393	0.6128
	7	1.0000	1.0000	0.9999	0.9994	0.9972	0.9905	0.9745	0.9427	0.8883	0.8062
	8	1.0000	1.0000	1.0000	0.9999	0.9996	0.9983	0.9944	0.9847	0.9644	0.9270
	9	1.0000	1.0000	1.0000	1.0000	1.0000	0.9998	0.9992	0.9972	0.9921	0.9807
	10	1.0000	1.0000	1.0000	1.0000	1.0000	1.0000	0.9999	0.9997	0.9989	0.9968
	11	1.0000	1.0000	1.0000	1.0000	1.0000	1.0000	1.0000	1.0000	0.9999	0.9998
13	0	0.5133	0.2542	0.1209	0.0550	0.0238	0.0097	0.0037	0.0013	0.0004	0.0001
	1	0.8646	0.6213	0.3983	0.2336	0.1267	0.0637	0.0296	0.0126	0.0049	0.0017
	2	0.9755	0.8661	0.6920	0.5017	0.3326	0.2025	0.1132	0.0579	0.0269	0.0112
	3	0.9969	0.9658	0.8820	0.7437	0.5843	0.4206	0.2783	0.1686	0.0929	0.0461
	4	0.9997	0.9935	0.9658	0.9009	0.7940	0.6543	0.5005	0.3530	0.2279	0.1334
	5	1.0000	0.9991	0.9925	0.9700	0.9198	0.8346	0.7159	0.5744	0.4268	0.2905
	6	1.0000	0.9999	0.9987	0.9930	0.9757	0.9376	0.8705	0.7712	0.6437	0.5000
	7	1.0000	1.0000	0.9998	0.9988	0.9944	0.9818	0.9538	0.9023	0.8212	0.7095
	8	1.0000	1.0000	1.0000	0.9998	0.9990	0.9960	0.9874	0.9679	0.9302	0.8666
	9	1.0000	1.0000	1.0000	1.0000	0.9999	0.9993	0.9975	0.9922	0.9797	0.9539
	10	1.0000	1.0000	1.0000	1.0000	1.0000	0.9999	0.9997	0.9987	0.9959	0.9888
	11	1.0000	1.0000	1.0000	1.0000	1.0000	1.0000	1.0000	0.9999	0.9995	0.9983
	12	1.0000	1.0000	1.0000	1.0000	1.0000	1.0000	1.0000	1.0000	1.0000	0.9999
14	0	0.4877	0.2288	0.1028	0.0440	0.0178	0.0068	0.0024	0.0008	0.0002	0.0001
	1	0.8470	0.5846	0.3567	0.1979	0.1010	0.0475	0.0205	0.0081	0.0029	0.0009

TABLE 1

Binomial Distribution Function (Continued)

n	x	0.05	0.10	0.15	0.20	0.25	0.30	0.35	0.40	0.45	0.50
14	2	0.9699	0.8416	0.6479	0.4481	0.2811	0.1608	0.0839	0.0398	0.0170	0.0065
	3	0.9958	0.9559	0.8535	0.6982	0.5213	0.3552	0.2205	0.1243	0.0632	0.0287
	4	0.9996	0.9908	0.9533	0.8702	0.7415	0.5842	0.4227	0.2793	0.1672	0.0898
	5	1.0000	0.9985	0.9885	0.9561	0.8883	0.7805	0.6405	0.4859	0.3373	0.2120
	6	1.0000	0.9998	0.9978	0.9884	0.9617	0.9067	0.8164	0.6925	0.5461	0.3953
	7	1.0000	1.0000	0.9997	0.9976	0.9897	0.9685	0.9247	0.8499	0.7414	0.6074
	8	1.0000	1.0000	1.0000	0.9996	0.9978	0.9917	0.9757	0.9417	0.8811	0.7880
	9	1.0000	1.0000	1.0000	1.0000	0.9997	0.9983	0.9940	0.9825	0.9574	0.9102
	10	1.0000	1.0000	1.0000	1.0000	1.0000	0.9989	0.9989	0.9961	0.9886	0.9713
	11	1.0000	1.0000	1.0000	1.0000	1.0000	1.0000	0.9999	0.9994	0.9978	0.9935
	12	1.0000	1.0000	1.0000	1.0000	1.0000	1.0000	1.0000	0.9999	0.9997	0.9991
	13	1.0000	1.0000	1.0000	1.0000	1.0000	1.0000	1.0000	1.0000	1.0000	0.9999
15	0	0.4633	0.2059	0.0874	0.0352	0.0134	0.0047	0.0016	0.0005	0.0001	0.0000
	1	0.8290	0.5490	0.3186	0.1671	0.0802	0.0353	0.0142	0.0052	0.0017	0.0005
	2	0.9638	0.8159	0.6042	0.3980	0.2361	0.1268	0.0617	0.0271	0.0107	0.0037
	3	0.9945	0.9444	0.8227	0.6482	0.4613	0.2969	0.1727	0.0905	0.0424	0.0176
	4	0.9994	0.9873	0.9383	0.8358	0.6865	0.5155	0.3519	0.2173	0.1204	0.0592
	5	0.9999	0.9978	0.9832	0.9389	0.8516	0.7216	0.5643	0.4032	0.2608	0.1509
	6	1.0000	0.9997	0.9964	0.9819	0.9434	0.8689	0.7548	0.6098	0.4522	0.3036
	7	1.0000	1.0000	0.9996	0.9958	0.9827	0.9500	0.8868	0.7869	0.6535	0.5000
	8	1.0000	1.0000	0.9999	0.9992	0.9958	0.9848	0.9578	0.9050	0.8182	0.6964
	9	1.0000	1.0000	1.0000	0.9999	0.9992	0.9963	0.9876	0.9662	0.9231	0.8491
	10	1.0000	1.0000	1.0000	1.0000	0.9999	0.9993	0.9972	0.9907	0.9745	0.9408
	11	1.0000	1.0000	1.0000	1.0000	1.0000	0.9999	0.9995	0.9981	0.9937	0.9824
	12	1.0000	1.0000	1.0000	1.0000	1.0000	1.0000	0.9999	0.9997	0.9989	0.9963
	13	1.0000	1.0000	1.0000	1.0000	1.0000	1.0000	1.0000	1.0000	0.9999	0.9995
	14	1.0000	1.0000	1.0000	1.0000	1.0000	1.0000	1.0000	1.0000	1.0000	1.0000
16	0	0.4401	0.1853	0.0743	0.0281	0.0100	0.0033	0.0010	0.0003	0.0001	0.0000
	1	0.8108	0.5147	0.2839	0.1407	0.0635	0.0261	0.0098	0.0033	0.0010	0.0003
	2	0.9571	0.7892	0.5614	0.3518	0.1971	0.0994	0.0451	0.0183	0.0066	0.0021
	3	0.9930	0.9316	0.7899	0.5981	0.4050	0.2459	0.1339	0.0651	0.0281	0.0106
	4	0.9991	0.9830	0.9209	0.7982	0.6302	0.4499	0.2892	0.1666	0.0853	0.0384
	5	0.9999	0.9967	0.9765	0.9183	0.8103	0.6598	0.4900	0.3288	0.1976	0.1051
	6	1.0000	0.9995	0.9944	0.9733	0.9204	0.8247	0.6881	0.5272	0.3660	0.2272
	7	1.0000	0.9999	0.9989	0.9930	0.9729	0.9256	0.8406	0.7161	0.5629	0.4018
	8	1.0000	1.0000	0.9998	0.9985	0.9925	0.9743	0.9329	0.8577	0.7441	0.5982
	9	1.0000	1.0000	1.0000	0.9998	0.9984	0.9929	0.9771	0.9417	0.8759	0.7728
	10	1.0000	1.0000	1.0000	1.0000	0.9997	0.9984	0.9938	0.9809	0.9514	0.8949
	11	1.0000	1.0000	1.0000	1.0000	1.0000	0.9997	0.9987	0.9951	0.9851	0.9616
	12	1.0000	1.0000	1.0000	1.0000	1.0000	1.0000	0.9998	0.9991	0.9965	0.9894
	13	1.0000	1.0000	1.0000	1.0000	1.0000	1.0000	1.0000	0.9999	0.9994	0.9979
	14	1.0000	1.0000	1.0000	1.0000	1.0000	1.0000	1.0000	1.0000	1.0000	0.9997
	15	1.0000	1.0000	1.0000	1.0000	1.0000	1.0000	1.0000	1.0000	1.0000	1.0000
17	0	0.4181	0.1668	0.0631	0.0225	0.0075	0.0023	0.0007	0.0002	0.0000	0.0000
	1	0.7922	0.4818	0.2525	0.1182	0.0501	0.0193	0.0067	0.0021	0.0006	0.0001
	2	0.9497	0.7618	0.5198	0.3096	0.1637	0.0774	0.0327	0.0123	0.0041	0.0012
	3	0.9912	0.9174	0.7556	0.5489	0.3530	0.2019	0.1028	0.0464	0.0184	0.0063
	4	0.9988	0.9779	0.9013	0.7582	0.5739	0.3887	0.2348	0.1260	0.0596	0.0245

TABLE 1

Binomial Distribution Function (Continued)

						p					
n	x	0.05	0.10	0.15	0.20	0.25	0.30	0.35	0.40	0.45	0.50
17	5	0.9999	0.9953	0.9681	0.8943	0.7653	0.5968	0.4197	0.2639	0.1471	0.0717
	6	1.0000	0.9992	0.9917	0.9623	0.8929	0.7752	0.6188	0.4478	0.2902	0.1662
	7	1.0000	0.9999	0.9983	0.9891	0.9598	0.8954	0.7872	0.6405	0.4743	0.3145
	8	1.0000	1.0000	0.9997	0.9974	0.9876	0.9597	0.9006	0.8011	0.6626	0.5000
	9	1.0000	1.0000	1.0000	0.9995	0.9969	0.9873	0.9617	0.9081	0.8166	0.6855
	10	1.0000	1.0000	1.0000	0.9999	0.9994	0.9968	0.9880	0.9652	0.9174	0.8338
	11	1.0000	1.0000	1.0000	1.0000	0.9999	0.9993	0.9970	0.9894	0.9699	0.9283
	12	1.0000	1.0000	1.0000	1.0000	1.0000	0.9999	0.9994	0.9975	0.9914	0.9755
	13	1.0000	1.0000	1.0000	1.0000	1.0000	1.0000	0.9999	0.9995	0.9981	0.9936
	14	1.0000	1.0000	1.0000	1.0000	1.0000	1.0000	1.0000	0.9999	0.9997	0.9988
	15	1.0000	1.0000	1.0000	1.0000	1.0000	1.0000	1.0000	1.0000	1.0000	0.9999
	16	1.0000	1.0000	1.0000	1.0000	1.0000	1.0000	1.0000	1.0000	1.0000	1.0000
18	0	0.3972	0.1501	0.0536	0.0180	0.0056	0.0016	0.0004	0.0001	0.0000	0.0000
	1	0.7735	0.4503	0.2241	0.0991	0.0395	0.0142	0.0046	0.0013	0.0003	0.0001
	2	0.9419	0.7338	0.4797	0.2713	0.1353	0.0600	0.0236	0.0082	0.0025	0.0007
	3	0.9891	0.9018	0.7202	0.5010	0.3057	0.1646	0.0783	0.0328	0.0120	0.0038
	4	0.9985	0.9718	0.8794	0.7164	0.5187	0.3327	0.1886	0.0942	0.0411	0.0154
	5	0.9998	0.9936	0.9581	0.8671	0.7175	0.5344	0.3550	0.2088	0.1077	0.0481
	6	1.0000	0.9988	0.9882	0.9487	0.8610	0.7217	0.5491	0.3743	0.2258	0.1189
	7	1.0000	0.9998	0.9973	0.9837	0.9431	0.8593	0.7283	0.5634	0.3915	0.2403
	8	1.0000	1.0000	0.9995	0.9957	0.9807	0.9404	0.8609	0.7368	0.5778	0.4073
	9	1.0000	1.0000	0.9999	0.9991	0.9946	0.9790	0.9403	0.8653	0.7473	0.5927
	10	1.0000	1.0000	1.0000	0.9998	0.9988	0.9939	0.9788	0.9424	0.8720	0.7597
	11	1.0000	1.0000	1.0000	1.0000	0.9998	0.9986	0.9938	0.9797	0.9463	0.8811
	12	1.0000	1.0000	1.0000	1.0000	1.0000	0.9997	0.9986	0.9942	0.9817	0.9519
	13	1.0000	1.0000	1.0000	1.0000	1.0000	1.0000	0.9997	0.9987	0.9951	0.9846
	14	1.0000	1.0000	1.0000	1.0000	1.0000	1.0000	1.0000	0.9998	0.9990	0.9962
	15	1.0000	1.0000	1.0000	1.0000	1.0000	1.0000	1.0000	1.0000	0.9999	0.9993
	16	1.0000	1.0000	1.0000	1.0000	1.0000	1.0000	1.0000	1.0000	1.0000	0.9999
19	0	0.3774	0.1351	0.0456	0.0144	0.0042	0.0011	0.0003	0.0001	0.0000	0.0000
	1	0.7547	0.4203	0.1985	0.0829	0.0310	0.0104	0.0031	0.0008	0.0002	0.0000
	2	0.9335	0.7054	0.4413	0.2369	0.1113	0.0462	0.0170	0.0055	0.0015	0.0004
	3	0.9868	0.8850	0.6841	0.4551	0.2630	0.1332	0.0591	0.0230	0.0077	0.0022
	4	0.9980	0.9648	0.8556	0.6733	0.4654	0.2822	0.1500	0.0696	0.0280	0.0096
	5	0.9998	0.9914	0.9463	0.8369	0.6678	0.4739	0.2968	0.1629	0.0777	0.0318
	6	1.0000	0.9983	0.9837	0.9324	0.8251	0.6655	0.4812	0.3081	0.1727	0.0835
	7	1.0000	0.9997	0.9959	0.9767	0.9225	0.8180	0.6656	0.4878	0.3169	0.1796
	8	1.0000	1.0000	0.9992	0.9933	0.9713	0.9161	0.8145	0.6675	0.4940	0.3238
	9	1.0000	1.0000	0.9999	0.9984	0.9911	0.9674	0.9125	0.8139	0.6710	0.5000
	10	1.0000	1.0000	1.0000	0.9997	0.9977	0.9895	0.9653	0.9115	0.8159	0.6762
	11	1.0000	1.0000	1.0000	1.0000	0.9995	0.9972	0.9886	0.9648	0.9129	0.8204
	12	1.0000	1.0000	1.0000	1.0000	0.9999	0.9994	0.9969	0.9884	0.9658	0.9165
	13	1.0000	1.0000	1.0000	1.0000	1.0000	0.9999	0.9993	0.9969	0.9891	0.9682
	14	1.0000	1.0000	1.0000	1.0000	1.0000	1.0000	0.9999	0.9994	0.9972	0.9904
	15	1.0000	1.0000	1.0000	1.0000	1.0000	1.0000	1.0000	0.9999	0.9995	0.9978
	16	1.0000	1.0000	1.0000	1.0000	1.0000	1.0000	1.0000	1.0000	0.9999	0.9996
	17	1.0000	1.0000	1.0000	1.0000	1.0000	1.0000	1.0000	1.0000	1.0000	1.0000

TABLE 1

Binomial Distribution Function (Continued)

n	x	0.05	0.10	0.15	0.20	0.25	0.30	0.35	0.40	0.45	0.50
20	0	0.3585	0.1216	0.0388	0.0115	0.0032	0.0008	0.0002	0.0000	0.0000	0.0000
	1	0.7358	0.3917	0.1756	0.0692	0.0243	0.0076	0.0021	0.0005	0.0001	0.0000
	2	0.9245	0.6769	0.4049	0.2061	0.0913	0.0355	0.0121	0.0036	0.0009	0.0002
	3	0.9841	0.8670	0.6477	0.4114	0.2252	0.1071	0.0444	0.0160	0.0049	0.0013
	4	0.9974	0.9568	0.8298	0.6296	0.4148	0.2375	0.1182	0.0510	0.0189	0.0059
	5	0.9997	0.9887	0.9327	0.8042	0.6172	0.4164	0.2454	0.1256	0.0553	0.0207
	6	1.0000	0.9976	0.9781	0.9133	0.7858	0.6080	0.4166	0.2500	0.1299	0.0577
	7	1.0000	0.9996	0.9941	0.9679	0.8982	0.7723	0.6010	0.4159	0.2520	0.1316
	8	1.0000	0.9999	0.9987	0.9900	0.9591	0.8867	0.7624	0.5956	0.4143	0.2517
	9	1.0000	1.0000	0.9998	0.9974	0.9861	0.9520	0.8782	0.7553	0.5914	0.4119
	10	1.0000	1.0000	1.0000	0.9994	0.9961	0.9829	0.9468	0.8725	0.7507	0.5881
	11	1.0000	1.0000	1.0000	0.9999	0.9991	0.9949	0.9804	0.9435	0.8692	0.7483
	12	1.0000	1.0000	1.0000	1.0000	0.9998	0.9987	0.9940	0.9790	0.9420	0.8684
	13	1.0000	1.0000	1.0000	1.0000	1.0000	0.9997	0.9985	0.9935	0.9786	0.9423
	14	1.0000	1.0000	1.0000	1.0000	1.0000	1.0000	0.9997	0.9984	0.9936	0.9793
	15	1.0000	1.0000	1.0000	1.0000	1.0000	1.0000	1.0000	0.9997	0.9985	0.9941
	16	1.0000	1.0000	1.0000	1.0000	1.0000	1.0000	1.0000	1.0000	0.9997	0.9987
	17	1.0000	1.0000	1.0000	1.0000	1.0000	1.0000	1.0000	1.0000	1.0000	0.9998
	18	1.0000	1.0000	1.0000	1.0000	1.0000	1.0000	1.0000	1.0000	1.0000	1.0000

TABLE 2

Poisson Distribution Function*

$$F(x; \lambda) = \sum_{k=0}^{x} e^{-\lambda} \frac{\lambda^k}{k!}$$

λ \ x	0	1	2	3	4	5	6	7	8	9
0.02	0.980	1.000								
0.04	0.961	0.999	1.000							
0.06	0.942	0.998	1.000							
0.08	0.923	0.997	1.000							
0.10	0.905	0.995	1.000							
0.15	0.861	0.990	0.999	1.000						
0.20	0.819	0.982	0.999	1.000						
0.25	0.779	0.974	0.998	1.000						
0.30	0.741	0.963	0.996	1.000						
0.35	0.705	0.951	0.994	1.000						
0.40	0.670	0.938	0.992	0.999	1.000					
0.45	0.638	0.925	0.989	0.999	1.000					
0.50	0.607	0.910	0.986	0.998	1.000					
0.55	0.577	0.894	0.982	0.998	1.000					
0.60	0.549	0.878	0.977	0.997	1.000					
0.65	0.522	0.861	0.972	0.996	0.999	1.000				
0.70	0.497	0.844	0.966	0.994	0.999	1.000				
0.75	0.472	0.827	0.959	0.993	0.999	1.000				
0.80	0.449	0.809	0.953	0.991	0.999	1.000				
0.85	0.427	0.791	0.945	0.989	0.998	1.000				
0.90	0.407	0.772	0.937	0.987	0.998	1.000				
0.95	0.387	0.754	0.929	0.984	0.997	1.000				
1.00	0.368	0.736	0.920	0.981	0.996	0.999	1.000			
1.1	0.333	0.699	0.900	0.974	0.995	0.999	1.000			
1.2	0.301	0.663	0.879	0.966	0.992	0.998	1.000			
1.3	0.273	0.627	0.857	0.957	0.989	0.998	1.000			
1.4	0.247	0.592	0.833	0.946	0.986	0.997	0.999	1.000		
1.5	0.223	0.558	0.809	0.934	0.981	0.996	0.999	1.000		
1.6	0.202	0.525	0.783	0.921	0.976	0.994	0.999	1.000		
1.7	0.183	0.493	0.757	0.907	0.970	0.992	0.998	1.000		
1.8	0.165	0.463	0.731	0.891	0.964	0.990	0.997	0.999	1.000	
1.9	0.150	0.434	0.704	0.875	0.956	0.987	0.997	0.999	1.000	
2.0	0.135	0.406	0.677	0.857	0.947	0.983	0.995	0.999	1.000	

*Reprinted by kind permission from E. C. Molina, *Poisson's Exponential Binomial Limit*, D. Van Nostrand Company, Inc., Princeton, N.J., 1947.

TABLE 2

Poisson Distribution Function (Continued)

λ \ x	0	1	2	3	4	5	6	7	8	9
2.2	0.111	0.355	0.623	0.819	0.928	0.975	0.993	0.998	1.000	
2.4	0.091	0.308	0.570	0.779	0.904	0.964	0.988	0.997	0.999	1.000
2.6	0.074	0.267	0.518	0.736	0.877	0.951	0.983	0.995	0.999	1.000
2.8	0.061	0.231	0.469	0.692	0.848	0.935	0.976	0.992	0.998	0.999
3.0	0.050	0.199	0.423	0.647	0.815	0.916	0.966	0.988	0.996	0.999
3.2	0.041	0.171	0.380	0.603	0.781	0.895	0.955	0.983	0.994	0.998
3.4	0.033	0.147	0.340	0.558	0.744	0.871	0.942	0.977	0.992	0.997
3.6	0.027	0.126	0.303	0.515	0.706	0.844	0.927	0.969	0.988	0.996
3.8	0.022	0.107	0.269	0.473	0.668	0.816	0.909	0.960	0.984	0.994
4.0	0.018	0.092	0.238	0.433	0.629	0.785	0.889	0.949	0.979	0.992
4.2	0.015	0.078	0.210	0.395	0.590	0.753	0.867	0.936	0.972	0.989
4.4	0.012	0.066	0.185	0.359	0.551	0.720	0.844	0.921	0.964	0.985
4.6	0.010	0.056	0.163	0.326	0.513	0.686	0.818	0.905	0.955	0.980
4.8	0.008	0.048	0.143	0.294	0.476	0.651	0.791	0.887	0.944	0.975
5.0	0.007	0.040	0.125	0.265	0.440	0.616	0.762	0.867	0.932	0.968
5.2	0.006	0.034	0.109	0.238	0.406	0.581	0.732	0.845	0.918	0.960
5.4	0.005	0.029	0.095	0.213	0.373	0.546	0.702	0.822	0.903	0.951
5.6	0.004	0.024	0.082	0.191	0.342	0.512	0.670	0.797	0.886	0.941
5.8	0.003	0.021	0.072	0.170	0.313	0.478	0.638	0.771	0.867	0.929
6.0	0.002	0.017	0.062	0.151	0.285	0.446	0.606	0.744	0.847	0.916

λ \ x	10	11	12	13	14	15	16
2.8	1.000						
3.0	1.000						
3.2	1.000						
3.4	0.999	1.000					
3.6	0.999	1.000					
3.8	0.998	0.999	1.000				
4.0	0.997	0.999	1.000				
4.2	0.996	0.999	1.000				
4.4	0.994	0.998	0.999	1.000			
4.6	0.992	0.997	0.999	1.000			
4.8	0.990	0.996	0.999	1.000			
5.0	0.986	0.995	0.998	0.999	1.000		
5.2	0.982	0.993	0.997	0.999	1.000		
5.4	0.977	0.990	0.996	0.999	1.000		
5.6	0.972	0.988	0.995	0.998	0.999	1.000	
5.8	0.965	0.984	0.993	0.997	0.999	1.000	
6.0	0.957	0.980	0.991	0.996	0.999	0.999	1.000

TABLE 2

Poisson Distribution Function (Continued)

λ \ x	0	1	2	3	4	5	6	7	8	9
6.2	0.002	0.015	0.054	0.134	0.259	0.414	0.574	0.716	0.826	0.902
6.4	0.002	0.012	0.046	0.119	0.235	0.384	0.542	0.687	0.803	0.886
6.6	0.001	0.010	0.040	0.105	0.213	0.355	0.511	0.658	0.780	0.869
6.8	0.001	0.009	0.034	0.093	0.192	0.327	0.480	0.628	0.755	0.850
7.0	0.001	0.007	0.030	0.082	0.173	0.301	0.450	0.599	0.729	0.830
7.2	0.001	0.006	0.025	0.072	0.156	0.276	0.420	0.569	0.703	0.810
7.4	0.001	0.005	0.022	0.063	0.140	0.253	0.392	0.539	0.676	0.788
7.6	0.001	0.004	0.019	0.055	0.125	0.231	0.365	0.510	0.648	0.765
7.8	0.000	0.004	0.016	0.048	0.112	0.210	0.338	0.481	0.620	0.741
8.0	0.000	0.003	0.014	0.042	0.100	0.191	0.313	0.453	0.593	0.717
8.5	0.000	0.002	0.009	0.030	0.074	0.150	0.256	0.386	0.523	0.653
9.0	0.000	0.001	0.006	0.021	0.055	0.116	0.207	0.324	0.456	0.587
9.5	0.000	0.001	0.004	0.015	0.040	0.089	0.165	0.269	0.392	0.522
10.0	0.000	0.000	0.003	0.010	0.029	0.067	0.130	0.220	0.333	0.458

λ \ x	10	11	12	13	14	15	16	17	18	19
6.2	0.949	0.975	0.989	0.995	0.998	0.999	1.000			
6.4	0.939	0.969	0.986	0.994	0.997	0.999	1.000			
6.6	0.927	0.963	0.982	0.992	0.997	0.999	0.999	1.000		
6.8	0.915	0.955	0.978	0.990	0.996	0.998	0.999	1.000		
7.0	0.901	0.947	0.973	0.987	0.994	0.998	0.999	1.000		
7.2	0.887	0.937	0.967	0.984	0.993	0.997	0.999	0.999	1.000	
7.4	0.871	0.926	0.961	0.980	0.991	0.996	0.998	0.999	1.000	
7.6	0.854	0.915	0.954	0.976	0.989	0.995	0.998	0.999	1.000	
7.8	0.835	0.902	0.945	0.971	0.986	0.993	0.997	0.999	1.000	
8.0	0.816	0.888	0.936	0.966	0.983	0.992	0.996	0.998	0.999	1.000
8.5	0.763	0.849	0.909	0.949	0.973	0.986	0.993	0.997	0.999	0.999
9.0	0.706	0.803	0.876	0.926	0.959	0.978	0.989	0.995	0.998	0.999
9.5	0.645	0.752	0.836	0.898	0.940	0.967	0.982	0.991	0.996	0.998
10.0	0.583	0.697	0.792	0.864	0.917	0.951	0.973	0.986	0.993	0.997

λ \ x	20	21	22
8.5	1.000		
9.0	1.000		
9.5	0.999	1.000	
10.0	0.998	0.999	1.000

TABLE 2

Poisson Distribution Function (Continued)

λ \ x	0	1	2	3	4	5	6	7	8	9
10.5	0.000	0.000	0.002	0.007	0.021	0.050	0.102	0.179	0.279	0.397
11.0	0.000	0.000	0.001	0.005	0.015	0.038	0.079	0.143	0.232	0.341
11.5	0.000	0.000	0.001	0.003	0.011	0.028	0.060	0.114	0.191	0.289
12.0	0.000	0.000	0.001	0.002	0.008	0.020	0.046	0.090	0.155	0.242
12.5	0.000	0.000	0.000	0.002	0.005	0.015	0.035	0.070	0.125	0.201
13.0	0.000	0.000	0.000	0.001	0.004	0.011	0.026	0.054	0.100	0.166
13.5	0.000	0.000	0.000	0.001	0.003	0.008	0.019	0.041	0.079	0.135
14.0	0.000	0.000	0.000	0.000	0.002	0.006	0.014	0.032	0.062	0.109
14.5	0.000	0.000	0.000	0.000	0.001	0.004	0.010	0.024	0.048	0.088
15.0	0.000	0.000	0.000	0.000	0.001	0.003	0.008	0.018	0.037	0.070

λ \ x	10	11	12	13	14	15	16	17	18	19
10.5	0.521	0.639	0.742	0.825	0.888	0.932	0.960	0.978	0.988	0.994
11.0	0.460	0.579	0.689	0.781	0.854	0.907	0.944	0.968	0.982	0.991
11.5	0.402	0.520	0.633	0.733	0.815	0.878	0.924	0.954	0.974	0.986
12.0	0.347	0.462	0.576	0.682	0.772	0.844	0.899	0.937	0.963	0.979
12.5	0.297	0.406	0.519	0.628	0.725	0.806	0.869	0.916	0.948	0.969
13.0	0.252	0.353	0.463	0.573	0.675	0.764	0.835	0.890	0.930	0.957
13.5	0.211	0.304	0.409	0.518	0.623	0.718	0.798	0.861	0.908	0.942
14.0	0.176	0.260	0.358	0.464	0.570	0.669	0.756	0.827	0.883	0.923
14.5	0.145	0.220	0.311	0.413	0.518	0.619	0.711	0.790	0.853	0.901
15.0	0.118	0.185	0.268	0.363	0.466	0.568	0.664	0.749	0.819	0.875

λ \ x	20	21	22	23	24	25	26	27	28	29
10.5	0.997	0.999	0.999	1.000						
11.0	0.995	0.998	0.999	1.000						
11.5	0.992	0.996	0.998	0.999	1.000					
12.0	0.988	0.994	0.997	0.999	0.999	1.000				
12.5	0.983	0.991	0.995	0.998	0.999	0.999	1.000			
13.0	0.975	0.986	0.992	0.996	0.998	0.999	1.000			
13.5	0.965	0.980	0.989	0.994	0.997	0.998	0.999	1.000		
14.0	0.952	0.971	0.983	0.991	0.995	0.997	0.999	0.999	1.000	
14.5	0.936	0.960	0.976	0.986	0.992	0.996	0.998	0.999	0.999	1.000
15.0	0.917	0.947	0.967	0.981	0.989	0.994	0.997	0.998	0.999	1.000

TABLE 2

Poisson Distribution Function (Continued)

λ \ x	4	5	6	7	8	9	10	11	12	13
16	0.000	0.001	0.004	0.010	0.022	0.043	0.077	0.127	0.193	0.275
17	0.000	0.001	0.002	0.005	0.013	0.026	0.049	0.085	0.135	0.201
18	0.000	0.000	0.001	0.003	0.007	0.015	0.030	0.055	0.092	0.143
19	0.000	0.000	0.001	0.002	0.004	0.009	0.018	0.035	0.061	0.098
20	0.000	0.000	0.000	0.001	0.002	0.005	0.011	0.021	0.039	0.066
21	0.000	0.000	0.000	0.000	0.001	0.003	0.006	0.013	0.025	0.043
22	0.000	0.000	0.000	0.000	0.001	0.002	0.004	0.008	0.015	0.028
23	0.000	0.000	0.000	0.000	0.000	0.001	0.002	0.004	0.009	0.017
24	0.000	0.000	0.000	0.000	0.000	0.000	0.001	0.003	0.005	0.011
25	0.000	0.000	0.000	0.000	0.000	0.000	0.001	0.001	0.003	0.006

λ \ x	14	15	16	17	18	19	20	21	22	23
16	0.368	0.467	0.566	0.659	0.742	0.812	0.868	0.911	0.942	0.963
17	0.281	0.371	0.468	0.564	0.655	0.736	0.805	0.861	0.905	0.937
18	0.208	0.287	0.375	0.469	0.562	0.651	0.731	0.799	0.855	0.899
19	0.150	0.215	0.292	0.378	0.469	0.561	0.647	0.725	0.793	0.849
20	0.105	0.157	0.221	0.297	0.381	0.470	0.559	0.644	0.721	0.787
21	0.072	0.111	0.163	0.227	0.302	0.384	0.471	0.558	0.640	0.716
22	0.048	0.077	0.117	0.169	0.232	0.306	0.387	0.472	0.556	0.637
23	0.031	0.052	0.082	0.123	0.175	0.238	0.310	0.389	0.472	0.555
24	0.020	0.034	0.056	0.087	0.128	0.180	0.243	0.314	0.392	0.473
25	0.012	0.022	0.038	0.060	0.092	0.134	0.185	0.247	0.318	0.394

λ \ x	24	25	26	27	28	29	30	31	32	33
16	0.978	0.987	0.993	0.996	0.998	0.999	0.999	1.000		
17	0.959	0.975	0.985	0.991	0.995	0.997	0.999	0.999	1.000	
18	0.932	0.955	0.972	0.983	0.990	0.994	0.997	0.998	0.999	1.000
19	0.893	0.927	0.951	0.969	0.980	0.988	0.993	0.996	0.998	0.999
20	0.843	0.888	0.922	0.948	0.966	0.978	0.987	0.992	0.995	0.997
21	0.782	0.838	0.883	0.917	0.944	0.963	0.976	0.985	0.991	0.994
22	0.712	0.777	0.832	0.877	0.913	0.940	0.959	0.973	0.983	0.989
23	0.635	0.708	0.772	0.827	0.873	0.908	0.936	0.956	0.971	0.981
24	0.554	0.632	0.704	0.768	0.823	0.868	0.904	0.932	0.953	0.969
25	0.473	0.553	0.629	0.700	0.763	0.818	0.863	0.900	0.929	0.950

λ \ x	34	35	36	37	38	39	40	41	42	43
19	0.999	1.000								
20	0.999	0.999	1.000							
21	0.997	0.998	0.999	0.999	1.000					
22	0.994	0.996	0.998	0.999	0.999	1.000				
23	0.998	0.993	0.996	0.997	0.999	0.999	1.000			
24	0.979	0.987	0.992	0.995	0.997	0.998	0.999	0.999		
25	0.966	0.978	0.985	0.991	0.994	0.997	0.998	0.999	1.000	

TABLE 3

Normal Distribution Function

$$F(z) = \frac{1}{\sqrt{2\pi}} \int_{-\infty}^{z} e^{-(1/2)t^2} \, dt$$

z	0.00	0.01	0.02	0.03	0.04	0.05	0.06	0.07	0.08	0.09
0.0	0.5000	0.5040	0.5080	0.5120	0.5160	0.5199	0.5239	0.5279	0.5319	0.5359
0.1	0.5398	0.5438	0.5478	0.5517	0.5557	0.5596	0.5636	0.5675	0.5714	0.5753
0.2	0.5793	0.5832	0.5871	0.5910	0.5948	0.5987	0.6026	0.6064	0.6103	0.6141
0.3	0.6179	0.6217	0.6255	0.6293	0.6331	0.6368	0.6406	0.6443	0.6480	0.6517
0.4	0.6554	0.6591	0.6628	0.6664	0.6700	0.6736	0.6772	0.6808	0.6844	0.6879
0.5	0.6915	0.6950	0.6985	0.7019	0.7054	0.7088	0.7123	0.7157	0.7190	0.7224
0.6	0.7257	0.7291	0.7324	0.7357	0.7389	0.7422	0.7454	0.7486	0.7517	0.7549
0.7	0.7580	0.7611	0.7642	0.7673	0.7704	0.7734	0.7764	0.7794	0.7823	0.7852
0.8	0.7881	0.7910	0.7939	0.7967	0.7995	0.8023	0.8051	0.8078	0.8106	0.8133
0.9	0.8159	0.8186	0.8212	0.8238	0.8264	0.8289	0.8315	0.8340	0.8365	0.8389
1.0	0.8413	0.8438	0.8461	0.8485	0.8508	0.8531	0.8554	0.8577	0.8599	0.8621
1.1	0.8643	0.8665	0.8686	0.8708	0.8729	0.8749	0.8770	0.8790	0.8810	0.8830
1.2	0.8849	0.8869	0.8888	0.8907	0.8925	0.8944	0.8962	0.8980	0.8997	0.9015
1.3	0.9032	0.9049	0.9066	0.9082	0.9099	0.9115	0.9131	0.9147	0.9162	0.9177
1.4	0.9192	0.9207	0.9222	0.9236	0.9251	0.9265	0.9279	0.9292	0.9306	0.9319
1.5	0.9332	0.9345	0.9357	0.9370	0.9382	0.9394	0.9406	0.9418	0.9429	0.9441
1.6	0.9452	0.9463	0.9474	0.9484	0.9495	0.9505	0.9515	0.9525	0.9535	0.9545
1.7	0.9554	0.9564	0.9573	0.9582	0.9591	0.9599	0.9608	0.9616	0.9625	0.9633
1.8	0.9641	0.9649	0.9656	0.9664	0.9671	0.9678	0.9686	0.9693	0.9699	0.9706
1.9	0.9713	0.9719	0.9726	0.9732	0.9738	0.9744	0.9750	0.9756	0.9761	0.9767
2.0	0.9772	0.9778	0.9783	0.9788	0.9793	0.9798	0.9803	0.9808	0.9812	0.9817
2.1	0.9821	0.9826	0.9830	0.9834	0.9838	0.9842	0.9846	0.9850	0.9854	0.9857
2.2	0.9861	0.9864	0.9868	0.9871	0.9875	0.9878	0.9881	0.9884	0.9887	0.9890
2.3	0.9893	0.9896	0.9898	0.9901	0.9904	0.9906	0.9909	0.9911	0.9913	0.9916
2.4	0.9918	0.9920	0.9922	0.9925	0.9927	0.9929	0.9931	0.9932	0.9934	0.9936
2.5	0.9938	0.9940	0.9941	0.9943	0.9945	0.9946	0.9948	0.9949	0.9951	0.9952
2.6	0.9953	0.9955	0.9956	0.9957	0.9959	0.9960	0.9961	0.9962	0.9963	0.9964
2.7	0.9965	0.9966	0.9967	0.9968	0.9969	0.9970	0.9971	0.9972	0.9973	0.9974
2.8	0.9974	0.9975	0.9976	0.9977	0.9977	0.9978	0.9979	0.9979	0.9980	0.9981
2.9	0.9981	0.9982	0.9982	0.9983	0.9984	0.9984	0.9985	0.9985	0.9986	0.9986
3.0	0.9987	0.9987	0.9987	0.9988	0.9988	0.9989	0.9989	0.9989	0.9990	0.9990
3.1	0.9990	0.9991	0.9991	0.9991	0.9992	0.9992	0.9992	0.9992	0.9993	0.9993
3.2	0.9993	0.9993	0.9994	0.9994	0.9994	0.9994	0.9994	0.9995	0.9995	0.9995
3.3	0.9995	0.9995	0.9995	0.9996	0.9996	0.9996	0.9996	0.9996	0.9996	0.9997
3.4	0.9997	0.9997	0.9997	0.9997	0.9997	0.9997	0.9997	0.9997	0.9997	0.9998
3.5	0.9998									
4.0	0.99997									
5.0	0.9999997									
6.0	0.999999999									